Vannessa Goodship (Ed.)

Injection Moulding

Also of interest

Chemical Product Technology.
Tylkowski, Wieszczycka, Jastrzab (Eds.), 2017
ISBN 978-3-11-046828-1, e-ISBN 978-3-11-046974-5

Polymeric Composites with Rice Hulls
An Introduction
Defonseka, 2019
ISBN 978-3-11-063968-1, e-ISBN 978-3-11-064320-6

Rubber.
Science and Technology
Princi, 2019
ISBN 978-3-11-064031-1, e-ISBN 978-3-11-064032-8

Smart Rubbers.
Synthesis and Applications
Polgar, von Essen, Pucci, Picchioni, 2019
ISBN 978-3-11-063892-9, e-ISBN 978-3-11-063901-8

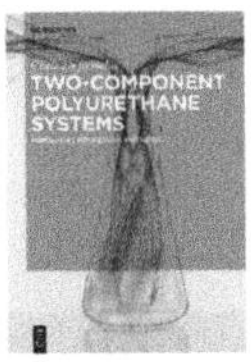

Two-Component Polyurethane Systems
Innovative Processing Methods
Defonseka, 2019
ISBN 978-3-11-063957-5, e-ISBN 978-3-11-064316-9

Injection Moulding

A Practical Guide

Edited by
Vannessa Goodship

3rd Edition

DE GRUYTER

Editor
Dr. Vannessa Goodship
University of Warwick
WMG
International Manufacturing Centre
Coventry CV4 7AL
United Kingdom
V.Goodship@warwick.ac.uk

ISBN 978-3-11-065302-1
e-ISBN (PDF) 978-3-11-065481-3
e-ISBN (EPUB) 978-3-11-065303-8

Library of Congress Control Number: 2019946034

Bibliographic information published by the Deutsche Nationalbibliothek
The Deutsche Nationalbibliothek lists this publication in the Deutsche Nationalbibliografie; detailed bibliographic data are available on the Internet at http://dnb.dnb.de.

Cover image: skodonnell / E+ / gettyimages
Typesetting: Integra Software Services Pvt. Ltd.
Printing and binding: CPI books GmbH, Leck

www.degruyter.com

This 3rd edition book is dedicated to the memory of my dad, who allowed me to dream, and to Deb, whose laughter and colour taught me to come back to earth again.

Dr. Vannessa Goodship
WMG
University of Warwick
2016

Preface

This 3rd edition book is designed as a guide to the injection moulding process. Whilst it is imagined to be primarily of interest to engineers and engineering students, many of the chapters should be accessible to non-specialist readers. It concentrates mainly on the moulding of plastics but also provides a basic guide to the moulding of ceramics and metal powders.

This book can be basically split into three sections: Chapters 1–4 form the underlying basics which are relevant to the rest of the book; Chapters 5–8 are the heart of the book, the processing guides; Chapter 9 is a troubleshooting guide and finally, Chapter 10 is an introduction to some of the more specialist injection moulding processes, such as multimaterial and gas-assisted moulding.

This book should equip the reader with the knowledge to understand the relationship of material, machine and mould tool. The materials covered include common thermoplastic and thermoset materials, as well as liquid silicone rubber, and provides information on the machinery required for successfully moulding these materials.

Dr. Vannessa Goodship
WMG
University of Warwick
2016

https://doi.org/10.1515/9783110654813-202

Acknowledgements

Thanks to Colin Tirel at ARBURG Ltd, UK, and all at ARBURG GmbH, for their assistance and help in gathering the material for this guide, and especially for the many people who contributed to the original source materials. This book is based on the ARBURG Plastics Technology Course KT1 and their associated technical literature.

https://doi.org/10.1515/9783110654813-203

Contents

1 Introduction

1.1 The big picture

When the Hyatt brothers, John and Isaiah, built and patented the first injection moulding machine (IJM) in 1872, it is doubtful they could possibly have imagined the impact this invention would have on the world. It spawned a worldwide industry employing millions of people.

In 2013, plastic manufacturers generated over 373 billion US dollars (USD) in the US alone with the plastic industry employing 900,000 people there. The worldwide consumption of plastic materials rose from 245 million metric tonnes in 2008, to 311 million metric tonnes in 2014. China uses over a quarter of the world's plastics, with this figure increasing year on year as manufacturing grows. Europe's manufacturing base is the second largest global market consuming around 20%. The Chinese market is increasingly prominent across the entire plastics sector and China is currently both the largest producer of machines in the world and the biggest consumer of them, which drives the need for a plastic raw materials market as well.

IJM sales in China had reached a market worth 25.5 billion Chinese yuan [renminbi (RMB)] in 2013, and was expected to reach RMB30 billion by 2016.

The packaging sector accounts for the majority of the world market, with the automotive industry and the electrical equipment industry also both heavy consumers of plastic and injection mouldings. Other important markets for plastic materials are the construction industry and consumer products.

A major factor for the entire injection moulding industry (and manufacturing in general) has been the globalisation of the market. Like other industries, successful IJM manufacturers have kept pace with changes in global demand and moved to lower cost production plants in the world's developing economies, such as China, India and Brazil. To adapt to the increased competition, the successful manufacturers are the ones that have been able to adapt, innovate and compete.

From these few facts and figures it can be seen that injection moulding is an extremely large and important manufacturing industry, but what is injection moulding?

1.2 Introduction to injection moulding

Injection moulding is one of the most commonly employed processes to produce plastic parts. It is a cyclic process of rapid mould filling followed by cooling and ejection. A variety of materials both plastic and non-plastic can be used as feedstock; however, the machine must be configured for the type of material used.

https://doi.org/10.1515/9783110654813-001

The material, which is generally available as grains or powder, is plasticised in an injection unit and injected into a clamped mould under high pressure (500–1,500 bar). The main advantage of injection moulding is it is a very economical method of mass production. Ready parts with tight tolerances can be produced in one step, often completely automatically and, in general, subsequent processing steps are not necessary. It is also possible to integrate different functions into one part to avoid the formation of different components that would be more expensive, and some of these more advanced and integrated processes are discussed in Chapter 10.

To guarantee a high-quality injection moulded part the following points have to be considered:

- The material has to be plasticised and injected carefully to avoid negative effects on the properties and characteristics of the material.
- The process settings (such as pressures and temperatures) concerning the machine and mould have to remain constant with regard to time and space.

An example of a commercially available IJM is shown in Figure 1.1 and the basic parts that make up a machine are shown in Figure 1.2.

The computer control unit allows the machine operator to control and monitor the process; control systems will be discussed in Chapter 5.

The purpose of the injection unit is to melt and meter the material into the tool; machinery will be discussed in Chapter 3.

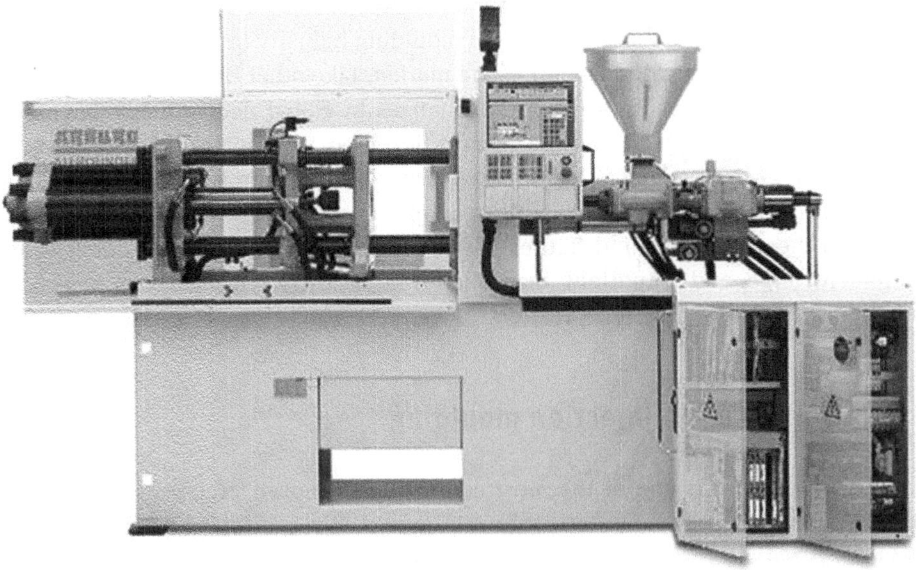

Figure 1.1: An injection moulding machine (IJM).

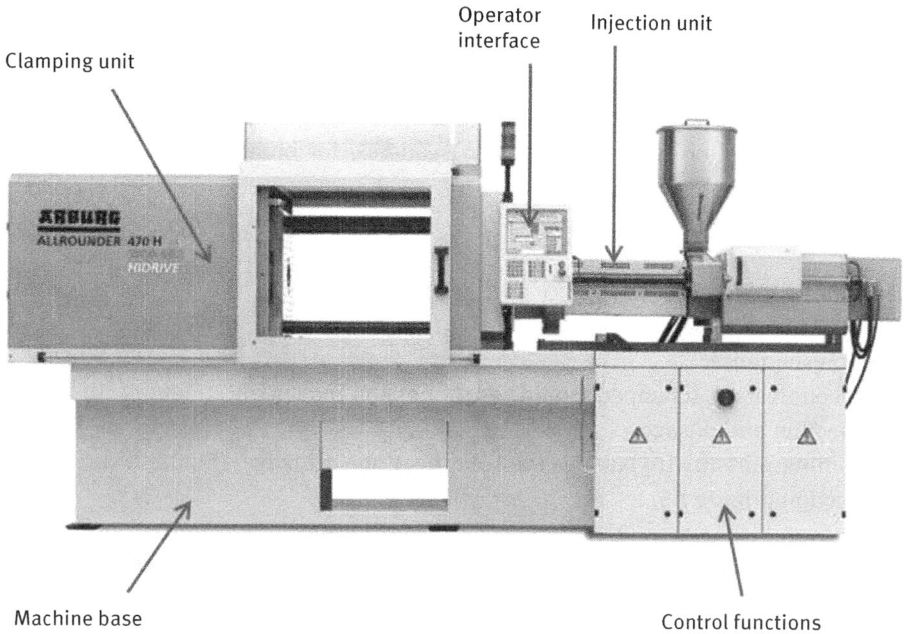

Figure 1.2: Basic components of an IJM.

The clamp unit holds the injection moulding tool and gives the required clamp force to hold the two sides of the mould tool together; tooling will be discussed in Chapter 4.

The machine base is powered by either electrical or hydraulic means, and houses the systems that drive the movement of the machine. Each machine has a certain 'footprint', which is the floor space it requires for operation.

1.3 The injection moulding process

During the injection moulding process, the mould and the plasticising area are separated from each other. The temperature of the plasticising area, i.e., the plasticising cylinder, is maintained at the same level as the processing temperature. The mould on the other hand, is kept cold enough for demoulding of the injection moulded part (thermoplastics), or warm enough for crosslinking (thermosets). The plasticised material is injected into the clamped mould. In an IJM, the clamping unit, which contains the mould and injection unit, is integrated. Completely automated production is possible if the mould is installed with a vertical parting line, which enables the parts to fall down and out of the mould after demoulding. IJM are more typically used for the processing of thermoplastics.

There are two types of injection unit available: a piston injection unit and a screw piston injection unit (reciprocating); the reciprocating screw method is the most common. Only screw piston machines can be used for the processing of thermosets as without the screw, the dwell time would be too long and the risk of early cross-linking would be too high. The injection sequence for both types of machine now follows.

1.3.1 Piston injection unit

Injection sequence:
1. Injection starts (clamped mould, start of piston movement) (Figure 1.3).
2. Injection and dosage.
3. Holding pressure (to balance the solidification shrinkage) (Figure 1.4).
4. Ejection (Figure 1.5)

In the piston injection unit, the material is transported step-by-step through the heated plasticating cylinder until the required temperature is reached. A dosage device replaces the consumed plastic by new material. As the dwell time in the

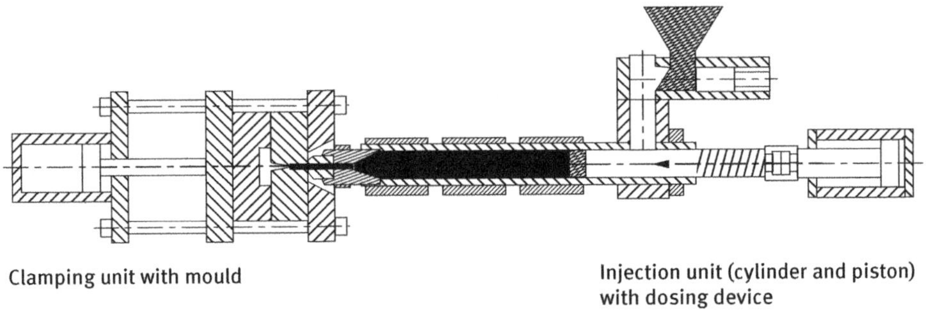

Figure 1.3: Piston injection unit (1).

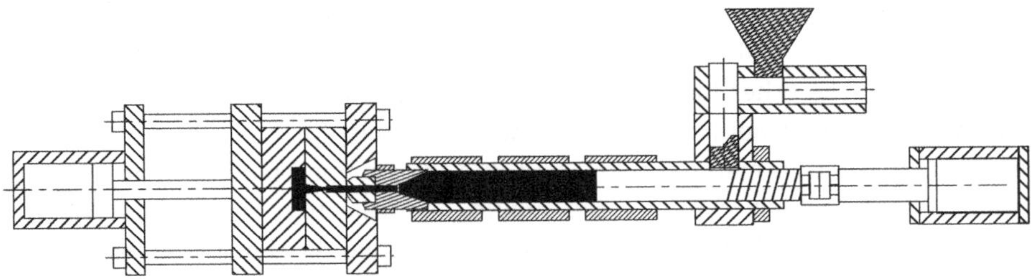

Figure 1.4: Piston injection unit (2).

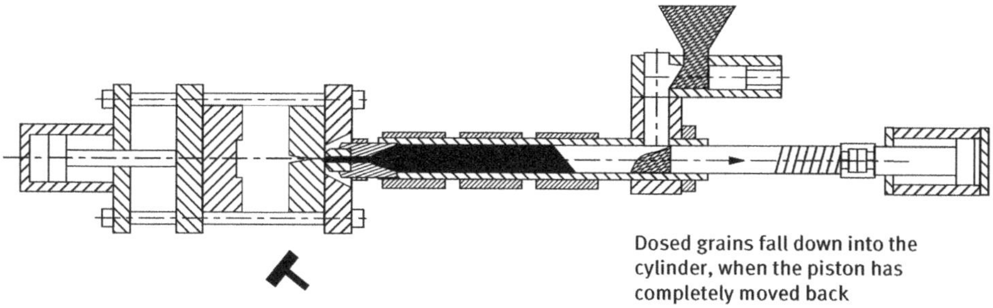

Figure 1.5: Piston injection unit (3).

cylinder is long, this method is not suitable for processing heat-sensitive materials such as rigid polyvinyl chloride and thermosets. Figures 1.3–1.5illustrate the machine cycle.

1.3.2 Reciprocating screw machine

For a reciprocating screw machine the process cycle can be split into five stages:

1. In stage 1, as shown in Figure 1.6, material is injected into the tool.
2. In stage 2 (Figure 1.7), the screw begins to turn and retract, metering a specified weight of molten material for the next shot. The previous shot is now cooling in the closed tool.
3. In stage 3, the injection unit moves back from the clamping unit as shown in Figure 1.8.
4. Stage 4 is shown in Figure 1.9. In this stage, the tool opens to reveal a cooled injection moulded component.
5. Stage 5 is ejection of the part as shown in Figure 1.10. The injection unit will then move forward to the clamp unit to start a fresh cycle as shown in stage 1.

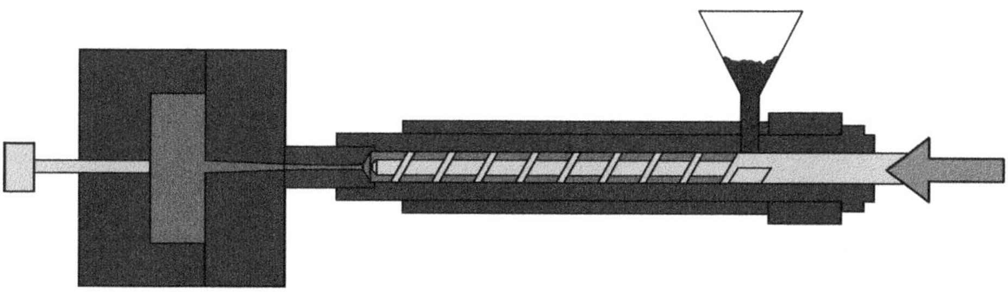

Figure 1.6: Injection moulding: injection.

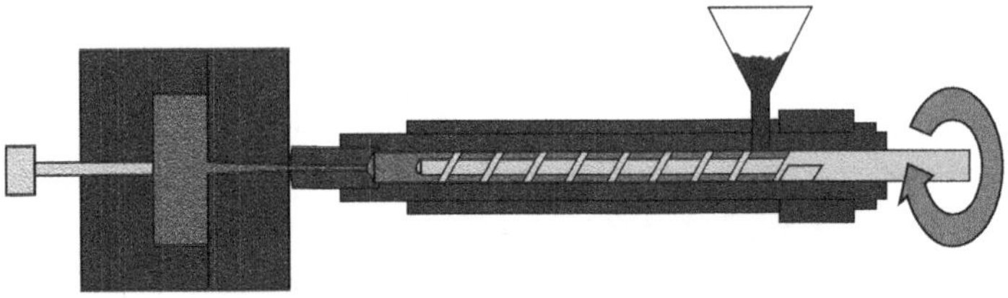

Figure 1.7: Injection moulding: metering.

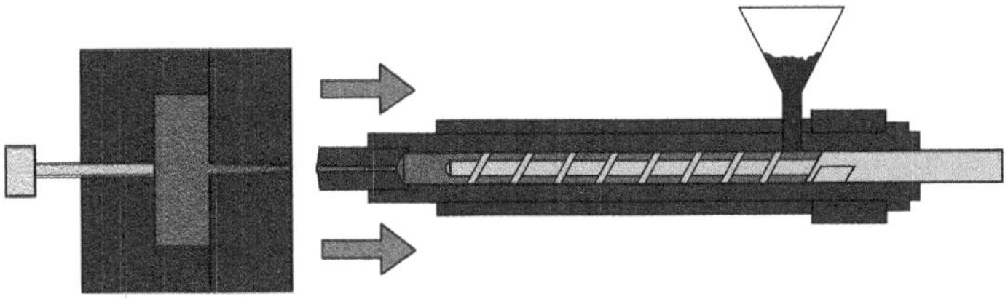

Figure 1.8: Injection moulding: injection unit retracts.

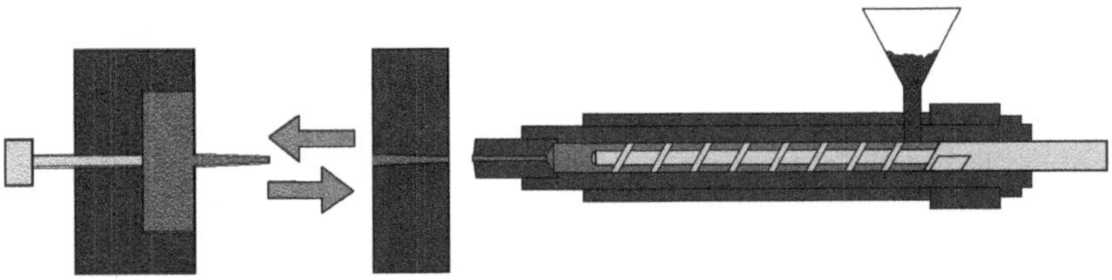

Figure 1.9: Injection moulding: mould open.

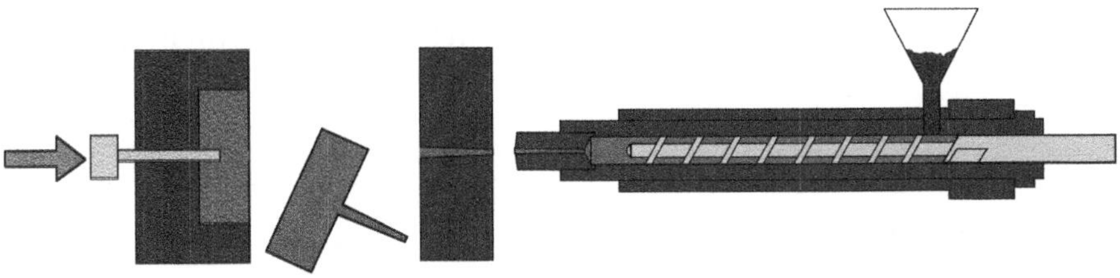

Figure 1.10: Injection moulding: ejection.

Thermoplastics as well as thermosets and classic elastomers can be processed with screw injection units. These material types will be discussed in more depth in Chapters 6 and 7.

Thermoplastics are injected into a cold mould. The temperature of the mould must be sufficiently below the melting temperature of the material for it to solidify, as solidification is a physical process.

Thermosets and classical elastomers are injected into a hot mould to make the crosslinking of the material possible, as crosslinking is a chemical process.

In the screw piston injection unit, the material is dosed and plasticised simultaneously. The material is kneaded thoroughly by a rotating, axially movable screw. It is heated up to the processing temperature by the heat transferred from the hot cylinder wall and by friction. The material is transported by the screw to the screw tip. As the nozzle opening (cylinder opening) is still closed, the screw moves backwards. As soon as enough material is in the area in front of the screw tip, the screw is stopped, which is controlled by a limit switch or by a stroke measure device. This is the end of the plasticising and dosage stage.

In screw piston injection machines, the material is plasticised more homogeneously and has to withstand less thermal stress than in piston injection units, as the plasticising itself happens just shortly after the injection.

1.3.3 Breaking down the injection moulding cycle

A single injection moulding cycle can be broken down into three distinct stages: plastication, mould filling and cooling with solidification.

1.3.3.1 Plastication

This stage is carried out in the injection unit and is similar to the extrusion process.

The polymer flow rate is governed by the material process conditions of the plastication stage and is a combination of material rheology, barrel temperature and shear, back pressure and screw speed. The basic aim is to produce a homogeneous melt for the next stage where the material enters the mould. Moulding parameters which control the plastication stage are cylinder temperature, screw back temperature and back pressure.

1.3.3.2 Filling

Here the injection unit delivers a preset amount of molten polymer to the mould tool.

The parameters of mould filling are of great importance to the end result, especially when considering factors such as warpage (orientation effects) and surface finish (skin formation). Filling dynamics are also thought to be the major factor in affecting the levels of residual stress. It is important that injection speeds are reproducible as slight changes can cause variations in the end product. Injection speeds that are too high can cause jetting and degradation and thus affect the mechanical properties of the material. A low speed may cause an increase in pressure requirements due to a thicker frozen layer and short shots (incomplete filling of the mould).

Thinner sections will generally need faster injection speeds than thick-walled parts, mainly due to the decreased importance of the relationship between mould filling time and cooling time with a thicker section. The important thing is that the speeds are reproducible from one shot to the next. Important moulding parameters for filling are the injection speed and injection pressure.

1.3.3.3 Packing and Solidification

Once the material is in the tool, filling must be completed (tool packing), the part cooled and finally ejected. The purpose of the packing stage is to add extra material to compensate for the shrinkage caused by the decreasing density of the solidifying polymer. If the additional polymer were not injected, the component would shrink and warp due to non-uniform cooling.

Ideally the packing and cooling stages should be such that the final product dimension is maintained as close as possible to design tolerances. Variables during this stage are packing pressure, packing time and the mould temperature. Bad mould design can lead to inconsistent cooling along the dimensions of the mould surface, which can cause increased residual stresses. Once the material has cooled sufficiently, the component can be injected and the injection cycle continues. The cycle does not occur sequentially, while one part is cooling, plastication of the next cycle has already begun.

A breakdown of the cycle and the relative time for each stage is shown in Figure 1.11.

The injection moulding process and parameters that affect the process will be further discussed in Chapter 8.

A brief introduction to injection moulding has now been given. Before moving on to consider it in greater depth, a quick summary of other commercial methods of polymer processing will now be considered for comparison.

1.4 Comparison with other moulding methods

1.4.1 Extrusion

Extrusion is used for the production of intermediate products such as profiles, boards and sheets, and in general, screw extruders are used. For the processing of thermosets, piston extruders are also suitable. The material is kneaded thoroughly and plasticised by friction and additional cylinder heating. It is fed

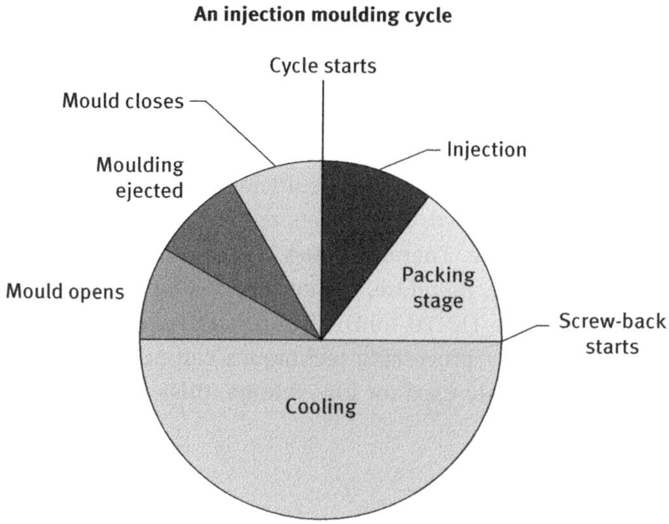

Figure 1.11: Breakdown of an injection moulding cycle.

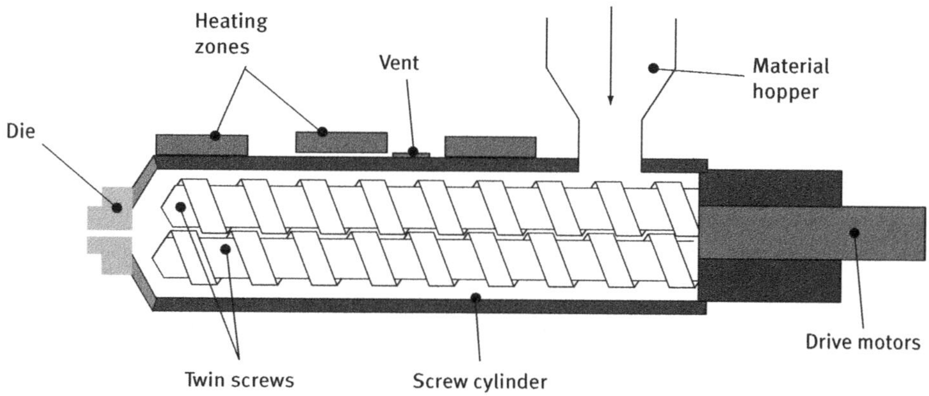

Figure 1.12: Extrusion.

through an open mould (die) with a diameter corresponding to the desired profile (Figure 1.12).

The extruder is very similar in design and purpose to an injection unit on an IJM; however, there are certain differences in both design and purpose. Whilst the material is simply fed through an extruder, an injection unit must also have the ability to both dose and rapidly inject the molten polymer into the mould tool cavity.

1.4.2 Compression moulding

During compression moulding a weighed amount of material is placed in an open mould tool and then compressed (Figure 1.13). During the clamping movement, the material is heated up to its processing temperature by heat transfer from the hot mould allowing it to form to the shape of the tool. After sufficient pressure and temperature has been applied, the tool is opened. Thermosets are demoulded from the hot mould when crosslinking is complete. Thermoplastics have to be cooled until the part is rigid enough for demoulding; therefore, the cycle times for thermoplastics and thermosets are both very long (12–20 min is not uncommon), in consequence, for thermoplastics, where other processing techniques can achieve much shorter cycle times, this method is mainly used for low volume, thick-walled parts or boards.

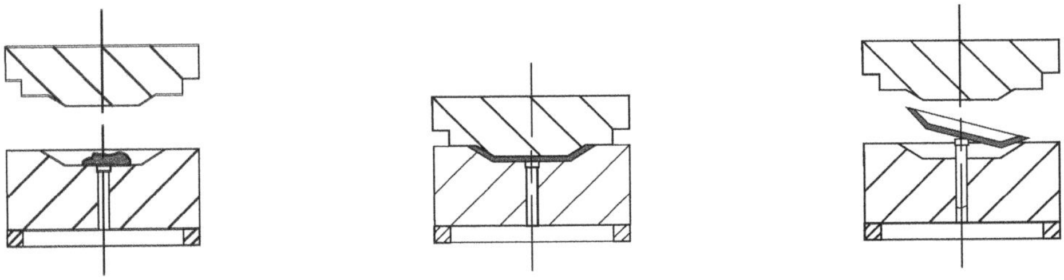

Figure 1.13: Compression moulding.

1.4.3 Transfer moulding

In a transfer mould, there is a special plasticising area (cylinder) as well as cavities.

With an open mould a weighed amount of material is placed into the plasticising area (Figure 1.14a). After mould clamping (Figure 1.14b), the material is pressed

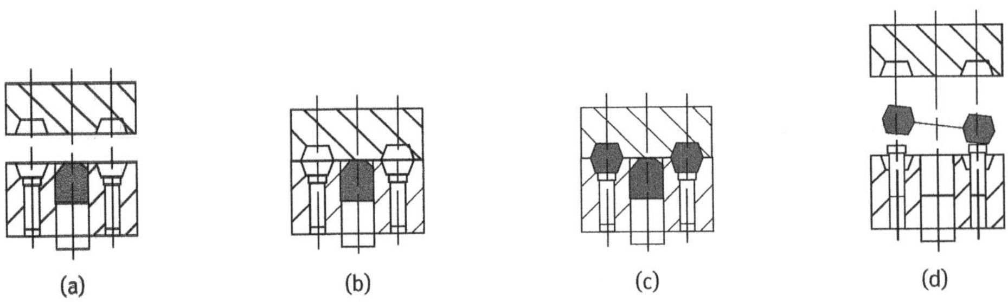

Figure 1.14: Transfer moulding.

through the sprue channels into the cavities by the piston (Figure 1.14c). The material is warmed by heat transfer from the hot mould. Frictional heat also warms the material, which occurs in the tool, especially when it passes though the sprue channels.

Practically, transfer moulding is used for the processing of thermosets only, as there are no benefits to the processing of thermoplastics compared with the injection moulding method.

1.4.4 Blow moulding

Blow moulding is the third most commercially important process for plastics production after extrusion and injection moulding, and is used to produce a range of hollow articles, for example, bottles, fuel tanks and other large containers. There are two main variations, injection blow moulding and extrusion blow moulding. The process sequence for both is the same:
- The material is either extruded or injected to produce a tube-like preform.
- The preform is blown out to the shape of the mould and then cooled.

Injection blow moulding is most commonly employed for the production of transparent soft drinks containers. However, extrusion blow moulding is the one most commonly employed for mouldings such as shampoos, detergents, plastic drums and milk bottles.

The material is fed through a transfer screw into a die head where the material is melted and passes out through a die as a tube-like extrudate termed a parison, as shown in Figure 1.15. This process can be either continuous or,

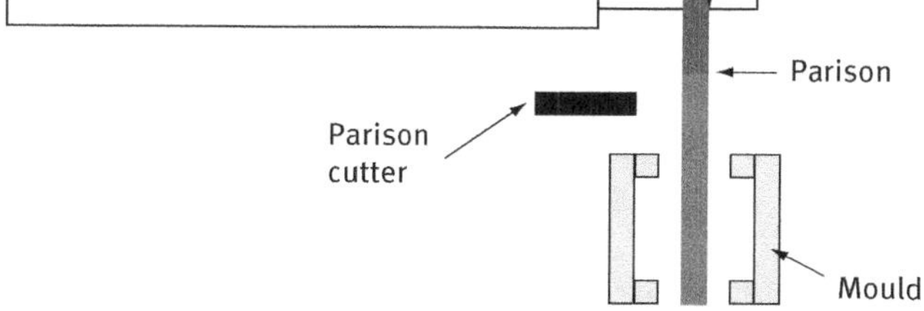

Figure 1.15: Blow moulding: stage 1.

with larger articles, intermittent. The parison extrudes down vertically and relies on the hot strength of the plastic to hold the parison weight in shape. For these reasons, blow moulding uses far more viscous materials than would normally be employed for the injection moulding process. A low viscosity material would simply pour out of the die onto the floor or split off before the parison had formed.

The parison will continue to extrude until it has reached the base of the mould. At this point the mould will close and the parison will be cut above the mould with a hot knife. The mould is now moved away from the parison, taking the cut-off slice of parison closed in the mould with it. The next parison continues to extrude. The inflation of the parison can now commence. A blow pin comes down into the top of the mould and blows air in to inflate the hot parison against the sides of the mould, as shown in Figure 1.16. The mould is cooled with water and this aids heat transfer to help solidify the newly formed article. The blow pin is removed and the tool opens to eject the part. The mould can then return to collect the next parison and start a new moulding cycle.

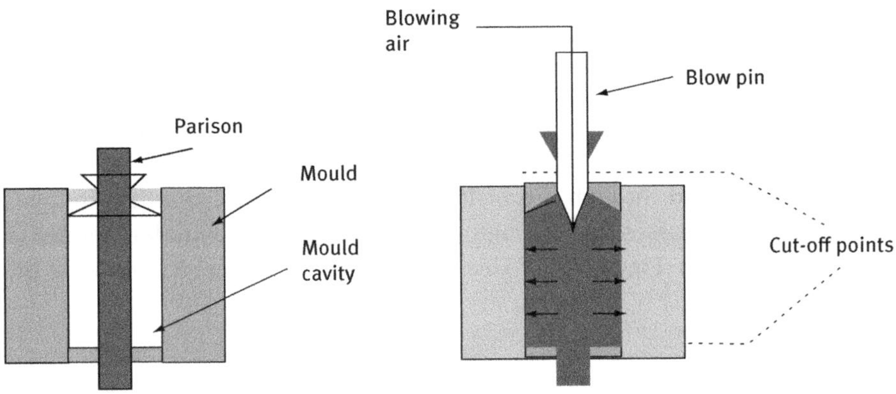

Figure 1.16: Blow moulding: stage 2.

The criteria for a blow moulding material are quite specific:
- Be of suitable viscosity.
- High melt strength.
- Be extendable when inflating, this is often given as the inflation ratio of the material.
- Be able to seal at the base of the moulding.

During inflation the material expands at a constant rate to give a constant wall thickness across the body of the moulding.

1.4.5 Film blowing

During film blowing the plastic material is fed through an extruder to an annular die opening. The cylindrical molten tube is inflated from the inside by blowing air, creating a bubble of material that can be fed and collected onto rollers. Cooling is achieved by blowing air through a cooling ring above the die. The process is shown in Figure 1.17.

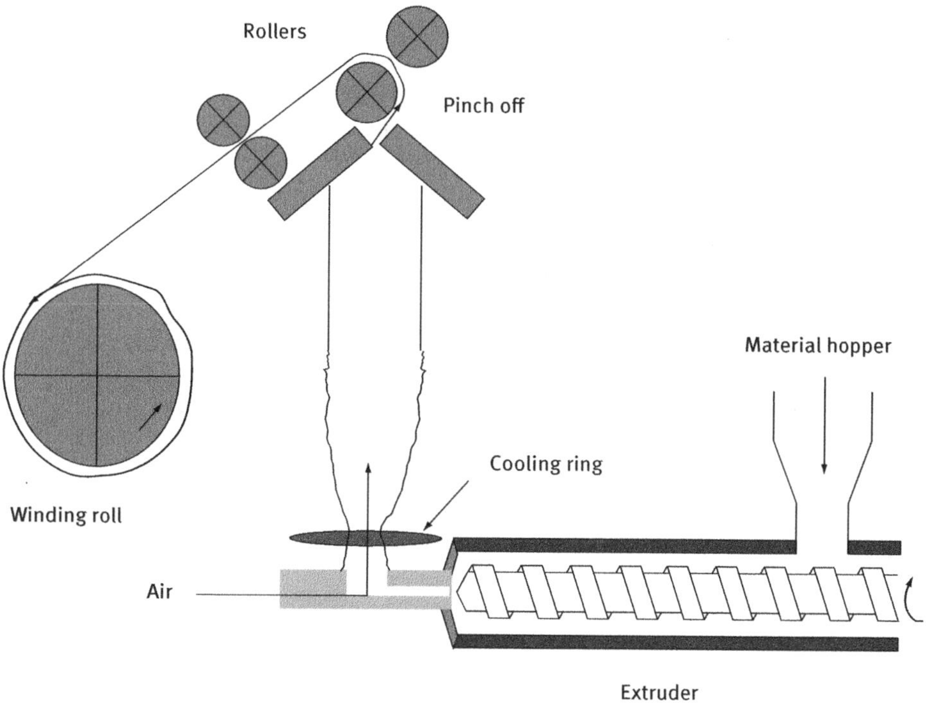

Figure 1.17: Film blowing.

A large amount of film scrap is available due to the short lifetime associated with packaging and industrial film materials such as carrier bags, dustbin liners and plastic sacks. A typical lifetime of products of this type is only two years.

100% recyclate material may be used in low-grade applications such as bin liners, other products such as carrier bags may incorporate scrap in with the virgin material for cost purposes.

Again, as with blow moulding, film blowing has specific material requirements in terms of melt strength, viscosity and inflation characteristics. Generally, film blowing is limited to polyolefin materials, the majority of usage being low-density polyethylene (LDPE), linear low-density polyethylene (LLDPE) and high-density polyethylene (HDPE).

1.4.6 Injection-compression

The injection-compression moulding process combines elements of both the injection moulding process and the compression moulding process.

Compared with injection moulding, there is a big reduction in the filling pressures required. There is also less orientation of the material, which can improve properties and reduce anisotropy. This can be of special importance in moulding thin-walled components where dimensional stability is important, and transparent mouldings where good optical properties are required.

Figure 1.18 shows the formation of an injection-compression moulded product. A preset amount of material is injected into a partly closed mould. The mould then closes to the size of the desired final component and squeezes the material by compression to complete cavity filling.

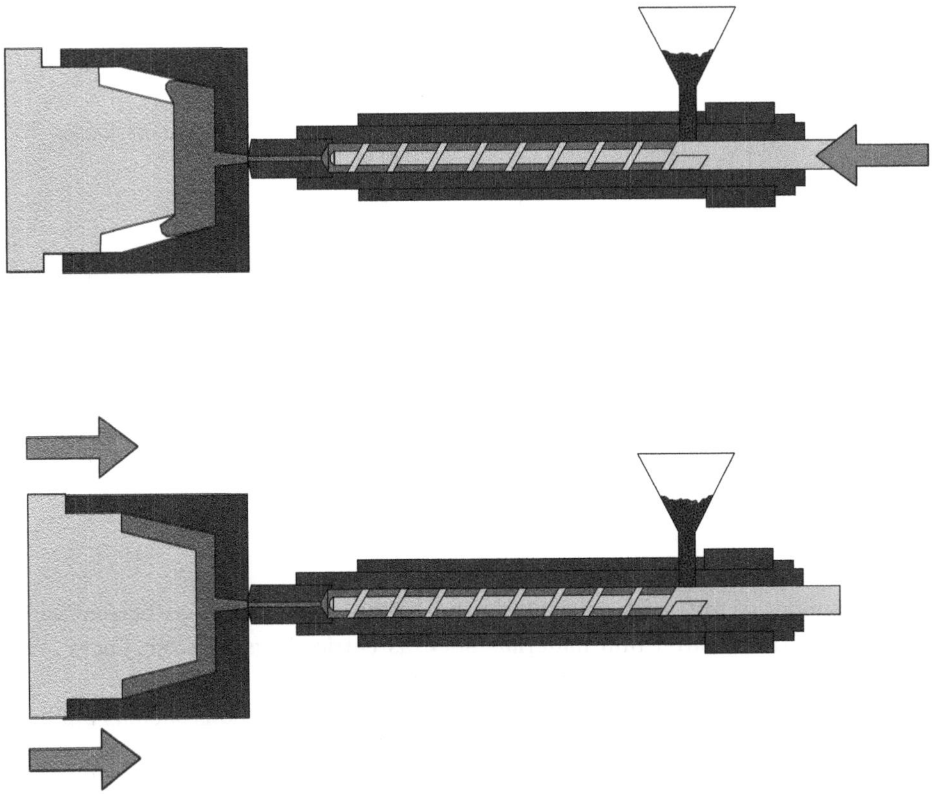

Figure 1.18: Injection-compression moulding.

It can be seen that injection-compression is a variation of the injection moulding process. Other variations on the injection moulding process such as this will be discussed in Chapter 10.

1.4.7 Additive manufacturing

The additive manufacturing (AM) process is also known as three-dimensional (3D) printing and is shown in Figure 1.19. Although it is a technology that has been around for a relatively long time, it has recently become more commercial and widely accessible as new technology and cheaper machines have reached technical maturity.

CAD* based, 3D model of part

↓

Create suitable file format (such as .STL file)

↓

Part split into sliced layers

↓

Fed into AM system for production

↓

Final part
(after scaffolding removal if needed)

**CAD is Computer Aided Design*

Figure 1.19: Additive manufacturing (AM).

The technology takes a 3D computer-generated design in a suitable format and translates it into a part build process, by slicing the component into numerous build layers. This enables AM machinery to accurately place and fuse parts bottom-up, piece-by-piece, layer-by-layer. No mould tool is required. This allows very intricate designs and actual moving parts to be built up during a single part build. The components are limited by the size of the build platform, the method of fusing materials and the types of materials that can be used in the process. Scaffolding materials can be integrated and later removed, to enable the production of parts that cannot support themselves during the build process. At present, several common plastic materials are not transferable to this process; however, it is a fast-growing market and rapid development is taking place. Hybrid processes are also developing around this technology, which will be discussed in Chapter 10. Compared with injection moulding, more intricate designs can be achieved, but the mechanical integrity is not currently as high as that produced with injection moulding. No tooling investment is required but build times per part are much slower than mass production processes and, at this point in time, it is not a high volume competitor to the traditional injection moulding market but a valuable niche market for customisable low volume production where complex designs (such as the additive manufactured part shown in Figure 1.19), which cannot be produced by more conventional production, can be built.

Other processes for thermoset and thermoplastics also exist which are not discussed here. The suggested reading at the end of this chapter provides sources of further information if required.

1.5 Conclusion

This chapter has briefly introduced the fascinating subject of injection moulding. A comparison with the various processing methods also discussed in this chapter is given in Table 1.1. For some components, there may be more than one suitable process route. The final choice of process will depend on both economics and the component design. For the remainder of this book however, only the injection moulding process will be discussed.

Table 1.1: Comparison of processes.

Process	Complexity of Parts	Example of parts	Forming action	Mould
Extrusion	Fairly simple profiles	Window profiles, tubes, granules	None	None
Injection moulding	Complex (solid)	TV housing, car handles	Injection	Closed
Blow moulding	Complex (hollow)	Bottles	Inflation	Closed
Film blowing	Very simple films	Carrier bags	Inflation	None
Compression moulding	Simple	Glass mat thermoplastic components, structural parts	Compression	Closed
Transfer moulding	Simple	Thermoset mouldings	Compression	Closed
Intrusion moulding	Simple	Low strength profiles and panels	Compression	Open
Injection-compression	Complex	Thin-walled mouldings	Compression	Closed
3D printing	Very complex	Hollow or solid, moving parts possible	Fusing	None

Suggested further reading

1. W. Michaeli in *Plastics Processing: An Introduction*, Carl Hanser Verlag, Munich, Germany, 1995.
2. V. Goodship in *The Instant Expert: Plastics, Processing and Properties*, Plastics Information Direct, Bristol, UK, 2010.

3. B. Strong in *Plastics: Materials and Processing*, Pearson Prentice Hall, New Jersey, NJ, USA, 2006.
4. D.V. Rosato, A.V. Rosato and D.P. Di Mattia in *Blow Molding Handbook*, Hanser Gardner Publications, Cincinnati, OH, USA, 2004.
5. C. Rauwendaal in *Polymer Extrusion*, 5th Edition, Carl Hanser Verlag GmbH & Company KG, Munich, Germany, 2014.
6. B.A. Davis in *Compression Molding*, Hanser Publishers, Cincinnati, OH, USA, 2003.
7. I. Gibson, D. Rosen and B. Stucker in *Additive Manufacturing Technologies*, 2nd Edition, Springer, New York, NY, USA, 2015.

2 Introduction to plastics

2.1 Introduction

Plastics were originally seen as substitute products for traditional materials such as metal and wood. However, now they have become as irreplaceable as the classic materials themselves. Plastics have managed this achievement due to their unique versatility and the ability to tailor their properties, which other materials cannot match. Our modern everyday life would be inconceivable without plastics. The use of plastics enables us to overcome problems that are insoluble with classic materials, whether it be in the fields of – to name only a few examples – electronics, light engineering, medical technology, space technology or machine and vehicle manufacture. Examples include squeezable bottles, products that are impossible to produce with glass, the use of plastic fuel tanks in cars, which enable designers to design tanks in shapes impossible to produce with metals, and biocompatible plastic-based materials are also used as internal lightweight medical implants that can be individually produced to size to fit each patient.

Plastics are made up of polymers and other materials, added to give the desired characteristics, and can be produced from petroleum (petroplastics) or natural sources (biobased). Natural polymeric materials such as rubber, shellac and gutta percha have a long history of being used as raw materials by mankind. The first thermoplastic, celluloid, was also manufactured from a natural product, cellulose, and today cellulose-based plastics, i.e., cellulose acetates (CA) are in common use. As a biobased cellulose plastic (1.18 kg of raw material wood chips can make 0.45 kg of plastic [1]), starch-based materials are also commonly found in packaging applications. Even today, there are still some cellulose-based plastics in common use. Cellulose is composed of the large molecules that are characteristic of plastics (macromolecules); however, to manufacture cellulose plastics, they still have to be 'prepared' with acids and anhydride chemicals. The first injection moulding machine was built and patented in 1872 in order to mould cellulose materials.

Today, the vast majority of plastics are manufactured artificially, i.e., the macromolecules are built up from smaller molecules (predominantly from carbon and hydrogen). Basically, plastics can also be manufactured from their basic constituents, carbon and hydrogen (coal or a biosource and water). For economic reasons, however, similar to petrol manufacturing, plastics are still predominately manufactured from products generated by the fractionated distillation of crude oil. This may look quite different in a 3rd edition if biobased materials expand into this market as predicted.

We can, therefore, divide plastics into:

https://doi.org/10.1515/9783110654813-002

1. Plastics made from natural substances, e.g., celluloid, CA, vulcanised fibre, casein plastics (galalith).
2. Artificial petroplastics, e.g., polypropylene (PP), polyethylene (PE), polystyrene (PS), polyamide (PA).
3. Artificial biobased plastics, e.g., PP.

This may look confusing but it should be remembered that plastics can be synthetically produced from both oil and natural resources. PP is not naturally occurring, but can be manufactured using sugar beet raw materials instead of crude oil. Waste petroplastics can also be used as raw materials to manufacture new synthetic biobased plastics.

However, the origin of plastics, whether obtained from naturally occurring large molecules or synthetically prepared from smaller molecules, makes no difference to the subsequent processing.

The first synthetic polymer, i.e., phenol formaldehyde, which is a thermoset phenolic resin, was developed in 1907 and called Bakelite after its inventor, Leo Baekeland. However, up until 1924, and the work of Herman Staudinger, there was no real understanding of the chemical structure of polymers. Staudinger proposed the concept of linear molecular chains and macromolecules, which once accepted by the scientific community (which actually took several more years), allowed the doors to open on the synthesis and development of new polymeric materials. This new understanding of the structure of polymers allowed the development of plastics such as polyvinyl chloride (PVC) and CA in the 1920s. The 1930s saw the introduction of PA, PS and acrylics as well as the introduction of single- and twin-screw extruders for polymer processing. New polymeric materials continued to be introduced, their development fuelled by World War II. The 1940s saw the introduction of epoxies, PE and acrylonitrile-butadiene-styrene (ABS) to name but three. The 1950s saw the birth of the PP industry as well as polyethylene terephthalate (PET) and polycarbonate (PC). American companies developed a number of engineering materials in the 1970s including polyphenylene sulfide and a number of fluoropolymers such as Du Pont's (USA) ethylene/tetrafluorethylene copolymer Tefzel and perfluoralkoxy plastics under the trade name Teflon PFA. In 1973, Dynamit Nobel (Germany) introduced their polyvinylidene fluoride Dyflon into the market and the 1980s saw the development of liquid crystalline polymers.

The development of new petropolymers has now slowed due to the expense and difficulty of synthesising new materials, however new plastics are still being developed by mixing existing materials together and are called polymer alloys and blends. An alloy has a single glass transition temperature (T_g) and generally has better properties than the individual components. A blend has more than one transition temperature and has properties between those of the original materials; an example of a commercially successful blend is ABS.

The last few decades have seen a sustained and growing interest in the environment, as a result there has been considerable development focused on the polymer industry and the production of both natural and petropolymer materials. There has also been environmental pressure and legislation on waste disposal and material recycling, and these developments are likely to continue.

2.2 Structure and typical properties of polymers

The word polymer derives from the Greek word *poli*, which means many and the word *meros*, which means parts, as polymers are made up of a number of smaller repeating units called monomers. The simplest and most commonly used monomer is ethylene, and chemically it consists of two carbon atoms (C) and four hydrogen atoms (H). It can be represented in the two ways shown in Figure 2.1. The lines in this diagram represent bonds that exist between the atoms to form a molecule.

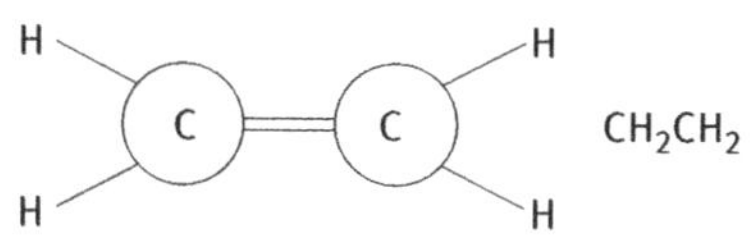

Figure 2.1: An ethylene molecule.

It is the existence of the double bond between the carbon atoms in ethylene, which allow the creation of PE. This happens when the monomers are combined by a process called polymerisation to form a chain such as the one shown in Figure 2.2. A useful polymer chain may consist of 200–2,000 of the monomers joined together; this particular type of polymerisation is called addition polymerisation.

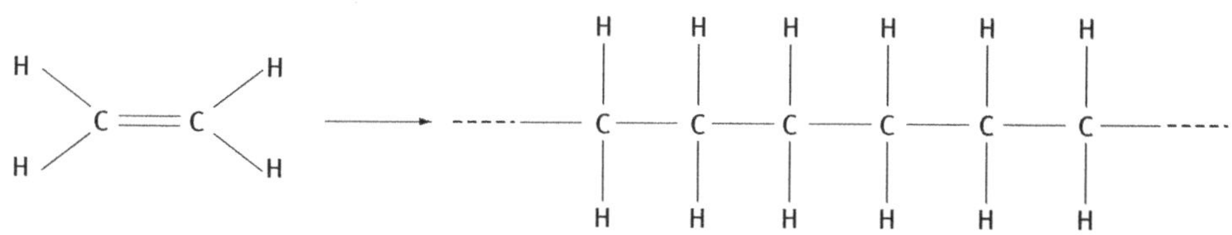

Figure 2.2: A polyethylene molecule.

It can be seen from Figure 2.2 that carbon atoms form the backbone of the polymer and many polymeric systems are made of long chains of carbon atoms such as this. However, polymers are not confined to carbon forms and injection moulding materials, such as liquid silicone rubbers, have different chemical structures as the backbone of the polymer chain. This will be illustrated later in Chapter 7. The characteristic structure of a number of common polymers is shown in Table 2.1.

Table 2.1: Characteristic structures of common polymers.

Polymer	Repeat unit(s)
PE	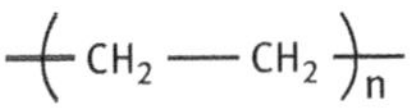
PP	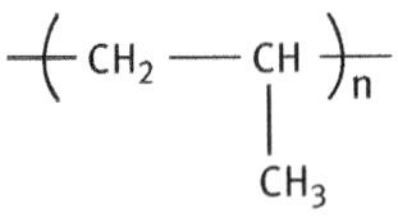
PS	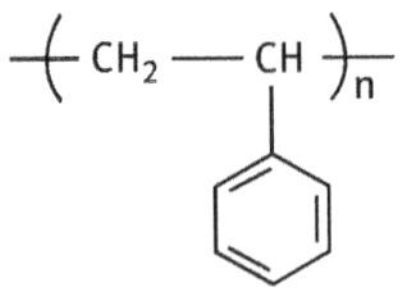
PA6	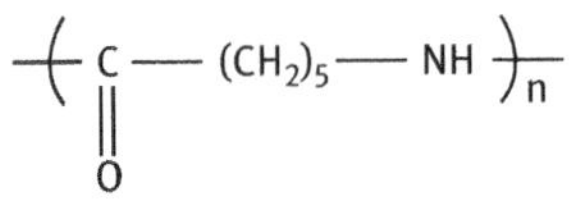
PA66	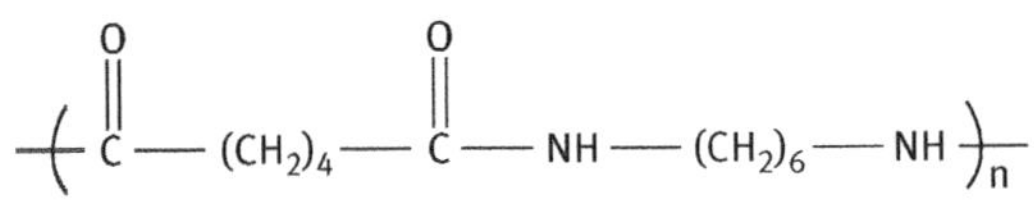
PET	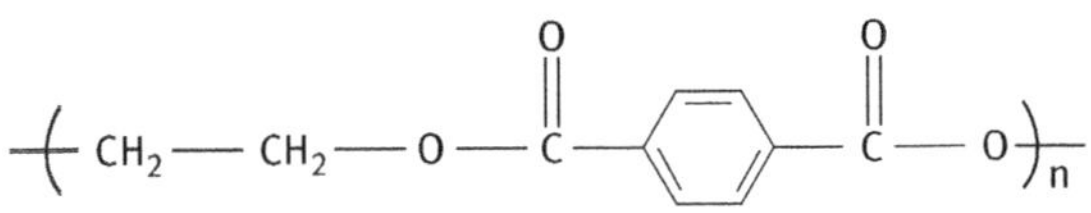
PVC	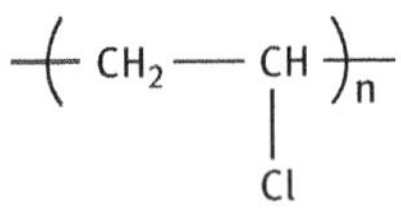
EP	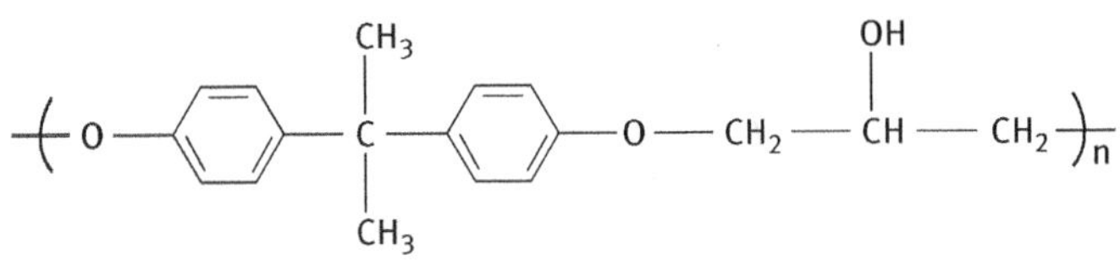

Table 2.1 (continued)

Polymer	Repeat unit(s)
Melamine formaldehyde resin	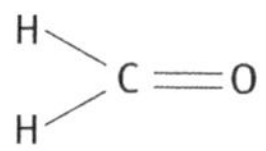 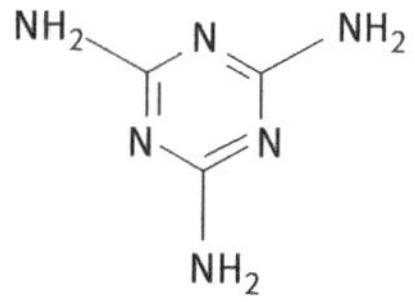

EP: Epoxy reins

2.3 Formation of macromolecules

There are three separate types of formation reactions for macromolecules. In polymerisation, the polymers merge in lines and form macromolecules without any kind of by-product being produced, which has already been illustrated in Figure 2.2. The number of monomers contained in a macromolecule is also referred to as the degree of polymerisation. If the macromolecules are made up of only one type of monomer, they are called homopolymers, one example being standard PS. If various monomers form part of the structure of a macromolecule, then they are called copolymers (mixed polymers); thus, for example, SB consists of styrene and butadiene monomers, and ABS comprises three different types of monomers: acrylonitrile, butadiene and styrene.

In polycondensation, macromolecules are formed from monomers under the splitting action of another substance, usually water. If polymerisation takes place at a temperature of over 100 °C, steam is formed. The macromolecules generated by polycondensation can be made up of one type of monomer (e.g., PA6) or two different monomers (e.g., PA6.6), depending on the shape of the monomers. Further examples of polycondensates are PC and linear polyesters such as PET.

During polyaddition, macromolecules are formed from monomers, without the generation of any cleavage products, and two different monomers are always required. The monomers undergo slight changes during this process and a few atoms change places between the different monomers; examples are polyurethane (PU) and EP.

The type of formation reaction plays no part in the subsequent injection moulding of thermoplastics. The molecules are already complete before injection moulding begins. They are merely melted, and then solidify in the mould to form the component.

During the processing of thermosets however, the type of formation reaction can affect the process/product if crosslinkage is brought about by polycondensation. Indeed, the locking pressure of the mould must then be great enough to prevent the mould halves being forced apart by steam production, thus making the component porous or blistered. Particular attention should be paid to this during the compression moulding or transfer moulding processing of thermosets. During injection moulding, the mould clamp force selected for the mould must be high enough to avoid any danger of it being forced open by steam, because of the high injection pressures required.

2.4 Molecular weight

A polymeric material may contain many macromolecular chains, all of various lengths or repeat units. The molecular weight distribution (MWD) is used to describe this variation and the average size of these chains determines the molecular weight (MW) of the polymer. As well as the backbone, polymers may also have side chains of varying lengths. This branching from the main chain affects the properties of the polymer, as branching restricts the ability of the polymer chains to pack together; therefore, branching affects the density of a polymer. For example with high-density polyethylene (HDPE) there is very little branching, hence the term high-density, in contrast, low-density polyethylene (LDPE) has many branches. An illustration of this effect on both density and melting temperature (T_m) can be seen in Table 2.2.

Table 2.2: Influence of branching on properties of PE.

PE	T_m (°C)	Density (g/cm^3)	Tensile strength (MPa)	Number/type of branches
LDPE	110–120	0.91–0.93	17–26	Long branches
LLDPE	122–124	0.92	13–27	10–35 short branches (per 1,000 carbon atoms)
HDPE	130–135	0.94–0.97	21–38	4–10 short branches (per 1,000 carbon atoms)

LLDPE: Linear low-density polyethylene

It can be seen that the properties of the polymer are strongly linked to MW and MWD and it is necessary to match the MW, material properties and flow characteristics in order to shape the material during processing and give the desired final product.

2.5 Plastics

Plastics are made up of polymers and other materials that are added to increase the functionality of the material. The actual polymer content within a plastic can vary widely from less than 20% to nearly 100%, with plastics consisting of virtually all polymers termed 'prime grades'. The level and type of the other additives used depends on the application for which the plastic is intended. There is a vast range of materials available on the market, even if just considering one plastic such as PP this would still be literally thousands of different custom grades. Multiply this by all the potential polymers available and the sheer and amazing versatility of plastic materials is very apparent.

Ignoring the source of the polymer (petro or bio), plastics can be subdivided into three main categories, thermoplastics, thermosets and elastomers. This distinction is based on both the molecular structure and the processing routes that can be applied. These three classes of materials will now be introduced.

2.5.1 Thermoplastic

These materials melt and flow when heated and solidify as they cool. On subsequent reheating they regain the ability to flow meaning they can be reprocessed and hence recycled *via* remelting them. Thermoplastics are used to make consumer items such as drinks containers, carrier bags and buckets. The most common thermoplastic materials and their applications are shown in Table 2.3.

Table 2.3: Common thermoplastics and their applications.

Thermoplastic polymer	Applications
HDPE	Packaging, pipes, tanks, bottles, crates
LDPE	Packaging, grocery bags, toys, lids
PP	Caps, yogurt pots, suitcases, tubes, buckets, rugs, battery casings
PS	Mass-produced transparent articles, yogurt pots, fast-food foamed packaging
PA	Bearings, gears, bolts, skate wheels, fishing lines
PET	Transparent carbonated drinks bottles, woven fibre
PVC	Food packaging, shoes, flooring

When thermoplastics solidify they can take one of two molecular structures: an amorphous structure or a semi-crystalline structure which are both illustrated in Figure 2.3.

When semi-crystalline materials are cooled the molecular structure tends to become highly ordered and crystals are formed. The size of these crystalline regions varies according to both the structure of the chains themselves and the cooling rate. These materials display sharp melt points unlike amorphous materials which soften. Semi-crystalline materials also tend to shrink more due to this molecular

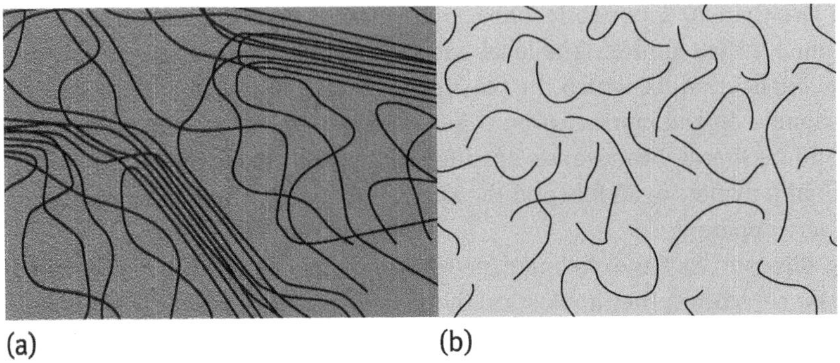

Figure 2.3: Semi-crystalline structure (a) and amorphous structure (b).

rearrangement, with some materials shrinking by as much as 20%. This shrinkage will be more in the direction of flow due to the molecular realignment caused by the process of injection moulding.

2.5.2 Thermosets

Thermoset injection moulding compounds change their structure when injected. Before injection moulding, they still consist of thread-shaped molecules similar to thermoplastics. However, during a process termed 'curing' the molecules crosslink forming a highly dense network of bonds, which makes the material stiff and brittle meaning that the thermoset moulded parts can then no longer be melted. Thermoset materials decompose before they can melt, therefore, they cannot be reprocessed in the same way as thermoplastics. The differences in the arrangement of molecules between thermoplastics and thermosets can be seen in Figure 2.4. Thermosets are often used where their strength and durability can be utilised, some common thermosets are shown in Table 2.4.

Different properties arise from the divergent forms of the molecules. When thermoplastic compounds are injection moulded, the structure of their molecules is not changed; melting only increases the mobility of the molecules. In contrast, once

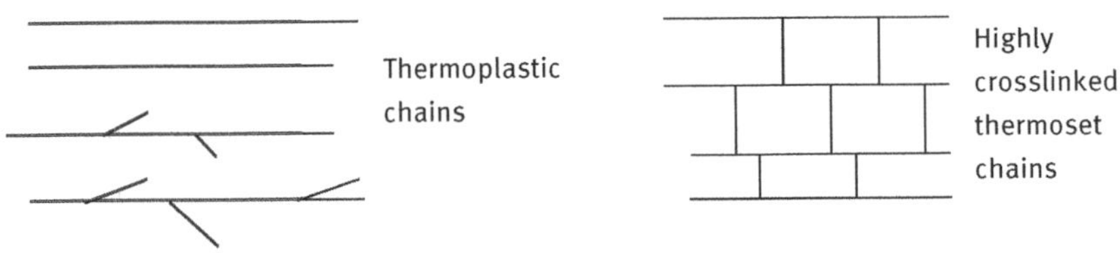

Figure 2.4: Arrangements of thermoplastic and thermoset molecular chains.

Table 2.4: Common thermoset materials.

Thermoset polymer	Application
Epoxy	Adhesives, electrical insulation
Melamine	Heat-resistant laminate surfaces, i.e., kitchen worktops
Phenolics	Heat-resistant handles for pans, irons, toasters
PU	Rigid or flexible foams for upholstery and insulation
Unsaturated polyesters	Partitions, toaster sides, satellite dishes, structural composites

thermoset plastics have crosslinked they can no longer be melted. Thermoset materials can thus be deformed only elastically and, in contrast, cannot be deformed plastically. The rigidity of thermoset materials depends on the width between the spaces of the crosslinking molecular networks. A material with a wide gap between the crosslinks can be deformed elastically to a larger degree; such thermosets are also known as elastomers.

2.5.3 Elastomers

According to the American Society for Testing and Materials (ASTM), a polymeric elastomer is defined as 'a material that, at room temperature, can be stretched repeatedly to at least twice its original length, and upon immediate release of the stretch, will return with force to its approximate original length'. Which means, to put it in layman's terms, it's rubbery!

2.5.3.1 Thermoplastic elastomers

For a long time elastomer materials were limited to thermoset-type materials with permanent crosslinks, such as polyisoprene (natural rubber). However, one major drawback with these materials was that they could not be recycled, which led to the development of thermoplastic elastomers (TPE). One common example is known as styrene-butadiene-styrene (SBS). Since styrene and butadiene do not mix, the styrene breaks up when SBS is heated allowing the material to be remelted. TPE offer a considerable advantage over thermoset-type elastomers such as:
- Lower part costs due to a lower material density.
- Faster cycle times.
- Recyclable scrap and parts.
- Can be coloured.

The market for TPE has also benefited from advanced processing techniques such as over-moulding (Chapter 10). The ability to use TPE either alone or in combination with other materials has enabled them to be used in applications such as:

- Gaskets, seals and stoppers.
- Shoe soles/heels, wrist straps.
- Handles, grips, knobs.

The properties of TPE that have made them so commercially successful are their lower modulus and flexibility. The ability to recover from stress and return to their original shape make them suitable for applications such as sealing rings and so on. Of course, thermosetting rubber materials have been available for a long time and it is these materials that TPE can replace in some applications.

Whilst newer materials have emerged over the last few years, primary TPE materials can be categorised into generic classes: block copolymers or thermoplastic/elastomeric blends and alloys. The latter are essentially two-phase systems, a hard thermoplastic phase is coupled mechanically or chemically with a soft elastomer phase, resulting in a TPE that has the combined properties of two phases.

TPE usage has increased significantly in recent years and is expected to rise annually from 5.6 million metric tonnes in 2015 to predicted values of 6–8% market growth annually. Some examples of common elastomers are shown in Table 2.5. As well as adhesion considerations, the suitability of a TPE will also be dependent on properties such as its hardness and compression ratio. Hardness can be defined as the resistance of a material to indentation and is usually measured on a durometer

Table 2.5: Examples of TPE and substrates.

Type	Elastomer description	Subgroup
TPE-O (or TP-O)	Thermoplastic olefinic elastomer Polyolefin blends of PP and EPDM rubber	PP/EPDM PP/EPDM crosslinked
TPE-V (or TP-V)	Thermoplastic vulcanisate Polyolefin alloys Vulcanised version of TPE-O and therefore with increased performance	Various
TPE-S	Styrenic block copolymers	SBS SEBS SEBS/PP
TPE-A	Thermoplastic amide elastomer Thermoplastic polyether block amides	PA PA12-based PA6-based
TPE-E	Thermoplastic polyester elastomer	Polyester ester
TPE-U (or TP-U)	Thermoplastic polyurethane elastomer	Polyester urethane Polyester ester urethane Polyether urethane
MPR	Melt processable rubber	–

EPDM: Ethylene propylene diene monomer
SEBS: Styrene-ethylene/butylene-styrene terblock copolymer

using a Shore hardness scale. TPE tend to be rated on a Shore A scale, the softest materials ranging from around 3 Shore A and the hardest to 95 Shore A. Out of interest, thermoplastic materials are measured using a different, Shore D scale, which gives an indication of the difference in properties. TPE material grades are available with a range of hardness levels, however it must be taken into consideration that hardness is a function of the thickness of the material and the substrate beneath. In a multi-shot application, the thinner the layer of elastomer on the substrate, the harder it will feel due to the effect of the harder substrate beneath it.

The compression set (CS) is often specified for sealing-type applications and is a measure of the deformation after compression for a specified deformation, time and temperature. The usual test method is ASTM D395 or International Organization for Standardization, ISO 815. Therefore, a CS is between 0 and 100% relative to an applied load, where 100% compression would represent a material that did not recover (non-elastic), and 0% a material that behaved completely elastically at that load.

2.5.4 The formulation of plastics

It was stated earlier in the chapter that plastics are mixtures of polymers and other materials. There are many additives commercially available that can be mixed with polymers, for example, glass- or carbon fibre reinforcement gives increased strength and flame retardants can be added for flame resistance. Plastics can be coloured with pigments for aesthetic or technical purposes, or they can be made more heat and light resistant for outdoor applications by the use of stabiliser additives. Those are just a few examples of the numerous possibilities. In addition, the levels of each of these additives can also be varied. It is easy to see how so many grades of plastic have come to exist. It is also hopefully now apparent how they have been able to compete so successfully with other materials as diverse as glass, metal and wood when all these possibilities are considered. A list of some of the more common plastic additives is given in Table 2.6.

Plastics which incorporate reinforcing materials such as glass fibre or clay are called polymer composites. Composites can be made from both thermoplastic and thermoset polymers, with materials such as glass and carbon fibres which increase the tensile stiffness and strength of the resultant materials enabling their use in structural composite applications. Polymer composites are widely used for a variety of applications where their high strength to weight ratio can be utilised. The automotive industry, in particular, has taken advantage of these properties to make weight savings over metal components to improve fuel efficiency. Nanomaterials can also be used as additives and newer additive materials such as graphene, carbon nanotubes and active agents, such as enzymes to aid the degradation of plastic, are all on or near the market.

Table 2.6: Common additives for plastics.

Additive	Purpose
Reinforcement, e.g., glass fibre	Increased strength and stiffness
Extenders, e.g., calcium carbonate	Cost reduction: much cheaper than polymers
Conductive fillers, e.g., aluminium powder, carbon black	Improved thermal and electrical conductivity
Flame retardant	Increased fire resistance
Light stabilisers	Increased resistance to degradation from daylight
Heat stabilisers	Increased resistance to degradation from elevated temperature exposure
Pigments	Give colour, increased aesthetic properties
Plasticisers	Improved flow properties and increased flexibility
Coupling agents	Improves interface bonding between the polymer and a reinforcing agent
Foaming agents	Lightness and stiffness
Mould release agents	Processing aids
Antistatic additives	Prevent dust build-up on consumer items

2.5.5 The binding structure of plastics

What is the relevance of the binding structure of polymers?

Well, the characteristic values of most immediate interest to the injection moulder are those which are needed in order to melt the injection moulding compound and inject it into the mould. It is important to know the temperature at which the individual plastic becomes molten, as different plastics react differently to heating and is dependent upon the holding power of the chains of molecules. We also need to know the decomposition temperature, where the molecule chains break apart into smaller chains or into the initial materials. This process is described as gasification and indicates severe damage to the material.

Generally, the forces which keep the molecules in the material compound are less powerful than the chemical bonds present between atoms. Therefore, materials with chemical bonds such as crosslinked thermosets are much harder to break down than thermoplastic materials. Polymers with a polar component (e.g., CN, NH, CO), generate greater intermolecular forces than those without; however, polar groups cause higher dielectric losses due to their dipole character.

2.6 The effects of processing on thermoplastics

In order to understand what happens to the polymers during injection moulding, the effects of processing on thermoplastic materials will now be considered. Pro-

cessing in its simplest form is the act of melting, forming and solidifying the melt. During this process three classes of properties need to be considered:
- Deformation processes, which enable the product to be formed.
- Heat and heat transfer, which enable the polymer to melt, flow and then solidify.
- Changes which occur to the structure and properties as a result of processing.

In order to study deformation processes it is necessary to consider the rheology of the material

2.6.1 Rheology

Rheology is the study of deformation and flow, consider the following flows:
1. Squeezing toothpaste through a tube.
2. Tipping honey from a spoon.
3. Pouring water into a glass.

The toothpaste, honey and water all have different viscosities. The toothpaste is the most viscous, and water the least. Viscosity represents the resistance to flow. In viscous flow, a material continues to deform as long as a stress is applied. To put it more simply, in the case of the toothpaste tube, we can squeeze out the toothpaste for as long as we squeeze the tube.

Consider the toothpaste tube again, when it is squeezed and then released it does not return to the original shape. A rubber ball however, does return to its original shape when released and likewise polymer melts will try to return to their original shape when stress is removed. So polymers combine both viscous and elastic properties and demonstrate what is called a viscoelastic response to stress.

The interaction between viscosity and elasticity frequently determines the success of any processing operation. Processing conditions must take into account not only how the polymers flow in their molten state but also how they change as the temperature goes up and down and the polymers melt and solidify. During injection moulding, the switch from mould packing to holding pressure causes the material viscosity to drastically change. It has been vividly described 'as changing from flowing like honey to flowing like modelling clay'. This explains why consistently switching at the same point can make the difference between producing consistent parts as opposed to producing rejects.

Polymer melts have viscosities in the range 2–300 Pa.S (for comparison the viscosity of water is 10^{-1} Pa.S). Two very common materials in commercial use are LDPE and PA.

As LDPE is heated, it turns from a solid, to a viscous gummy liquid and then to a mobile fluid as temperature is increased. PA on the other hand turns quite suddenly from a solid to a low viscosity (watery) fluid. It is important to understand that polymers may react differently to the heat and stress applied to them, in order to find the most suitable processing conditions for moulding.

2.6.2 Heat and heat transfer

A thermoplastic material cannot flow when it is in its solid state. To enable it to flow it needs to be heated to either:

- Above its T_m, if it is a crystalline material; or
- To its T_g, if it is an amorphous thermoplastic.

Some common T_m are shown in Table 2.7. T_m is also sometimes referred to as a flow temperature or a crystalline T_m for amorphous and crystalline polymers, respectively. Amorphous materials do not have a clearly defined T_m like crystalline materials, but tend to soften and flow instead. At T_m, which may be a fairly sharp point or a range, the polymer behaves as a viscous liquid and flows as the molecules are free to slide over one another. Highly crosslinked polymers do not have a melting temperature as the chemical bonds between the chains form a rigid structure that do not allow flow movement.

Table 2.7: T_m of common polymers.

Structural unit	Polymer	T_m (°C)
$-CH_2-CH_{2-}$	PE	135
$-CH_2-CH-CH_3$	PP	170
$-HOOC-C_6H_4-COOH-$	PET	245
$-CO-(CH_{2-})_5NH-$	PA6	233

From these examples, a wide range of Tm can be clearly seen. Once above the Tm, the viscosity will decrease as the temperature increases however, the rate of this change varies according to the particular type of material. Eventually the polymers will reach a point where they become thermally unstable and start to degrade. Generally, this shows itself as a discolouration of the melt which turns yellow or brown. Upon exposure to excessive heat, polymers burn or decompose.

2.6.3 Physical and chemical change

As well as the effects of heat on the polymer, the effects of shear also need to be considered. Shear is a type of force that involves exerting stress across the surface of the polymer, while in effect, the base of the polymer is untouched; this action is shown in Figure 2.5. The action of shear, like temperature, causes a decrease in viscosity, which is due to the shearing action causing mechanical damage and breakage of the polymer chains. At very high levels of shear this leads to degradation.

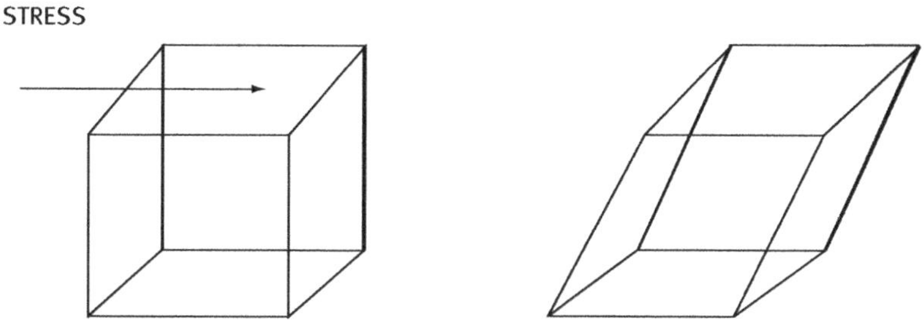

Figure 2.5: Shearing flow.

With this is mind, good thermal stability is a requirement for most polymer processing operations, as the combined action of heat and shear can produce degradation, the effect of which is chain breakage. The length of the chain is related to the MW, and the MW to the properties; therefore any change in chain length will effect:
- MW and viscosity.
- Tensile and impact properties.

Combined heat and shear can also result in:
- Change in colour (yellow or brown discolouration).
- Reduced fibre length in glass-filled materials.

Consideration must also be given to the residence time of the material in the processing machine, as a long residence time may cause thermal deterioration, especially in heat-sensitive materials such as PVC.

2.6.4 Fountain flow

Before injection the material in the injection unit is plasticated. Once the material is molten and homogeneous it is ready for injection. In order for injection to occur, the pressure in the nozzle must be high enough to overcome the flow resistance of the melt. The flow is then pressure driven as the material moves from a region of high pressure to a region of low pressure. The larger the pressure and pressure gradient, the faster the material will flow. In order to mould with an increased flow length, an increased entrance pressure is required. The flow length of a material under any particular processing conditions for a specific wall thickness will depend on the thermal properties and shear properties of the material.

During the process of injection moulding, polymers come under both shear stress and to a lesser extent elongational stresses. Elongational flow occurs when the melt encounters abrupt dimensional changes such as in the gate region, however, it is shear flow that dominates the mould-filling process as injection moulding tools fill *via* a fountain flow effect, as shown in Figure 2.6. The first material to enter the mould is under very little pressure as it flows into the channel, but as it proceeds towards the extremities of the mould, it is progressively cooled. The material that follows is hotter and begins to pressurise material already there. As the mould fills, a skin of solid plastic forms on the wall, which causes a reduction in the effective gap through which later molten material passes and so increases the shear rate. The extra shear stress causes orientation of the molecular structure and the melt is generally too viscous to allow relaxation of all the stress. As a result, it is often found that shrinkage in the direction of flow during filling is larger than that in the circumferential direction, which is due to orientation effects. Material will be orientated in the direction of flow and will shrink more in the direction of flow than perpendicular to it. It is not always possible to prevent it from occurring but it can be alleviated by differential cooling of the mould tool.

The parameters of mould filling, packing and cooling have been shown to be more important to final part quality than the plastication stage. All of these processes will be described in greater detail in Chapter 8.

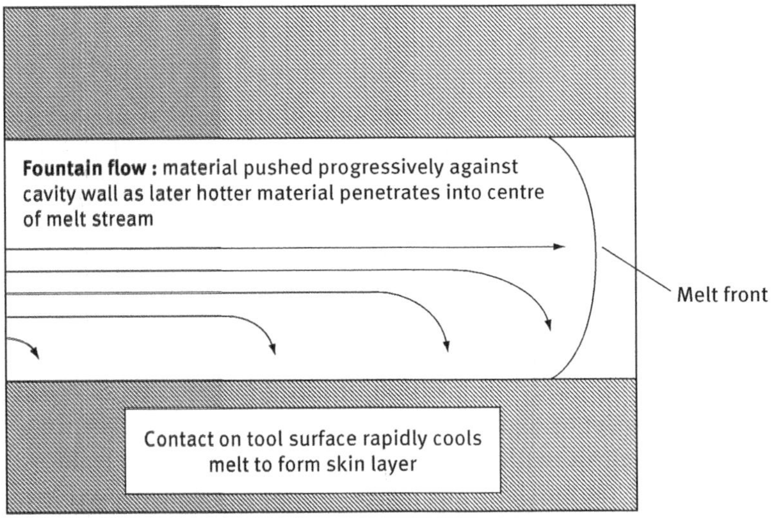

Figure 2.6: Schematic of fountain flow.

2.7 Conclusion

This chapter has presented a general introduction to polymers and plastics and some of their important properties that relate to injection moulding. There are more detailed chapters on thermoplastics (Chapter 6) and thermosets (Chapter 7) later in this book.

Reference

1. *Tenite™ Cellulosic Plastics: From Trees to Plastic*, Eastman Chemical Company, Kingsport, TN, USA, 2014. http://www.eastman.com/literature_center/p/ppc100.pdf.

3 Injection moulding machinery

In Chapter 1, the basic parts of an injection moulding machine (IJM) were identified as:

- Injection unit.
- Machine base with hydraulic and or/electrics.
- Control unit.
- Clamping unit with mould.

This chapter begins with a breakdown of the purpose and components of an injection unit. A glossary of common injection moulding machinery is now given.

Clamping unit	The part of the machine in which the mould is mounted. It provides both force and movement.
Daylight	The maximum distance between the stationary and moving platens.
Ejector	Ejects the moulded parts from the tool. The mechanism is activated through the clamping unit. The ejection force may be hydraulic, pneumatic or mechanical.
Fixed platen (stationary platen)	The inner face of the clamping unit that does not move during mould closing. It contains mould mounting holes.
(Full) hydraulic clamp	A clamping unit triggered by a hydraulic cylinder which is directly connected to the moving platen. The mould is opened and closed by the application of direct hydraulic fluid pressure. This also provides the clamping force required to keep the mould closed during injection.
Injection cylinder	Part of the injection unit which includes the screw, nozzle, hopper and heaters.
Injection screw	Transports, mixes, plasticises and injects the material from the hopper into the machine nozzle.
Injection unit	The part of the machine which feeds, melts and injects the material into the mould.
Mould (tool)	Contains the cavity which material is injected into and consists of two halves, a stationary and moving half which attach onto the platens. Halves are linked *via* the tie-bars.

https://doi.org/10.1515/9783110654813-003

(continued)

Moving platen	The inner face of the clamping unit that moves during mould closing. The moving half of the mould is bolted onto this platen, which also contains the ejection mechanism and mould mounting holes.
Plunger unit	Injects and plasticises material by heating it in a chamber between the mould and a plunger. Heating is achieved *via* conduction. The plunger forces material into the chamber which in turn forces the melt present in the chamber into the mould.
Tie-bars (tie-rods, tie-beams)	Bars which link and align the stationary and moving platens together.
Toggle clamp	A clamping mechanism with a toggle directly connected to the moving platen.
Reciprocating screw	A screw which both plasticises and injects material.

3.1 Injection units

The first aim of the plastication stage is to produce a homogeneous melt for the next stage, i.e., where the material enters the mould. A second important function of the injection unit is the actual injection into the mould. Here, it is important that injection speeds are reproducible as slight changes can cause variations in the end product. An injection unit is shown in Figure 3.1.

Basically, two different injection units are available, hydraulic driven machines as shown in Figure 3.1a, or an electrically driven unit (shown removed from the assembly) as shown in Figure 3.1b. There is little difference, technically speaking, between one type and the other, the main difference is the energy consumption, as the injection unit consumes the highest amount of energy during the injection moulding process. Energy savings may be made by the use of electrical machines if electric costs or the reduction of carbon dioxide (CO_2) emissions are an issue.

3.2 Piston (plunger) injection unit

The design of this unit was originally based on a method used to mould rubber. Material is metered by a dosage device and transported through the heated plasticising cylinder until it is in front of the plunger at the correct temperature. The material residence time in the cylinder is very long, making this method unsuitable for heat-sensitive materials such as rigid polyvinyl chloride (PVC) and thermosets. This type of machine is not widely used as it was replaced by the reciprocating screw-piston injection unit.

Figure 3.1: A hydraulically powered injection unit assembly (a) and an electric heating unit (b).

3.3 Reciprocating screw injection unit

This is the most common type of unit and will be the basis for further discussion inthis section. Thermoplastics as well as thermosets and classical elastomers can be processed using screw injection units *via* the process cycle described in Chapter 1.

In the screw injection unit, the material is simultaneously plasticised and dosed as previously described in Section 1.4.2. The design of a plasticising screw has several advantages over a piston-type injection unit mainly as a result of the ability to produce a homogeneous melt *via* mixing. The flow of the material is also improved as shear from the screw lowers the viscosity of the material. The long residence times of piston-type machines are eliminated allowing heat-sensitive materials such as PVC to be processed. The screw is also easier to purge and less prone to degradation or material hang-ups. Important parameters for these screws are:

- The diameter of the screw and its ratio to the length (L/D ratio) (see Section 3.5).
- Shot capacity.

- Plasticising rate.
- Plasticising capacity (the amount of material that can be melted at any one time).

3.3.1 Shot capacity

The shot capacity is the full amount of material, as a weight or volume, injected from the screw during moulding. This is usually given in machine specifications as a shot capacity for polystyrene (PS), and will vary with material. The shot size is the amount of material required to fully fill a moulding tool and therefore varies as it is dependent on the mould tool volume.

3.3.2 Plasticising capacity

This is the maximum rate at which the injection unit can deliver polymer melt. During extrusion, this is a continuous process; however, it should be remembered that injection is an intermittent process, therefore the plasticising rate will be lower. To calculate the melting rate consideration should be given to the overall cycle.

The effectiveness of plastication depends on the shot size, cylinder capacity, screw design, screw speed and heater band power – it will also vary from material to material.

3.3.3 The feeding hopper

Material is placed into the hopper prior to plastication and the location of the hopper is shown in Figure 3.1a. It must be designed to avoid material bridging in the throat so as to let gravity feed the material. (Alternative additional stuffer units are available for materials that do not feed by gravity.) Areas of material hold up must be avoided to allow a consistent feed and dose. Additives, especially when they are different weights to the polymer, may tend to accumulate and be fed inconsistently, which should be avoided where possible as it can lead to variations in melt quality. The hopper may also contain magnets to collect metal contamination in order to prevent it from entering the feed system and causing damage. It may also contain grids to prevent large particulates from entering and blocking the feeding system, which is especially important if using recyclate materials. Keeping the feed system cool is also important, in order to prevent material melting in the throat of the feeding system and sticking to the sides of the throat, which in extreme cases, could block the machine completely.

3.3.4 The injection cylinder

Once the material has passed through the hopper, it enters the injection barrel. The barrel consists of a number of separately controlled heating zones as can be seen in Figure 3.2. Heat is generated *via* conduction from the cylinder and by the shearing action of the screw on the material feedstock. Polymers are not particularly good conductors of heat; therefore, the polymer thickness in any section of the screw tends to be kept low. The amount of shear is material dependent, mainly viscosity related and controlled by the machine screw back and back pressure.

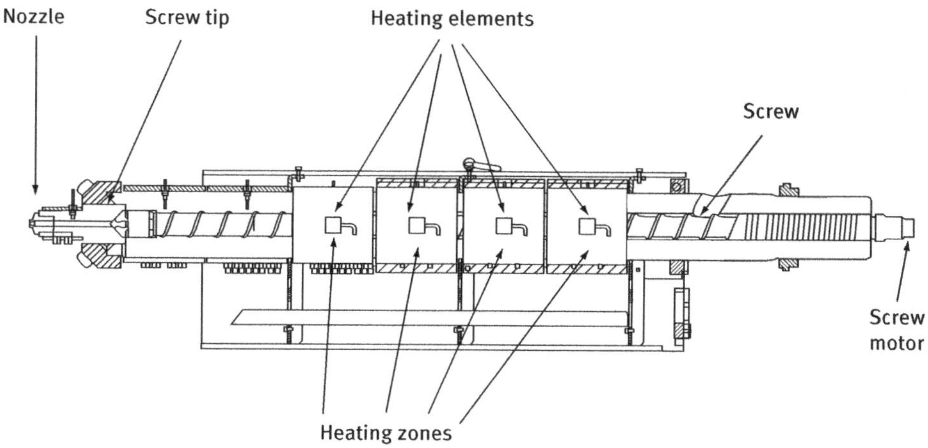

Figure 3.2: Schematic of an injection unit.

3.4 Selection of the injection unit

To select the correct configuration of injection unit for a particular material or material range, consideration must be given to the following factors:
- The selection of the correct nozzle type:
  - Flat?
  - Radius?
  - Shut-off nozzle?

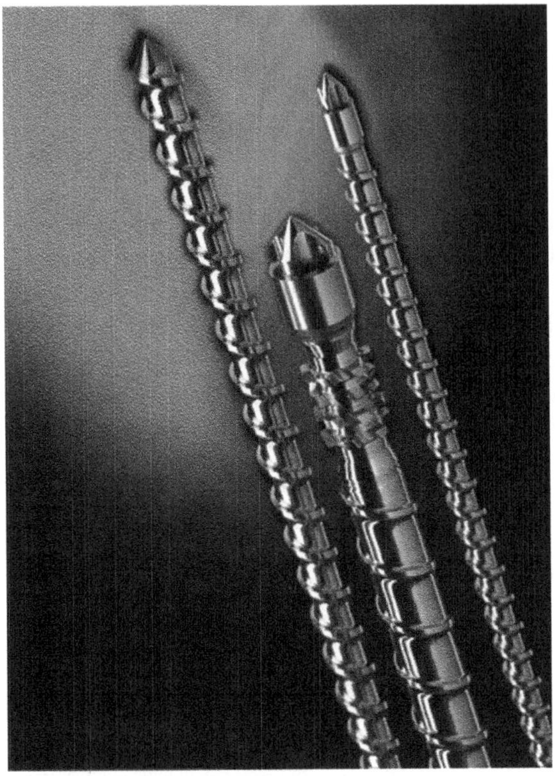

Figure 3.3: Various injection screws.

- Screw and cylinder outfitting must be adapted to suit the raw material being processed:
  - The geometry of the screw must be correct (Figure 3.3 shows some of the variety which are available).
  - The screw is suitably corrosion resistant [e.g., nitrided, Arbid (boronising heat treated), bimetallic].
- The dosage volume should be approximately 20–80% of the total shot capacity and the residence time should not be excessive for sensitive materials.
- Dosage capacity and melt capacity must be sufficient for processing requirements.
- Whether a mixing cylinder (screw) is required for use with colour additives.
- Whether a hydraulic accumulator is necessary for moulding with long lines of flow.

Some typical designs for injection cylinders are shown in Figure 3.4.

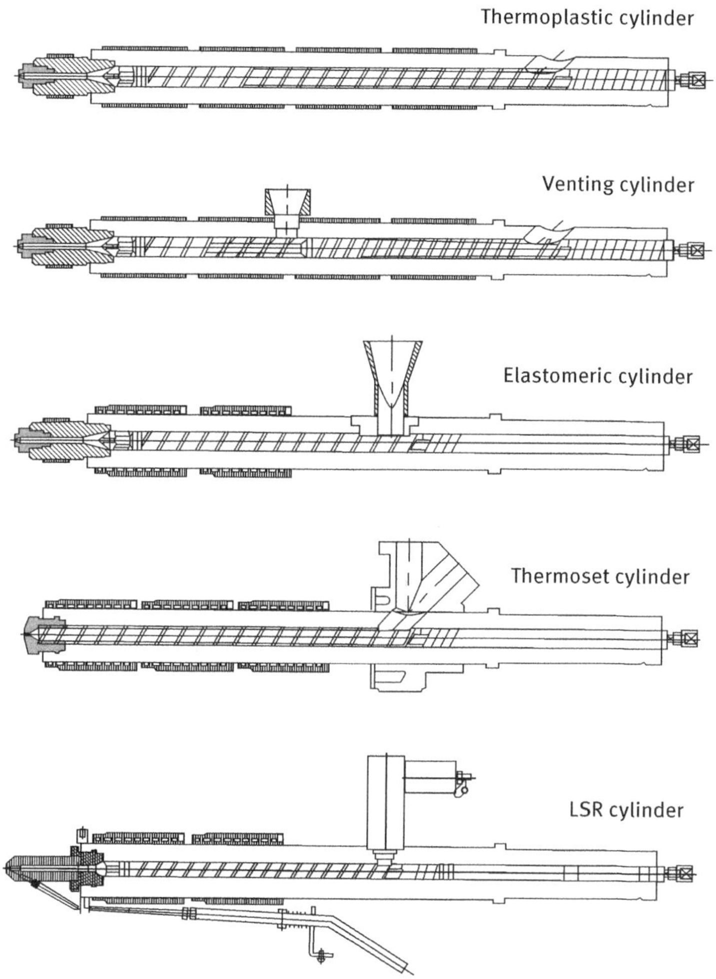

Figure 3.4: Various injection cylinders.
(LSR: liquid silicone rubber)

The design of the screw is very important for plastication. Some typical screw configurations will be detailed in the next section.

3.5 Length/diameter ratio

Perhaps the most important specification for the screw is the L/D ratio, which is the ratio of the length of the screw (the flighted length) to its diameter. For thermoplastics this ratio will be a minimum of about 20:1; whereas for thermosets, elastomers and LSR the L/D is approximately 14:1, and for extended plasticising screws the L/D may be 24:1. The higher L/D ratio is usually used for thermoplastics with colour additives, especially with polypropylene (PP) and polyethylene (PE) and enables better mixing of the colourant. On fast cycle machines with increased capacity, a higher L/D ratio may also be beneficial.

The correct ratio has also been found to improve melt performance with compact disc manufacturing machines. More information on mixing can be found in the next section.

From Figure 3.5 it can be seen that the design of the screw along its length is not constant but varies. Generally, screws are designed with three distinct regions: 1) a feed section, 2) a melting transition region (2-compression region) and 3) a metering section. The size of these regions will vary depending on the characteristics of the material it was designed for. A passive screw has low shear and compression and acts only to melt and transfer the material from one end to the other; it is sometimes termed a transfer screw. A standard screw has regions of shear mixing and compression.

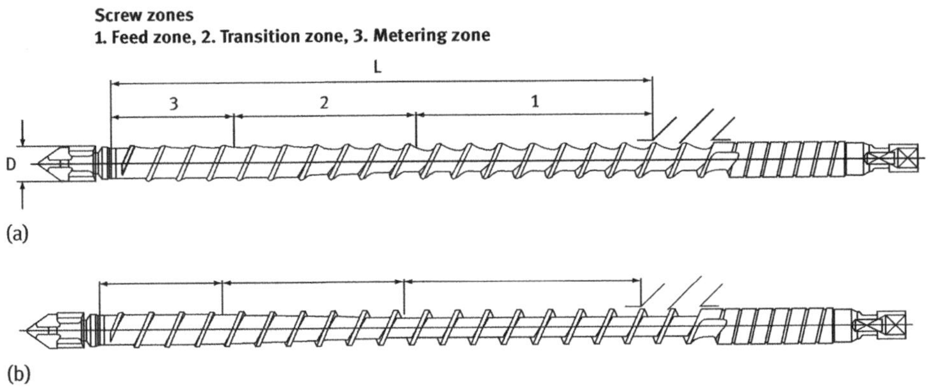

Figure 3.5: Sections of an injection screw: (a) passive screw and (b) standard screw.

The compression ratio can be defined as the ratio of the flight depth in the feed section to that in the metering section. As an example, a screw for a polyamide (PA) material may have a compression ratio of 3:1 on a 20 L/D screw with a 30 mm diameter.

A standard compression ratio is roughly 2:1 for thermoplastics, and for sensitive thermoplastics such as PVC or containing metal/ceramic-powder, this may drop to 1.6:1. Compressionless screws with a compression ratio of 1 are used for processing materials such as elastomers and LSR. The design of screws for various materials is illustrated in Figure 3.6.

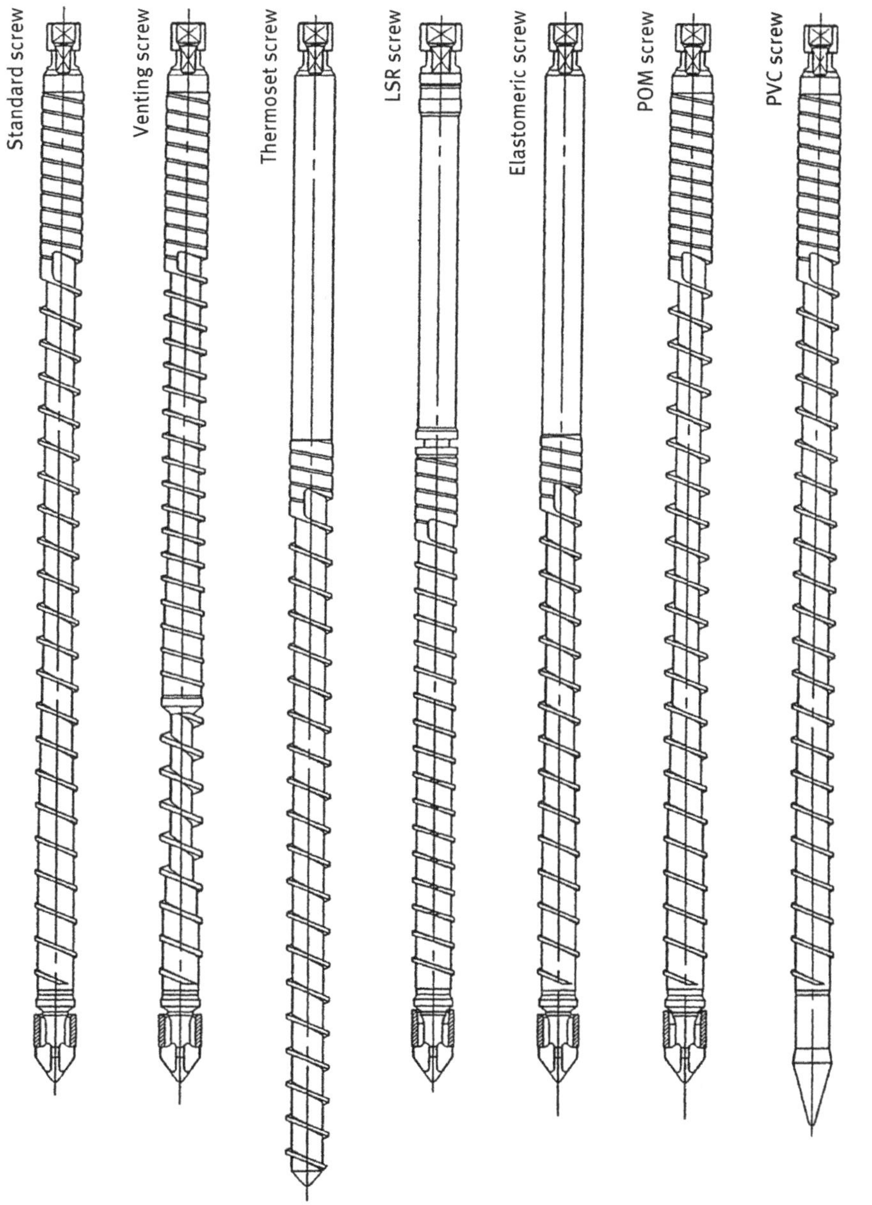

Figure 3.6: Injection moulding screws.
(POM: polyoxymethylene)

3.5.1 Mixing screws for additives and fast cycling machines

In cases where an increased mixing ability of the screw is required, there are two potential solutions: to add mixing elements or to use an extended screw design.

3.5.1.1 Mixing elements

Where higher levels of mixing are desired, mixing elements help create a more homogeneous melt as far as the distribution of temperature and filler is concerned.

The quality of the melt depends above all on two criteria:
- The retention time of the melt in the mixing element.
- The number of melt partitions.

The loss of pressure in mixing elements should be kept as small as possible, as any loss of pressure leads to a reduction in plasticising quality. A mixing element must therefore be designed as follows:
- The free volume of the raw material in the mixing element must be as large as possible.
- The melt deflection must be compulsory.
- The flow cross-section must be as wide as possible.

One solution is to use Rhombus mixers, which are similar to slot disc mixing elements that are used in extrusion techniques however, the Rhombus elements also have a specific conveying effect. An important advantage of these mixing elements is the compulsory deflection of the melt. On its way to the screw tip, the melt flows through several cut-offs in the slot discs. Mixing elements with one or more channels and flights with cut-offs do not have compulsory deflection, as a certain percentage of the melt flows along the screw channels without being impeded; hence, there is only a limited mixing effect.

3.5.1.2 Extended plasticising cylinder

An alternative is to use an extended plasticising cylinder, which is suitable in the following circumstances:
1. If a high melting capacity is required (e.g., raw material with a high specific thermal capacity, i.e., PE, PP, PA), leading to:
   - Short cycle times.
   - Medium cycle times and high product weight (dosage volume over 50%).
   - Moulding compounds that tend to shear (leading to an increase in the heating energy/overall heat energy ratio).
2. If a masterbatch is used for colouring purposes and the homogeneity of the colours is of great importance.

3. If the temperature level during the moulding process must be reduced or peak temperatures must be eliminated resulting in:
   - Reduction of cycle time (shorter cooling time).
   - Avoiding decomposition of thermal-sensitive raw material.

The combination of cylinder and screw depends on the specific case it is needed for, for example:
- Case 1: Extended screw with standard geometry (the separate zones of the screw are extended according to a constant factor).
- Case 2: Standard screw + Rhombus mixing element + metering zone extension.
- The cylinder is heated with an additional heater band or extended heater bands with increased heating power – this is shown in Figure 3.7.

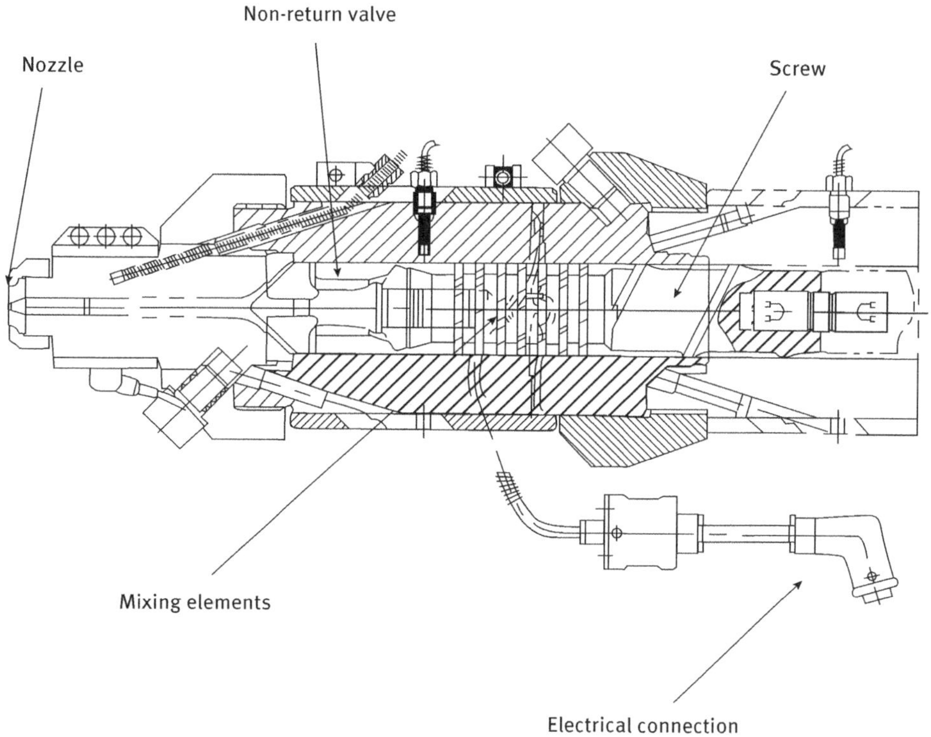

Figure 3.7: Cylinder extension module with extended plastication, additional heating circuit and mixing port.

3.5.2 Non-return valve

Many materials require the use of a valve with a check ring fitted to the end of the screw to prevent backflow and this also helps to ensure a constant cavity pressure is maintained. The most important design consideration is that they should avoid flow restrictions or hold-up of the melt flow. Non-return valves are more prone to wear than other components, hence suitably toughened materials are used for their manufacture. Various designs of non-return valves are shown in Figure 3.8.

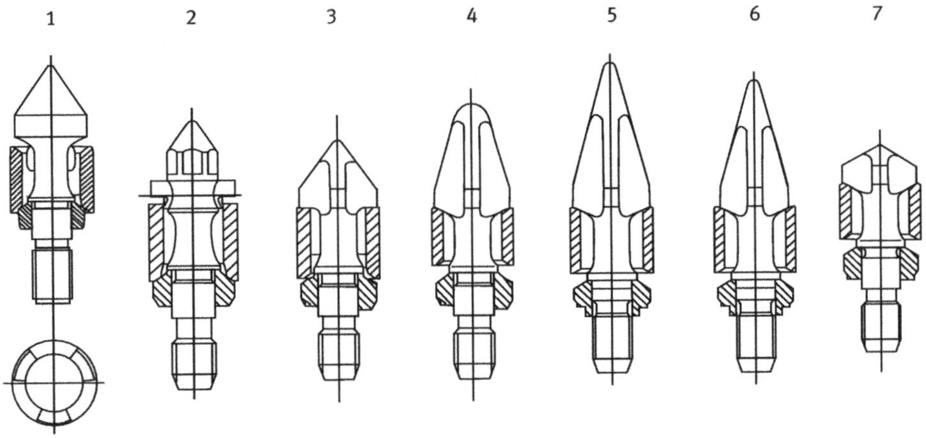

Figure 3.8: Various designs for non-return valves. 1) For polyester paste, 2) low abrasion type with M-C-V series machines, 3) standard type on M-C-V series machines, 4) standard type on Arburg ALLROUNDER® machines, 5) old type only on hydronical unit from 35 mm upwards, 6) old type with smaller flow cross-section and 7) spare part tips for the combination of a new screw with an old nozzle.

3.6 General information on wear and tear

There are several causes of wear and tear which include:

- Incorrect adjustment of process parameters, e.g., back pressure is too high, dosage speed is too high, no dosage delay, incorrect adjustment and setting of temperatures for the plasticating cylinder and feed yoke.
- Wear and tear generated by raw materials, e.g., mechanical wear caused by fibreglass, glass spheres, carbon fibre, stone powder, metallic powder, ceramic powder.
- Chemical corrosion, e.g., with additives, flame-resistant materials, materials containing fluorides. There are several ways to determine any mechanical wear and/or chemical corrosion. Mechanical wear can be seen by grooves and surface abrasion in one direction. Chemical corrosion can leave holes of varying sizes in different areas and directions as well as surface deposits. To determine the wear of the screw and barrel, consideration must be given to the original

heat treatment method used, i.e., using nitride and Arbid methods, the surface thickness can be measured, with bimetal outfitting, the surface can be examined. Generally, if the heat-treated surface has worn down this signifies to the manufacturer that the units have worn out.

Measurements of the thickness can be carried out according to the barrel and screw dimensions:
- By measurement with a micrometer or callipers and so on, e.g., manufacturer's dimensions.
- Barrel: 30 mm (30.00–30.05 mm).
- Screw: 30 mm (29.85–30.15 mm).
- Bushing: 30 mm (29.97 ± 0.01 mm).

Desired gap between barrel and bushing:
- Standard: 0.04 mm.
- LSR: 0.02 mm.

There are ways to minimise wear and tear on the injection units, including properly adjusting the process parameters, the correct selection of barrel and screw for the job, and suitable heat treatment outfitting (see hot combinations and applications).

3.7 Unit hardening treatment

Examples of the resistance of various plasticising screw and cylinder outfitting options are shown in Table 3.1. Table 3.2 lists the classifications of the various treatments and Table 3.3 gives recommendations for the type of outfitting for some common materials. From these three tables it should be relatively straightforward to ensure the correct equipment is being used and wear and tear is minimised.

3.8 The nozzle

The nozzle provides the connection between the injection cylinder and the mould tool. Its job is to convey the material with minimal pressure or heat change. There are two common types of nozzle:
1. An open nozzle.
2. Nozzle shut-off valve.

Various designs of nozzle are shown in Figures 3.9–3.11. The nozzle itself may not necessarily be made of just one piece. A tip that is screwed into the nozzle body can be replaced or repaired and may need to be an abrasion- and corrosion-resistant

Table 3.1: Plasticising screw and cylinder outfitting.

Metal treatment type	Plasticising		Resistant to abrasion	Resistant to corrosion	Thickness (mm)	Hardness	Temperature range up to
	Cylinder	Screw					
Nitride (low abrasion)	+	+			0.30	800 V	250 °C
Arbid (Abrasion proof)	+	+	*	*	0.12	1,800 V	450 °C
BMA	+		**	*	2.5–3.0	63 HRC	450 °C
PM		+	**	*	2.00	57–58 HRC	450 °C
PKV		+	*	**	2.00	54–56 HRC	450 °C

*Resistant
**Very resistant
Generally, all nozzles are in nitride execution with different steel types.
Non-return valves are in nitride execution on screws with Arbid.
BMA: Ultralow abrasion-bimetallic alloy
PM: High-chromium, powder-metallurgical steel
PKV: Highly wear resistant

Table 3.2: Abrasion classifications.

Abrasion classification	Cylinder/screw
Low abrasion	Nitride/nitride
Abrasion proof	Arbid/Arbid BMA/Arbid[1]
	VSX/Arbid[2]
Ultralow abrasion	BMA/PH
Corrosion proof	BMK/PK

[1] Only V-series from 370 V up, 470/520 °C and aggregate 50.
[2] Only 270 V.

tool steel tip. To ensure optimum flow conditions, there must be no material hang-ups and therefore, the flow must be streamlined. The land length is generally kept to a minimum and is dictated by the strength requirements. For high pressure applications an increased flange diameter may be required.

It is essential that the temperature of the nozzle be controlled. The location of the heating and control is equally important otherwise material degradation or premature material freezing (cold slugs) may occur.

Table 3.3: Recommendations for use.

Abrasion classification	Material types
Low abrasion	Thermoplastics without corrosive or abrasive components (LDPE, HDPE, PP, PS, CA)
Abrasion proof	Thermoplastics with corrosive and/or abrasive components (ABS, PVC, POM, PBT, PET, PA); low to middle content of filling material (e.g., glass, French chalk, TiO_2)
Ultralow abrasion	Thermoplastics with strongly abrasive and corrosive components, high temperature material (PC, PBT, PET, PA, POM, PE with high share of glass, PES, PSU, PEI, PEEK, PPS, PPA)
Corrosion proof	Thermoplastics with highly corrosive components (CPVC, FEP, ETFE, PFA, E/CTFE, PVDF), also flame-protected types

ABS: Acrylonitrile-butadiene-styrene
CA: Cellulose acetate
CPVC: Chlorinated polyvinyl chloride
ECTFE: Ethylene chlorotrifluoroethylene
ETFE: Ethylene-tetrafluoroethylene
FEP: Tetrafluoroethylene hexafluoropropylene
HDPE: High-density polyethylene
LDPE: Low-density polyethylene
PBT: Polybutylene terephthalate
PC: Polycarbonate
PEEK: Polyether ether ketone
PEI: Polyetherimide
PES: Polyether sulfone
PET: Polyethylene terephthalate
PFA: Perfluoroalkoxyalkane
PPA: Polyphthalamide
PPS: Polyphenylene sulfide
PSU: Polysulfone
PVDF: Polyvinylidene fluoride

A thermocouple can be used to close the gate and heater. Thermocouples may extend into the melt rather than measuring the temperature of the nozzle; some measurement locations are shown in Figure 3.12.

3.9 Clamping units

The clamping units of injection machines are described and rated separately to the injection unit. The clamping units are required to enable mounting and holding of the two mould halves and must also provide sufficient clamping force during injection and cooling to enable effective moulding. The mould halves must also open and close accurately and smoothly to enable part injection and begin the next

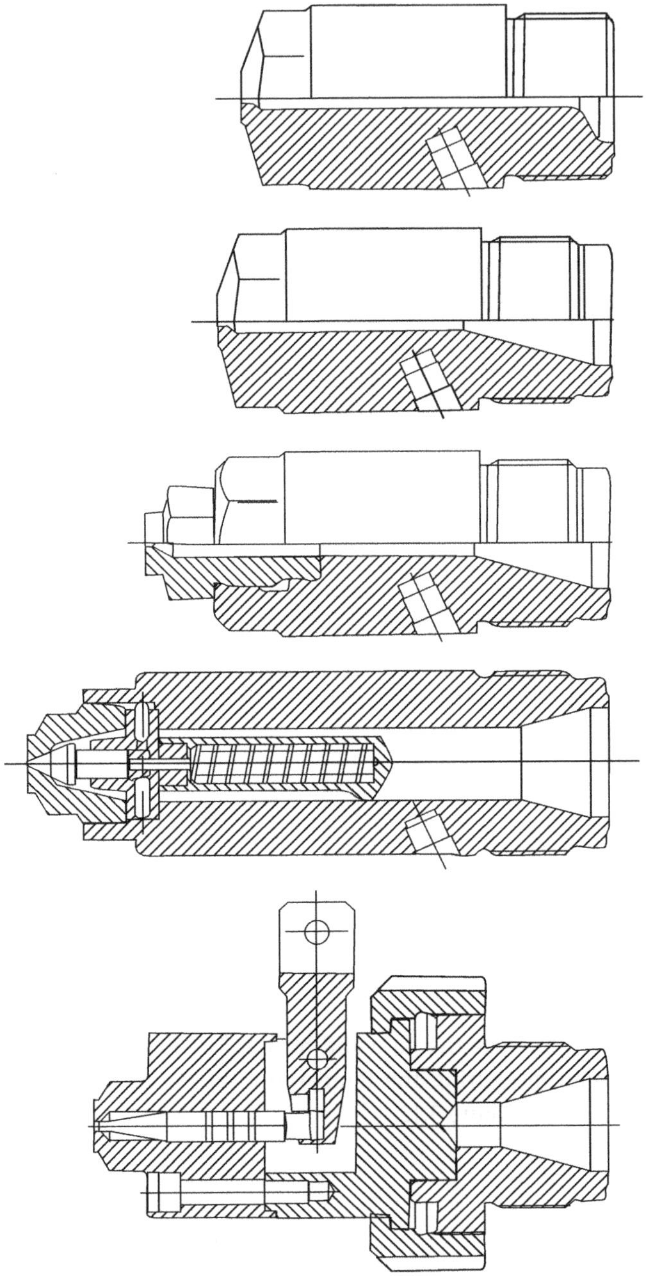

Figure 3.9: Various nozzle designs.

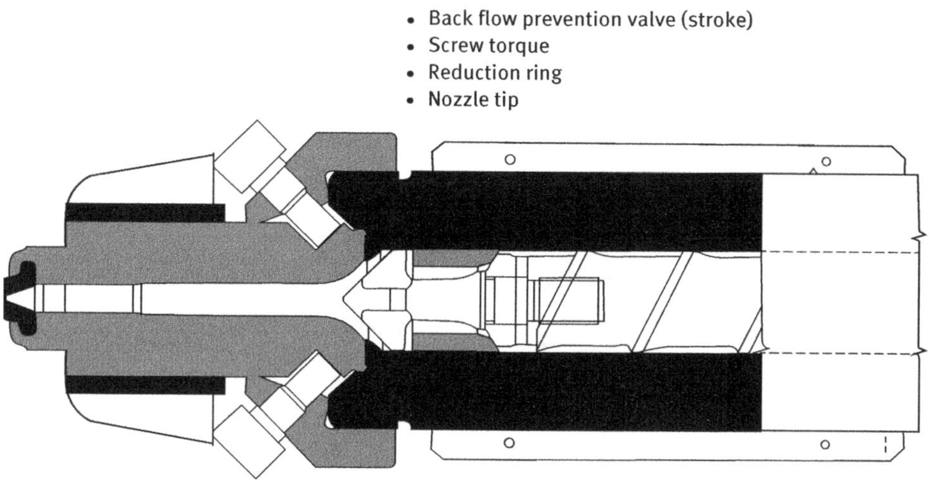

Figure 3.10: Nozzle attachment.

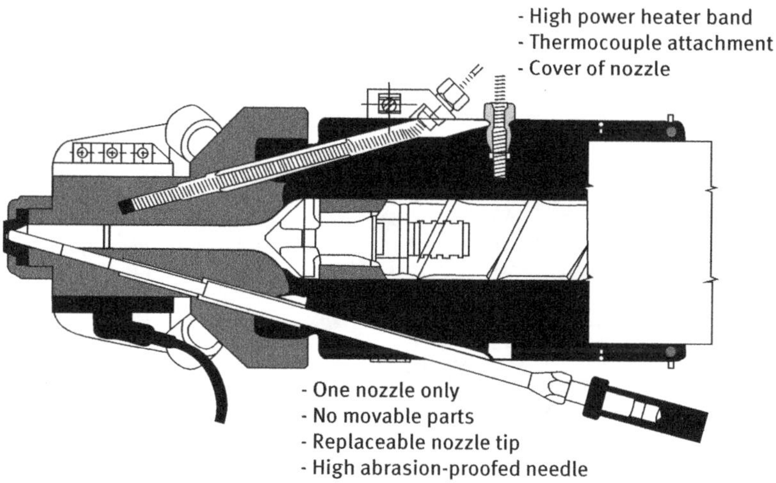

Figure 3.11: Replaceable hydraulic nozzle.

process cycle. The clamping units on injection machines can be run by hydraulics, a hydraulic and toggle combination or by electrical power.

Figure 3.13 shows a clamping unit. The stationary platen is attached to the machine with four tie-rods connecting it to the movable platen. Figure 3.14 shows a direct hydraulic clamping system. The clamp ram moves the moving platen until it reaches the stationary platen and the pressure begins to build up. The ejectors are fitted onto the moving platen and can be activated once the tool is opened and the moving platen retracted.

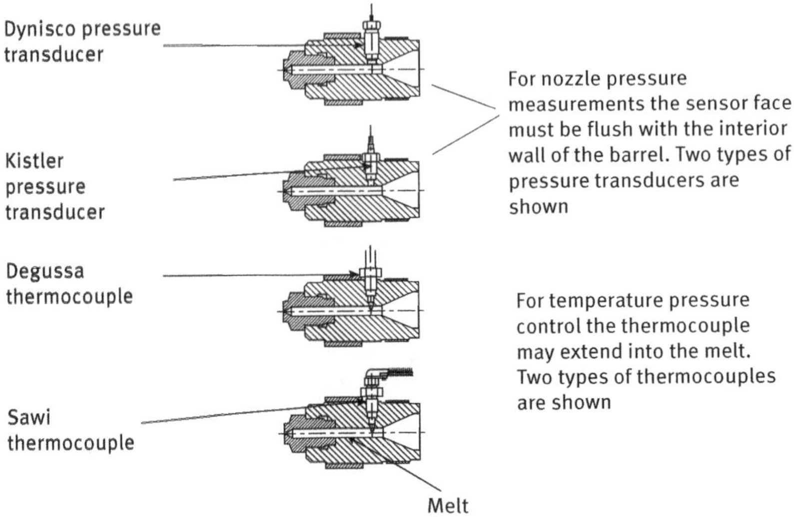

Figure 3.12: Measurement locations in nozzle body.

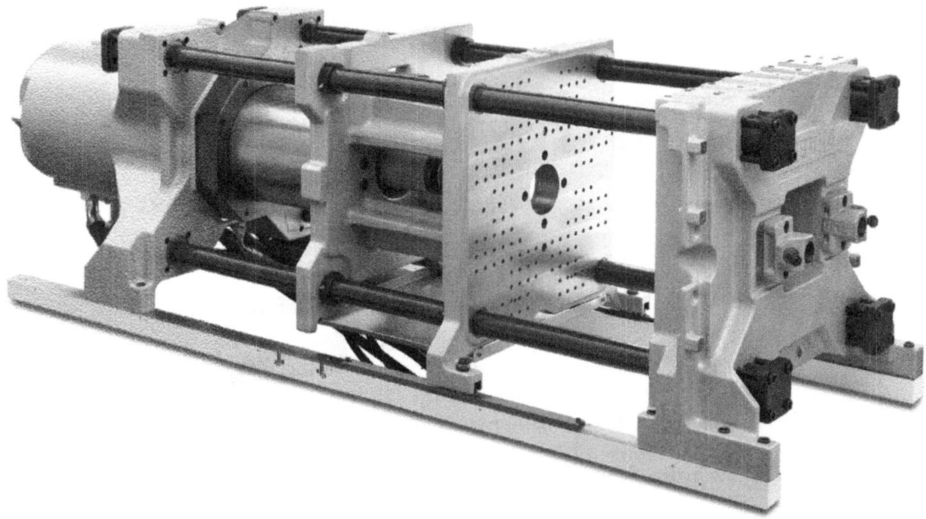

Figure 3.13: A clamp unit.

A toggle-type injection unit is shown in Figure 3.15 and this design enables the force to be amplified. Basically, it is two metal bars attached by a pivot; one end is attached to the stationary plate, the other to the movable platen. When open, it forms a distinctive 'V' configuration and when closed the bars form a straight line. The advantage of this design is that a much smaller force from the hydraulic

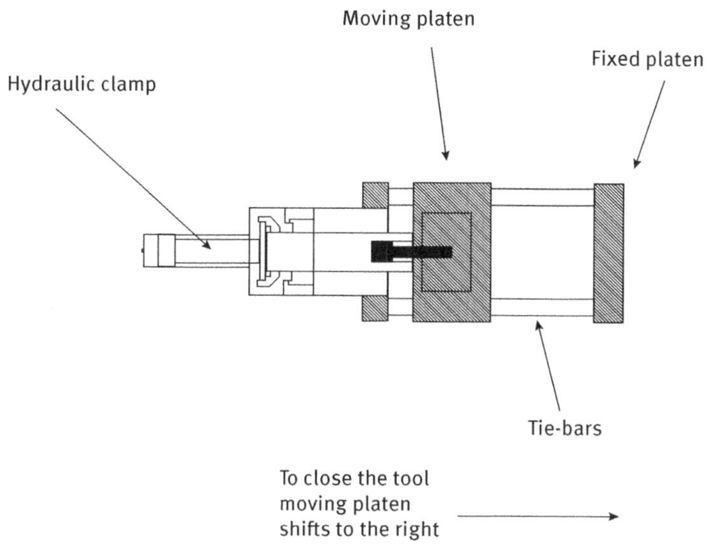

Figure 3.14: Direct hydraulic clamping unit.

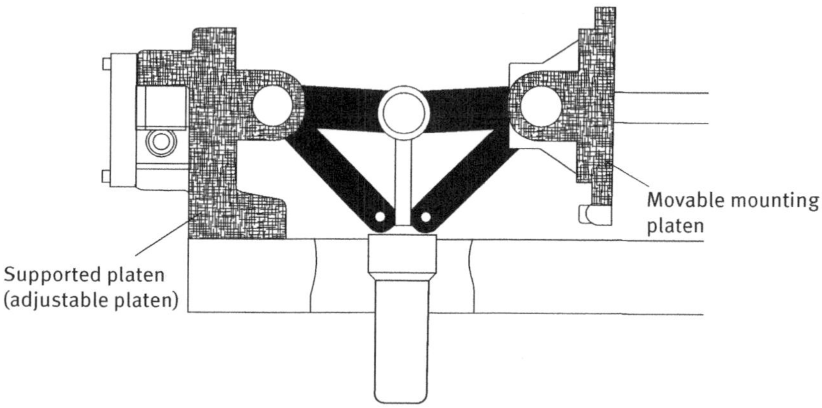

Figure 3.15: Toggle-type clamping unit.

cylinder is required, the size of this advantage varies but can be as high as 50:1. A further advantage of toggle machines is that once extended the toggles remain there until retracted, making them self-locking. The hydraulic system on the other hand requires the application of constant line pressure. The disadvantages are that it is more difficult to control the speed and force of a toggle mechanism. It must also be adjusted for different depths of mould tool to ensure the toggle is fully extended.

3.9.1 Differential piston system

A minimal volume of oil is required for the opening and closing movements of the mould tool. The oil volume required from the pump results from the differential surface and stroke of the piston (approximately 7% of the clamping cylinder volume). The rest of the oil flows through the borings in the main piston as a result of the piston stroke. The pressure cycle is shown in Figures 3.16–3.18.

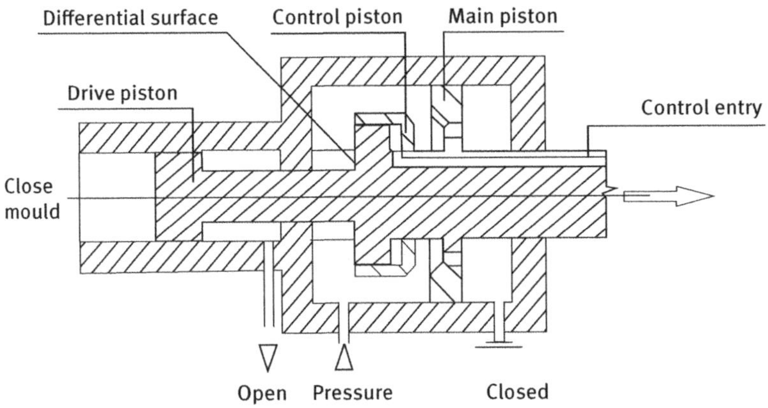

Figure 3.16: Mould closing.

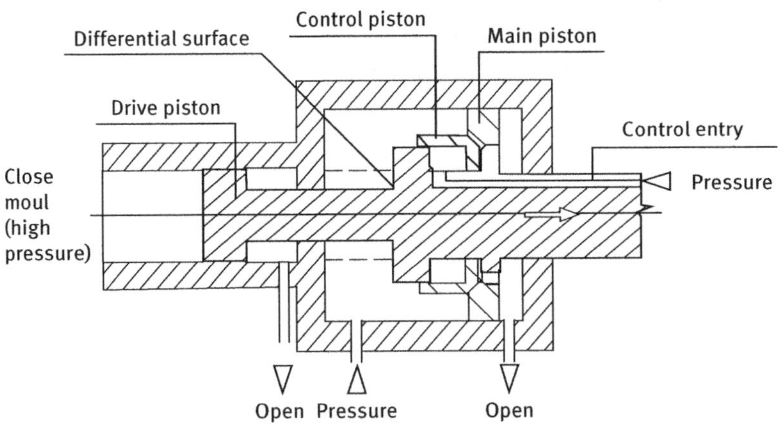

Figure 3.17: Mould clamping (high pressure).

When opening with increased opening force (high pressure opening), the control piston closes the main piston. The main piston and opening piston now open the mould with 50 bar pressure.

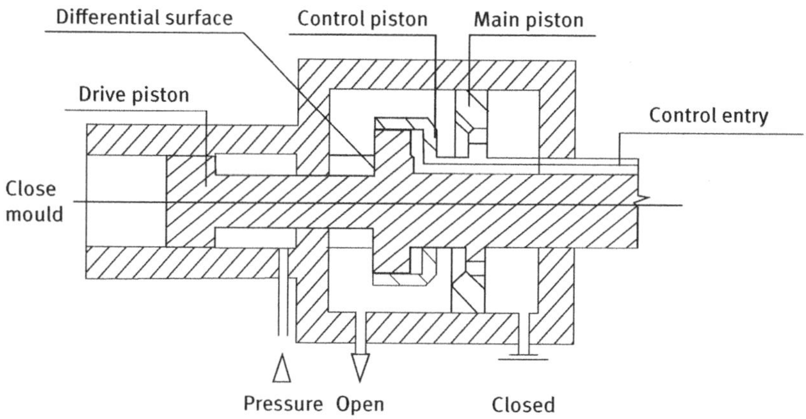

Figure 3.18: Opening of mould.

When the injection unit is in the vertical position and 'braking' is selected, the borings in the main piston are closed shortly before the end of the opening motion; this ensures an exact positioning of the movable platen in the lowest end position. Sinkage of the movable platen on an idle machine is also avoided.

3.9.2 Mould weights

Each machine will have a maximum permitted mould weight for the movable mould halves. These values should not be exceeded for any reason as production problems and premature wear would result; examples of some commercially available machines from the manufacturer Arburg are shown in Table 3.4, the type of supports are shown in Figure 3.19.

3.10 Selection of the clamping unit

To select a clamping unit, consideration must be given to the following factors:

- Injection mould size:
  - Dimensions.
  - Centering.
  - Ejector coupling.
  - Mould weight.
- Projected surface area:
  - Amorphous $cm^2 \times 4 = kN$.
  - Semi-crystalline $cm^2 \times 6–7 = kN$.

Table 3.4: Maximum mould weight.

Machine model	Of moving mould half kg/lbs	Of moving mould half on index unit (multi-comp.) kg/lbs	Of mould kg/lbs
270 A	200 / 441	–	300 / 661
370 A	360 / 794	360 / 794	540 / 1,190
470 A	760 / 1,676	600 / 1,323	1,140 / 2,513
520 A	1,000 / 2,205	600 / 1,323	1,500 / 3,307
570 A	1,300 / 2,866	750 / 1,655	2,000 / 4,409
630 A	2,100 / 4,630	1,000 / 2,205	3,200 / 7,055
720 A	2,900 / 6,393	2,000 / 4,409	4,300 / 9,480
820 A	4,000 / 8,818	2,000 / 4,409	6,000 / 13,228
920 A	6,000 / 13,228	3,000 / 6,614	9,000 / 1,9842
370 E	360 / 794	–	540 / 1,190
470 E	760 / 1,676	–	1,140 / 2,513
520 E	1,000 / 2,205	–	1,500 / 3,307
570 E	1,300 / 2,866	–	2,000 / 4,409
370 E GOLDEN ELECTRIC	360 / 794	–	540 / 1,190
470 E GOLDEN ELECTRIC	760 / 1,676	–	1,140 / 2,513
520 E GOLDEN ELECTRIC	1,000 / 2,205	–	1,500 / 3,307
570 E GOLDEN ELECTRIC	1,300 / 2,866	–	2,000 / 4,409
270 H	200 / 441	–	300 / 661
370 H	360 / 794	–	540 / 1,190
470 H	760 / 1,676	–	1,140 / 2,513
520 H	1,000 / 2,205	–	1,500 / 3,307
570 H	1,300 / 2,866	–	2,000 / 4,409
630 H	2,100 / 4,630	–	3,200 / 7,055
720 H	2,900 / 6,393	–	4,300 / 9,480
820 H	4,000 / 8,818	–	6,000 / 13,228
920 H	6,000 / 1,3228	–	9,000 / 19,842
270 C GOLDEN EDITION	180/396	–	400 / 882
320 C GOLDEN EDITION	180/396	–	400 / 882
420 C GOLDEN EDITION	600 / 1,323	–	900 / 1,984
470 C GOLDEN EDITION	800 / 1,764	–	1,200 / 2,646
570 C GOLDEN EDITION	1,500 / 3,307	–	2,250 / 4,960
170 S	36/79–55/121 [1)]	–	85 / 187
270 S	135/298–200/441 [1)]	200 / 441	300 / 661
370 S	360 / 794	360 / 794	540 / 1,190
470 S	760 / 1,676	600 / 1,323	1,140 / 2,513
520 S	1,050 / 2,315	600 / 1,323	1,600 / 3,527
570 S	1,400 / 3,086	750 / 1,655	2,100 / 4,630
630 S	2,500 / 5,512	1,000 / 2,205	3,800 / 8,378
720 S	3,600 / 7,937	2,000 / 4,409	5,300 / 11,684
820 S	5,000 / 11,023	2,000 / 4,409	7,500 / 16,535
920 S	8,000 / 17,637	3,500 / 7,716	12,000 / 26,455

Table 3.4 (continued)

Machine model	Of moving mould half kg/lbs	On rotary table kg/lbs	Of mould kg/lbs
175 V	36 / 79 [2] [3]	60 / 132	55 / 121 [3]
275 V	110 / 243 [2] [3]	120 / 265	165 / 364 [3]
375 V	230 / 507 [2] [3]	240 / 529	340 / 750 [3]
1200 T	400 / 882	750 / 1,655	–
1500 T	500 / 1,102	1,050 / 2,315	–
1800 T	800 / 1,764	1,650 / 3,638	–

[1] With optional support of moving mould mounting platen

[2] Upper mould half

[3] Restriction valid only for vertical injection unit Reproduced with permission from ARBURG.

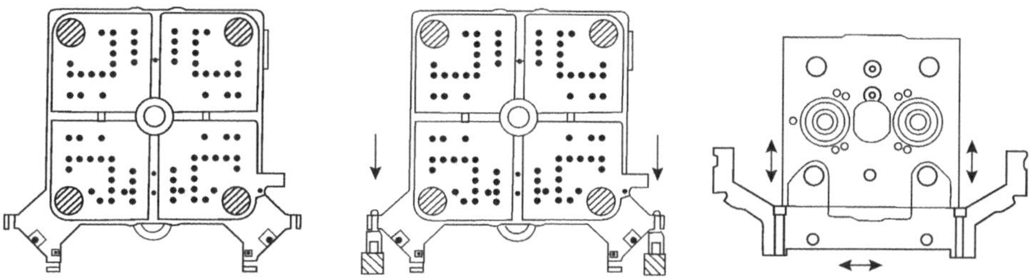

Figure 3.19: Support of the movable platen.

- Select higher clamping force with split fallower moulds:
  - With thin-walled parts and thermoset materials.

The next section will consider issues relating to mould clamping force requirements.

3.11 Mould clamping force

3.11.1 Mould clamping force level

The mould clamping force must be set high enough to prevent flash, which is caused by the swelling of the mould due to the force exerted during the initial material injection, resulting in the material coming out of the mould cavity. The required mould clamping force depends on the size of the moulded component surface which is projected onto the parting plane, and on the internal mould pressure. When the machine contains a new mould and is started-up, the internal mould pressure required is not yet known. Naturally, from experience, it may be possible to extrapolate a figure, which depends on the type of plastic and the component format. However, practice has continually shown that it is initially quite sufficient to use a guide value of 2.5–5 kN/cm^2 to calculate a projected moulded component surface. For settings during the start-up period, the calculations are performed using the upper limiting value, even with easy-flowing compounds, such as PS or PE, for which 2.5–3.5 kN/cm^2 is adequate throughout.

For example:
The size of the component moulded surface projected onto the parting plane = 52 cm. The guide value for mould clamping force = 52 × 5 = 260 kN. During the course of the test injections, the mould clamping force can be reduced until just above the point at which leaks begin. The mould clamping force should be as low as possible, i.e., only as high as necessary to save energy used on the toggle clamping units to keep wear as low as possible. Another reason to keep the mould clamping force as low as possible is to reduce the air extraction from the mould cavity required during the initial injection.

3.11.2 Mould clamping force and mould rigidity

In a high majority of cases flash (see Chapter 9 for a description) is not caused by inadequate mould clamping force or locking pressure, but by an insufficient rigidity of the mould in areas of concern.

The mould plates carrying the moulding nest can become thoroughly bent in an area of the central bores and, most critically, in the area of the ejector system, under the effect of the internal mould pressure. Thus, the mould plates must have sufficiently high flexural strength, if the formation of scratches is to be avoided. This danger is especially great for the mould plate on the ejector side hence, if possible, additional support columns, with an oversize of 0.03–0.05 mm in comparison with the external supports, should be incorporated.

If the moulding nests are directly built into the mould plate, there is a further measure available to reduce the tendency of leak and/or scratch formation. Only one sealing edge is present around the moulding cavities, and the remaining surface of the parting plane is free ground to approximately 0.05 mm. This means that, for the same mould clamping pressure, the sealing pressure increases as a result of a reduction in the sealing faces. The free grinding also favours air extraction from the mould cavity during filling. Rigid moulds thus require mould clamping forces which are not so high, meaning the machine can be improved and energy saved and, under certain circumstances, it can even mean using a smaller machine.

3.11.3 Setting mould closing and clamping force

Before setting mould closing, the mould protection functions need to be set to prevent any damage as a result of the moulds hitting one another. A low value is generally sufficient, i.e., 5.0 kN. If this is not sufficient to close the mould, the force should be increased little-by-little, however, first check if there is anything hindering smooth closing. The desired closing force can then be set, i.e., 20 kN.

3.11.3.1 The clamping force

The height of the clamping force depends on the moulded part's projected area in the parting line. The projected area is measured in cm^2 or square inches (1 sq in = 6.451 cm^2). Table 3.5 relates the specific clamping force needed per cm^2 to various plastic materials.

Table 3.5: Clamping force requirements per cm^2.

Material	**PS, SB, SAN, ABS, CA, CAB, PVC soft**	**PMMA, modified PPO, PC, PSU/PES, PVC hard**
	PE soft (LDPE), PE rigid (HDPE)	**PP, PA, POM, PET, PBT, PPS**
Specific clamping force	2 5 to 5.0 kN/cm^2 16.1 to 32.3 kN/sq in	5.0 to 7.0 kN/cm^2 32.3 to 45.2 kN/sq in

CAB: Cellulose acetate butyrate
PMMA: Polymethyl methacrylate
PPO: Polyphenylene oxide
SAN: Styrene-acrylonitrile
SB: Styrene-butadiene

To calculate the projected area of the moulded part, multiply projected area × specific clamping force = clamping force.

Example:

Material: SAN.

Specific clamping force: 2.5 to 5 kN/cm^2, average value 3.5 kN/cm^2. Moulded part's projected area: 90 cm^2.

Calculation of clamping force: $3.5 \times 90 = 315$ (kN).

Enter the calculated clamping force for the 'mould closing' phase.

3.11.3.2 Second clamping force

At the end of the holding pressure phase the clamping force can be reduced and generally, this will be half of the value calculated previously.

The second clamping force should last to the end of the holding pressure phase.

3.12 Data for mould closing

Table 3.6 gives recommended closing force values (clamping force).

Table 3.6: Recommended closing force values (clamping force).

Injection material	Recommended values for specific closing force (kN/cm²)	Practical values for the mould cavity pressure, based on the recommended closing force values (bar)
PS	1.5–3.5	150–350
SB	2.0–4.0	200–400
SAN	2.5–4.5	250–450
ABS	3.0–5.5	300–550
PVC rigid	2.5–5.0	250–500
PVC soft	1.5–3.0	150–300
CA	2.5–4.5	250–450
CP	2.0–3.5	200–350
PMMA	3.5–5.5	350–550
Modified PPE (modified PPO)	3.5–6.0	350–600
PC	3.5–6.5	350–650
PSU/PES	4.0–6.0	400–600
PEI	3.5–6.5	350–650
PE soft	2.0–6.0	200–600
PP	3.0–6.5	300–650
PA4.6	4.5–7.5	450–750
PA6	3.5–5.5	350–550
PA6.6	4.5–7.5	450–750
PA6.10	3.0–5.0	300–500
PA11, PA12	3.5–5.5	350–550
PA amorphous	3.5–4.5	350–450
POM	5.5–10.5	550–1,050
PET	4.5–7.5	450–750
PBT	4.0–7.0	400–700
PPS	3.5–6.5	350–650
FEP	3.0–6.0	300–500
PAA	3.0–7.0	300–700
LCP	3.0–8.0	300–800
	Thermosets/Elastomers	
Duroplaste	2.0–6.0	200–600
PE-U	2.0–4.5	200–450
LSR	0.8–2.5	80–250

CP: Cellulose propionate
LCP: Liquid crystal polymer
PAA: Polyacrylic acid
PE-U: Thermoplastic elastomer (urethane grade)
PPE: Polyphenylene ether

The clamping force (closing force) to be set can be calculated from the specific closing force and the projected injected part surface in the parting line.

Example:
Specific closing force = 3.5 kN/cm^2 = average value for SAN from the table above.
A projected injection part surface of:
- 20 cm^2 results in a closing force of = 3.5 × 20 cm^2 = 70 kN.
- 90 cm^2 results in a closing force of = 3.5 × 90 cm^2 = 315 kN.

The closing forces (clamping forces) in the fill phase are higher with thin-walled parts:

$$\text{Closing force (clamping force)} = z \times A_{proj} \times pw$$

Where:
z: number of cavities.
A_{proj}: projected parting line part surface.
pw: mould cavity pressure according to Figure 3.20, dependent upon the flow-path/wall thickness relationship (L/s) and the wall thickness (s).

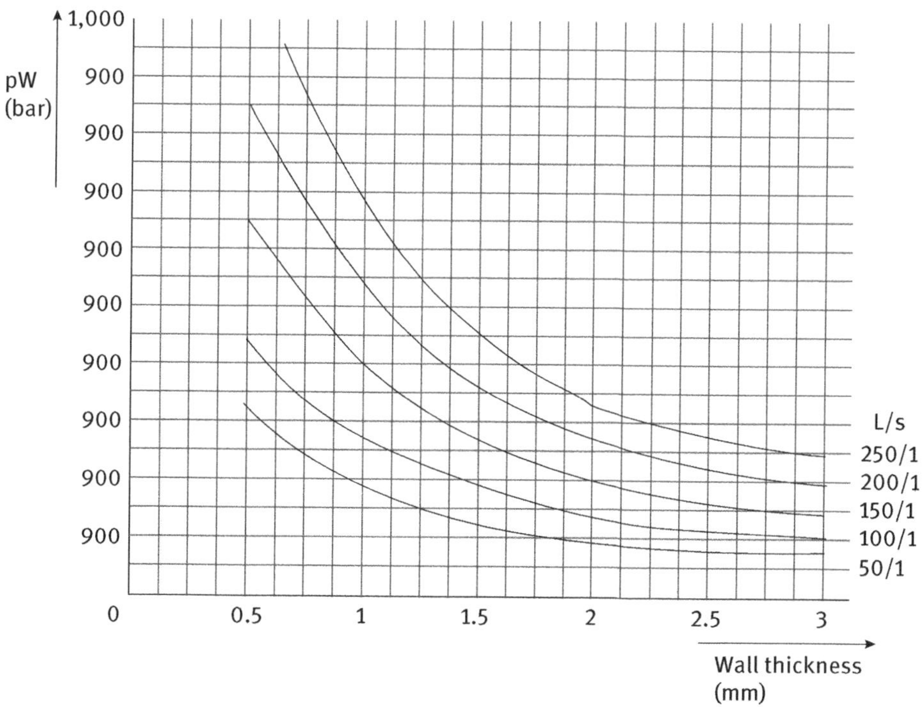

Figure 3.20: Mould cavity pressure and wall thickness.

Example 1: Coffee cup

Flowpath (L) = 80 mm.
Wall thickness (s) = 0.5 mm. L/s = 80/0.5 = 160.
Cavity count (z) = 1.
A_{proj} = 51 cm^2.
pw (according to Figure 3.20) = 650 bar ≈ 6,500 N/cm^2.
Closing (clamping force) = 1 × 51 × 6,500 = 331,500 N = 332 kN.

Example 2: Marmalade container

Flowpath (L) = 85 mm.
Wall thickness (s) = 0.5 mm. L/s = 85/0.5 = 170.
Cavity count (z) = 1.
A_{proj} = 71 cm^2.
pw (according to Figure 3.20) = 750 bar ≈ 7,500 N/cm^2.
Closing (clamping force) = 1 × 71 × 7,500 = 532,500 N = 533 kN.

Example 3: Flower pot

Flowpath (L) = 120 mm.
Wall thickness (s) = 0.55 mm. L/s = 120/0.55 = 218.
Cavity count (z) = 1.
A_{proj} = 113 cm^2.
pw (according to Figure 3.20) = 800 bar = 8,000 N/cm^2.
Closing (clamping force) = 1 × 113 × 8,000 = 904,000 N = 904 kN.

It should be noted in connection with Figure 3.20 that higher cavity pressures occur with high viscosity materials.

IMPORTANT: The relationship between flowpath and wall thickness should not exceed 250/1 with high viscosity materials; multiple injections are required if this is the case.

3.13 Other considerations

The type of clamp and the clamping force are the main specifications of a clamping unit: however, there are other design features which also need consideration:

Maximum daylight.
- Space between tie-bars.
- Clamp stroke.
- Clamp speed.
- Knockout stroke.

The first two of these points concern the size of the mould tool that can be physically fitted into the machine. The daylight is the maximum distance between the stationary and moving platen, obviously there must be room in the tool to open and eject the part. The tie-bars restrict the length and width of the tool that can be placed on the platen as they restrict entry. With this in mind, there are tie-barless machines on the market which become particularly pertinent with multi-shot moulding (see Chapter 10).

The clamp stroke is the maximum distance the moving platen will move. The clamp speed is the maximum speed at which it will do this, which can affect the overall cycle time and therefore process efficiency and so can be of great importance.

The maximum knockout stroke determines the movement available for ejection.

3.14 International standard for the designation of injection moulding machines

Now that the basic machinery components have been discussed, the way in which IJM are designated can now be considered. IJM are classified by the machinery that they are composed of as set out in the International Standard.

(As of January 1983) standard designation, e.g., 700–210.

1st designation = clamping force in kN (1 kN ≈ 0.1 Mp ≈ 0.11 ton)
2nd designation = shot capacity of inj. unit

$$= \frac{\text{max.inj.volume}(\text{cm}^3) \times \text{max.inj.pressure(bar)}}{1{,}000}$$

Machines with two injection units bear the designation for the second injection unit next to that of the first unit, e.g., 700–210/210.

The shot capacity of the injection unit does not depend on the size of the plasticising cylinder used:

Injection volume = screw-piston surface × stroke
Injection pressure = injection force/screw-piston surface
That is:

$$\frac{\text{inj.volume} \times \text{inj.pressure}}{1{,}000} = \frac{\text{screw} - \text{piston surface} \times \text{stroke} \times \text{inj.force}}{1{,}000\,\text{screw} - \text{piston surface}}$$

$$= \frac{\text{stroke} \times \text{inj.force}}{1{,}000}$$

Examples of machine designation:

1. ALLROUNDER® 305:
   - Clamping force – 700 kN.
   - Maximum injection volume with 35 θ screw – 140 cm^3.
   - Maximum injection pressure with 35 θ screw – 1,500 bar.
   - Shot capacity of injection unit $= \frac{140 \times 1{,}500}{1{,}000} = 210$
   - International Standard Designation = 700–210.

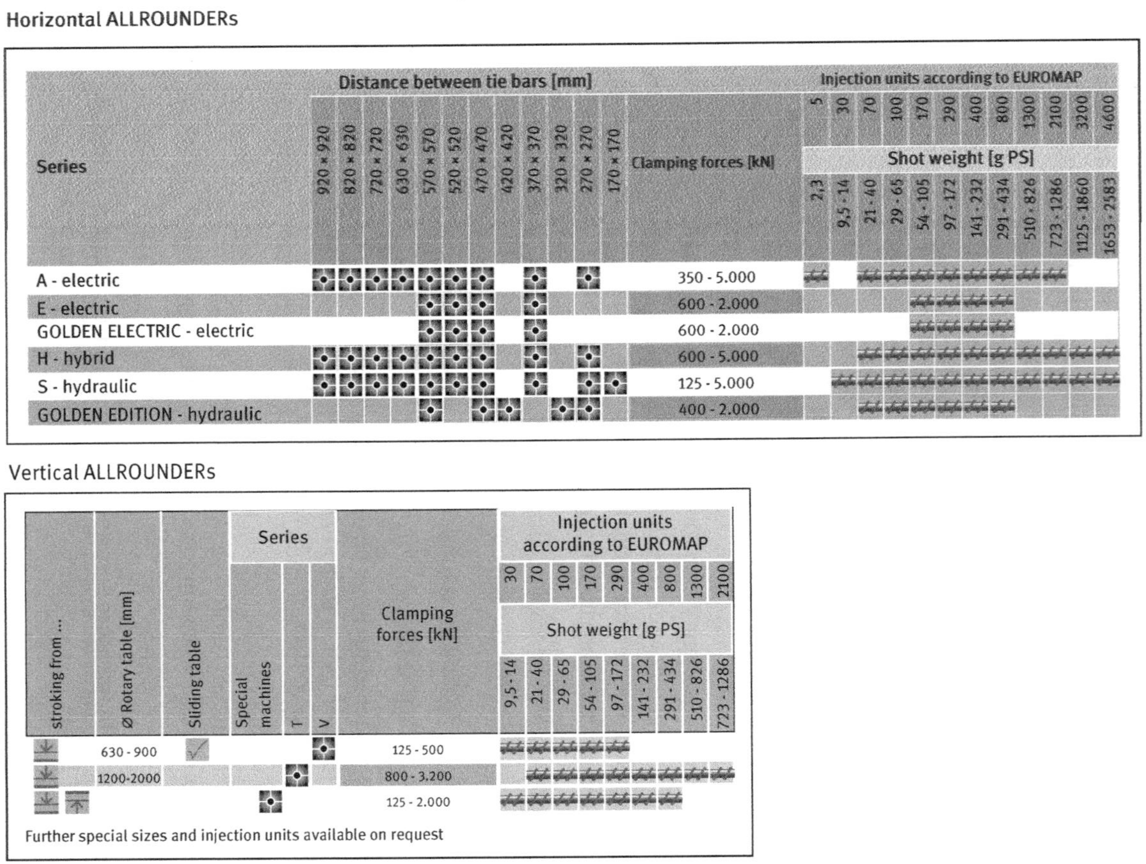

Figure 3.21: ALLROUNDER® S machine range offered by ARBURG.

2. Two-colour ALLROUNDER® 270
   - Clamping force: 500 kN.
   - Shot capacity of first injection unit: 90.
   - Shot capacity of second injection unit: 210.
   - International Standard Designation: 500–90/210.

IMPORTANT: until January 1983, the international standard designation was detailed as follows:
- 1st designation – shot capacity of injection unit (now 2nd).
- 2nd designation – clamping force in Mp (now clamping force in kN as 1st designation).

3.15 Injection moulding machine: drive technology

When buying an IJM as well considering the clamping force requirement, injection unit and tool area (as shown in a machine range offered by ARBURG in Figure 3.21), there may also be a need to consider whether all hydraulic, all electric or hybrid combination drive units are necessary.

There is considerable debate in the industry over whether electrical or hydraulic drives are better for various applications, as each can bring benefits dependent on

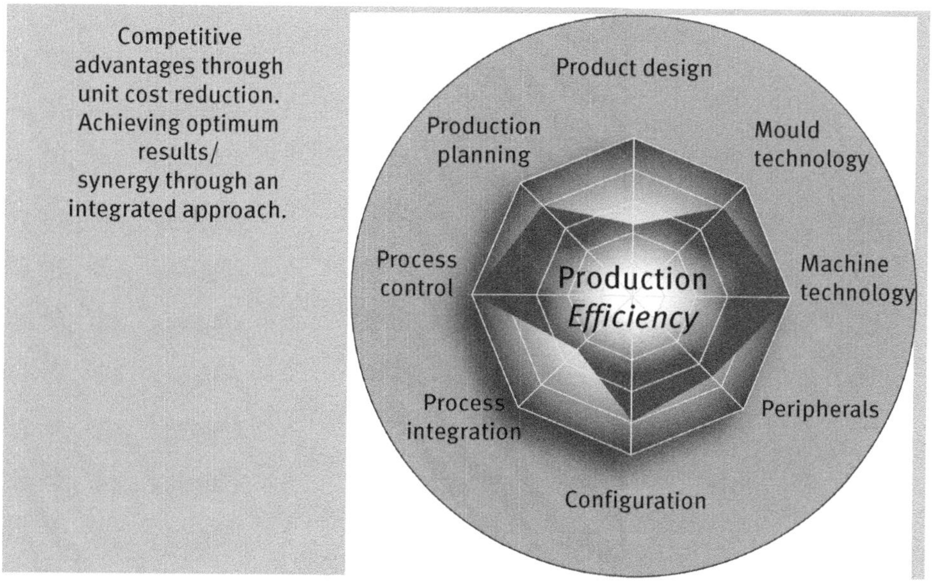

Figure 3.22: An integrated approach to injection moulding offered by ARBURG.

the specific application. As a result, most machine manufacturers, but not all, still offer all three and examples of manufacturers and machine builders offering a range of drives include: Arburg, Absolute Haitian, Battenfeld, Boy Machines, Chen Hsong, Demag Dima, Engel, Fortune, Husky, Kawaguchi, Krauss-Maffei Meiki, MHI (Mitsubishi), Milacron Negri Bossi, Nissei, Netstal, Sandretto, Toshiba, Toyo and Ube.

Energy consumption can certainly be one driver towards using an electric screw design, as the plastication unit, which includes the energy intensive screw recovery stage, consumes a much larger percentage of energy than any of the injection, clamping and ejection stages. However, all electrical machines currently require more expensive initial capital investment. Modern hybrid machines on the other hand combine hydraulics with the 'on demand' advantages of an electric servo motor and therefore pump oil when required in hydraulic accumulators. As machine technology advances many of the perceived differences in the capability of each technology may disappear, at this stage however it appears that there is considerable disagreement across the industry on which technology can claim to be 'the best' in all cases. As shown in Figure 3.22, an integrated approach to the entire production process allows optimum outcomes.

4 Injection mould tooling basics

4.1 Types of moulds

An injection mould tool has two major purposes:

- It is the cavity into which the molten plastic is injected.
- The surface of the tool acts as a heat exchanger (as the material solidifies on contact).

Injection mould designs differ depending on the type of material and component being moulded. Mould tool design and component design are equally important considerations for success. Component design is beyond the scope of this book but the various tooling, gating, temperature control and ejection systems that make up the mould tool will be considered here.

The standard design for injection mould tools is the two-plate design.

4.1.1 Two-plate mould

This is the simplest mould design. Mould cavities are formed in one plate only with the stationary half of the mould blank. A central sprue bushing can be placed into the stationary half of the mould or it is possible to have a direct runner system to a multi-impression mould. The moving half of the mould contains the ejection mechanism. This is illustrated in Figure 4.1.

4.1.2 Stripper mould

A stripper mould is very similar to the standard two-plate mould except for the ejection system. This design has a stripper plate for ejection, whereas the standard mould has pins or sleeves as the ejectors. This is illustrated in Figure 4.2. The advantage of a stripper plate is the increased surface area for ejection that it offers.

4.1.3 Slide mould

Like the two previous designs, this is also a two-plate mould; however, it has slides and cam pins for additional lateral movement, as shown in Figure 4.3. This type of design is suitable for producing parts with undercuts or external threads.

https://doi.org/10.1515/9783110654813-004

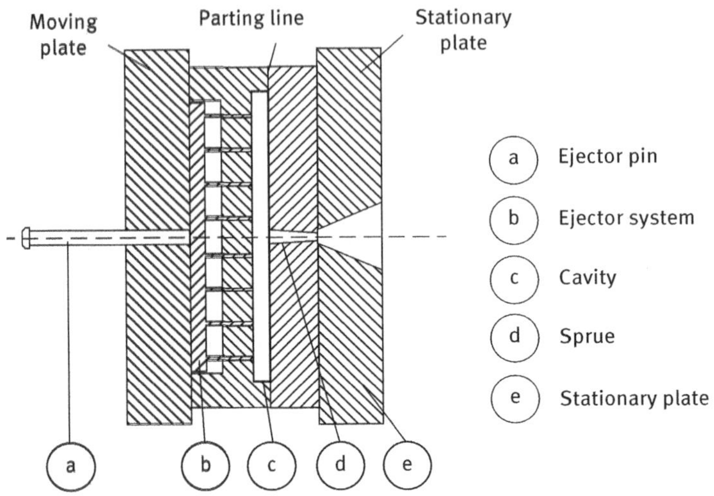

Figure 4.1: Two-plate injection mould.

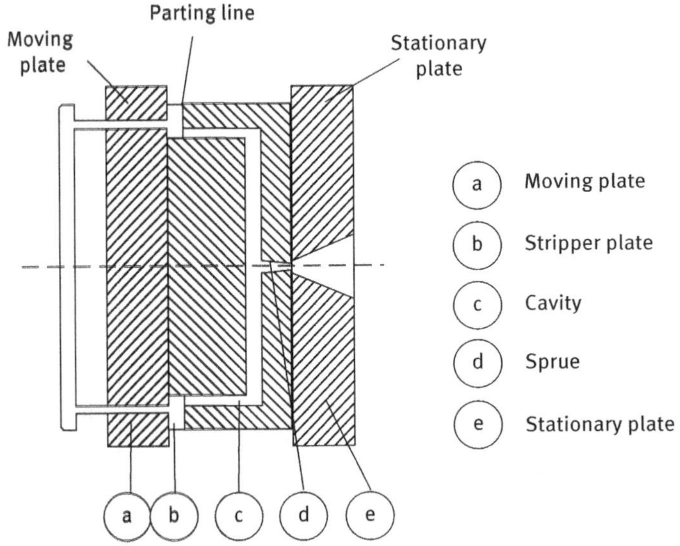

Figure 4.2: Mould with stripper plate.

4.1.4 Three-plate mould

These are normally used when multicavities are involved and semi or fully automatic working is required. This type of mould, as its name suggests, has an extra

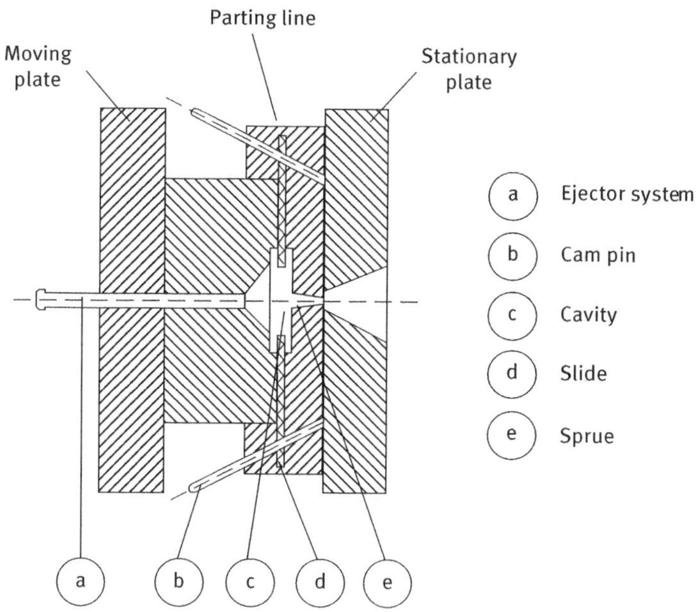

Figure 4.3: Slide mould.

plate (Figure 4.4), and usually continues the gate on one of its sides with the complete runner system, preferably trapezoidal. The opposite side of the plate carries part of the mould form (usually the female part).

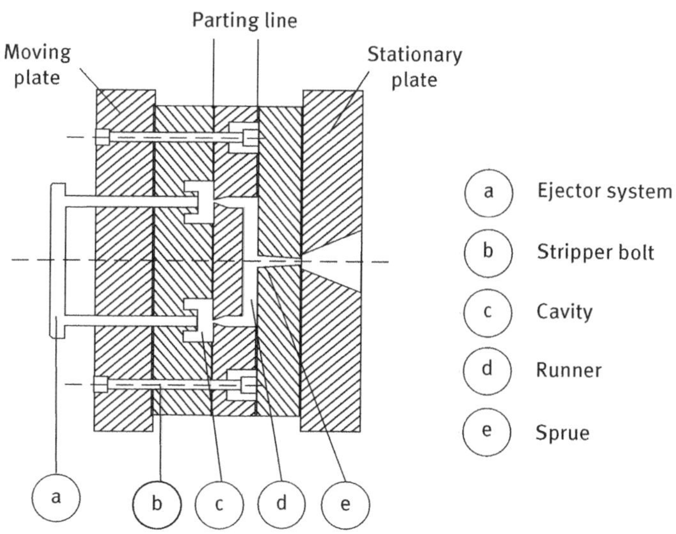

Figure 4.4: Three-plate mould.

When the mould opens, the plate is separated by means of a delayed action mechanism (e.g., chains or length bolts), hence breaking the restricted gate. The mouldings are then ejected from one daylight and the sprue and runner system are ejected from the other.

The successful ejection of mouldings relies on clean separation of the moulding and gate at the parting line – Figure 4.5 illustrates the sequence. With this method of tooling, restricted gates of the correct design must be used.

Multiplate moulds are usually more expensive than two-plate moulds and can lead to slower production if an operator has to remove the sprue and runner system when the mould is open; this can usually be avoided by providing automatic ejection of the sprue and runner. The distance travelled by the plates is governed by the length of the chain or the length of the bolts used to separate them.

4.2 The feed system

4.2.1 Introduction

The feed system accommodates the molten polymer coming from the barrel and guides it into the mould cavity. Its configuration, dimensions and connection with the moulding greatly affect the mould filling process and subsequently, the quality of the product. A design that is based primarily on economic viewpoints, (rapid solidification and short cycles) is mostly incompatible with quality demands. The two main areas that need to be considered are the runner system and the gate.

4.2.2 The runner

When designing runner systems the three primary considerations are:
1. The shape of the runner.
2. The runner layout.
3. The runner dimensions.

More specific demands of the runner design could include the following points:
1. The cavity should fill with a minimum of weld lines.
2. The cavities fill at the same time.
3. Restrictions to flow should be as low as possible.
4. Share of the total shot weight should be as low as possible.
5. Should be easily demoulded.
6. Appearance of the product should be unaffected.

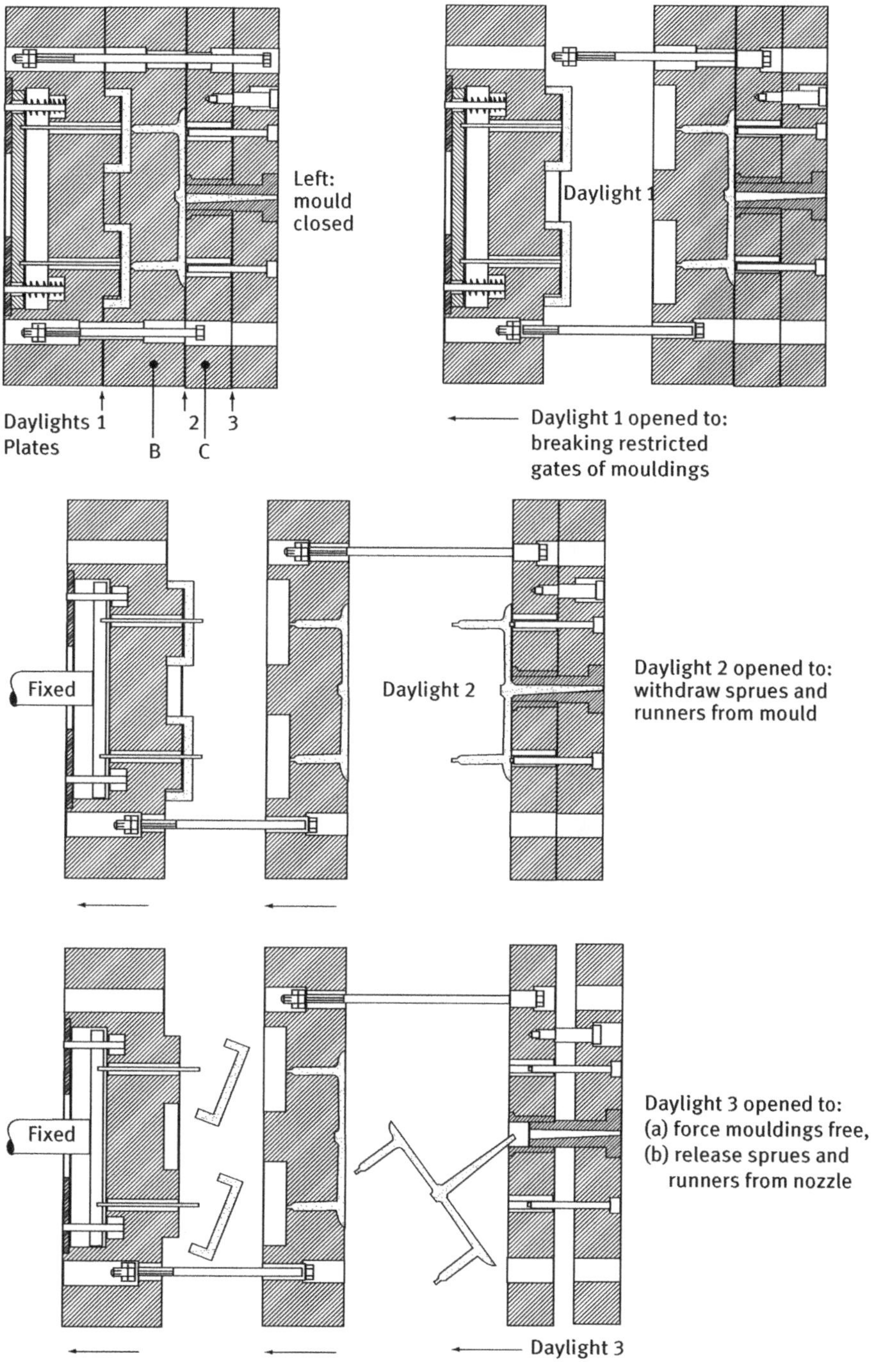

Figure 4.5: General assembly and operation of a typical three-plate mould.

7. Length as short as is technically feasible to reduce losses in temperature and pressure and keep scrap to a minimum.
8. Cross-section as large as required to allow a longer or equal freezing time to that of the component (to allow effective packing of the part).

4.2.3 Runner shape

The cross-sectional shape of the runner used in a mould is usually one of four forms as shown in Figure 4.6.

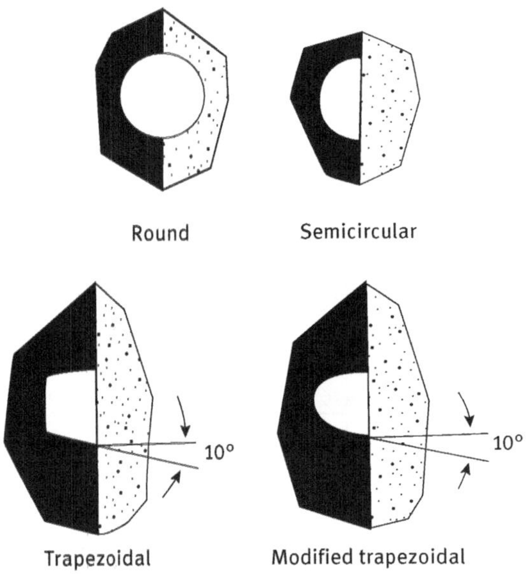

Figure 4.6: Cross-sectional shapes of runners.

The criteria for efficient runner design are that the runner should provide a maximum cross-sectional area and minimum contact on the periphery, when taking heat transfer into account.

As can be seen, the round and modified trapezoidal types of runners are the two most satisfactory designs, whereas the ratios exhibited by the semicircular and trapezoidal systems make their use less desirable.

As the plastic melt progresses through the runner, the melt touching the cold mould surface will rapidly decrease in temperature and solidify. The material which follows will pass through the centre of this solidified material and, because of the low conductivity that thermoplastics possess, the solidified material acts as insulation and maintains the temperature in the central melt flow.

Ideally, the gate should be positioned in line with the central melt flow and a fully round runner will meet this requirement. The trapezoidal runners are not as satisfactory in this respect since the gate cannot usually be positioned in line with the central flow stream.

The main objection to the fully round runner is that the runner is formed from two semicircular channels cut into each half of the mould. It is essential that these channels are accurately matched to prevent an undesirable and inefficient runner system being developed. Due the requirement of extra machining and so on, a fully round runner adds to the cost of the mould, but only marginally. A modified trapezoidal design is preferred if the runner is to be machined in only one half of the mould.

The choice of runner section is also influenced by the question of whether positive ejection of the runner system is possible. In a simple two-plate mould this does not generally pose any problems.

For multiplate moulds, however, positive ejection of the runner system is not practical. Therefore, the basic trapezoidal runner is always specified, the runner channel being machined into the fixed half from which it is pulled as the mould opens. In this way, the runner is free to fall under gravity between the mould plates.

4.2.4 Runner layout

The layouts of the runner will depend upon certain factors:
- The number of impressions.
- The shape of the components.
- The type of mould.
- The type and positioning of the gate.

The runner length should be kept as short as possible to reduce material wastage and to reduce pressure losses. The runner should also be balanced, i.e., the distance the material has to travel from the sprue to the gate should be the same for each cavity. This is illustrated in Figure 4.7.

Sometimes however, it is not always practical to have a balanced runner system (family moulds and so on). In these cases, uniform filling of the cavities can be achieved by varying the runner and gate dimensions as shown in Figure 4.8.

Runner systems can be divided into three main types: standard-runner, cold-runner and hot-runner.

4.2.4.1 Standard-runner

Standard-runner systems are machined straight into the mould plates, their temperature is therefore that of the mould temperature (i.e., usually 20 to 120 °C). The material passes through the runner to the cavities, which are filled and packed by the

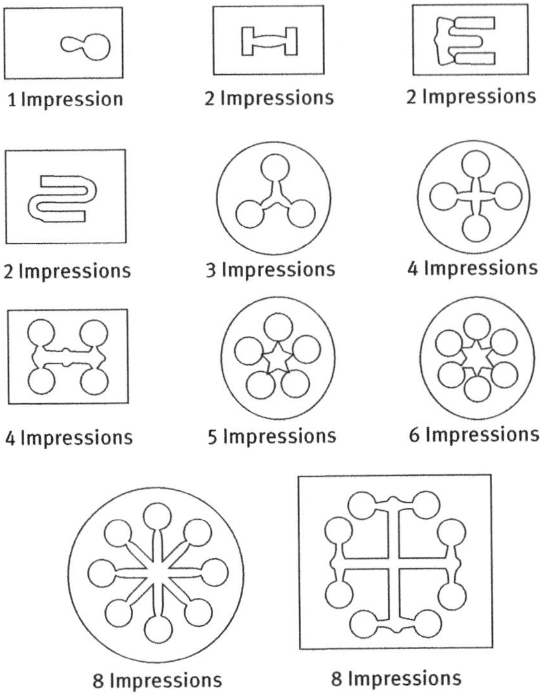

Figure 4.7: Balanced runner layouts: the melt has the same length of travel for all impressions.

holding pressure, and then the molten material in the runner freezes with the rest of the component during cooling.

4.2.4.2 Cold-runner

For reactive materials, such as thermosetting materials or rubber, a cold-runner is required to prevent a premature reaction of the materials in the runner.

4.2.4.3 Hot-runner

The use of the hot-runner technique for feeding multi-impression and large area mouldings is now firmly established. The advantages of hot-runner mouldings are as follows:

- Melt enters the cavities in a more controlled manner than with a sprue and runner system, as temperature control in the hot-runner is adjustable to finer limits.
- A possible reduction in post-moulding finishing operations to remove large sprue gate witness marks.
- The elimination of cold sprues and runners in multi-impression moulds, which would normally be scrapped or reworked.

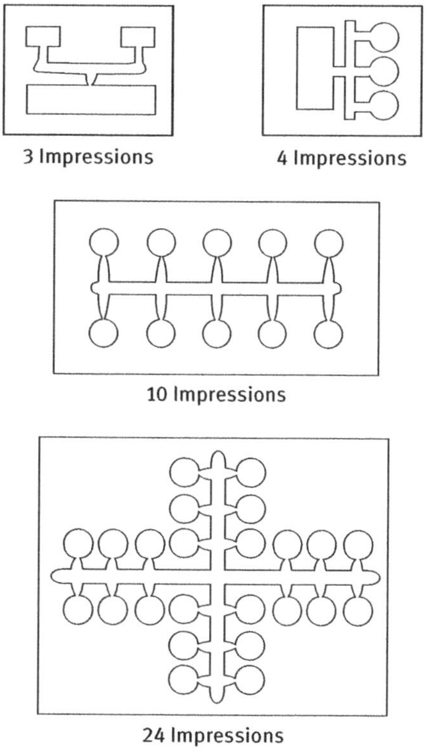

Figure 4.8: Runner layouts which require balanced gating.

- Hot-runners enable single impression, large area mouldings to be edge-gated, whilst keeping the moulding in the centre of the machine platen (Figure 4.9).
- Effective increase in the shot capacity of the machine as, once the hot-runner is filled, the injection capacity can be fully concentrated into the cavities.

In designing hot-runner moulds, the following important points should be observed:
- Provide adequate heating for the hot-runner manifold (1.8 or 30 W/in3) and nozzle (approximately 300 W).
- Make provision for closely controlling the temperature of the manifold and nozzles with suitable instruments.
- Insulate the hot-runner manifold and nozzles from the machine platen or mould cavities by air or compressed temperature-resistant sheeting.
- Provide adequate runner channels in the heated manifold, i.e., minimum 12 mm diameter.
- Make the machine nozzle orifice diameter of similar size to the channels in the hot- runner manifold.
- Ensure that the runner channels are devoid of any sharp corners or blind spots where melt could become trapped and consequently degraded.

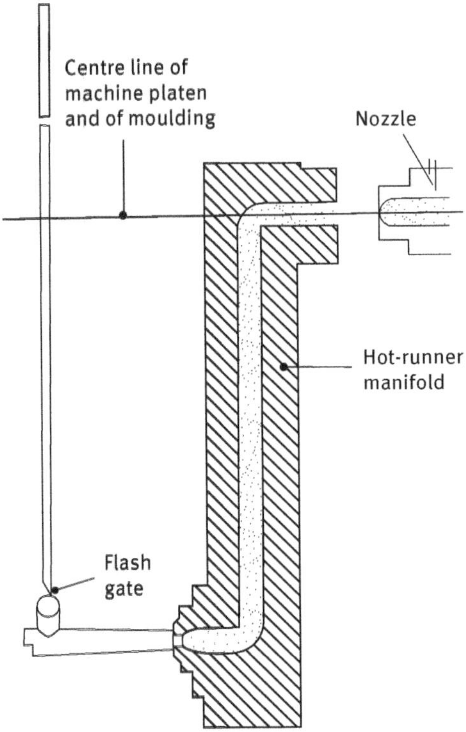

Figure 4.9: Hot-runner layout permitting the mould to be placed centrally on the platen.

Typical hot-runner systems for a thermoplastic have a heated manifold. A hot-runner system has a temperature which is in the range of the melt temperature of the thermoplastic and therefore, significantly higher than the standard-runner system. This ensures the material does not freeze-off prematurely in the runner. Less raw material is required because the runner content does not need to be demoulded and is available for the next shot.

However, the disadvantages of the hot-runner system are:
- More rejects can be expected especially during start-up.
- Higher costs from the purchase and installation of auxiliary equipment, such as heaters, temperature controllers and sensors.
- Long flowpaths and high shear velocity can increase the likelihood of material degradation.

The uneven melt temperature distribution can result in non-uniform filling.

Further various designs and components of hot-runner systems can be seen in Figures 4.10–4.13.

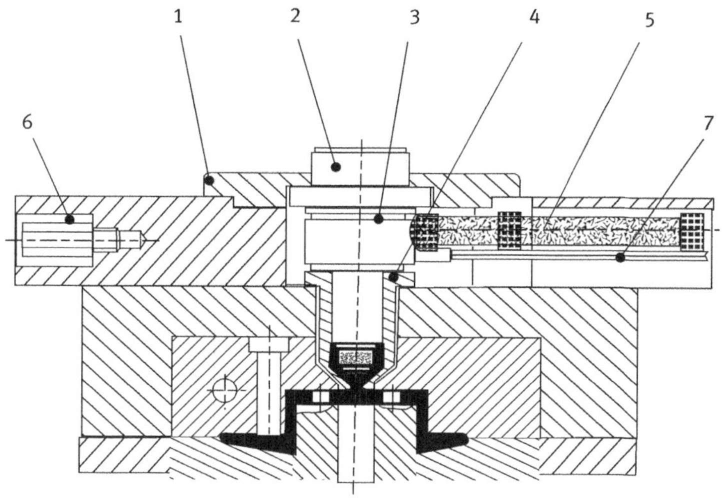

1. Locating ring
2. Machine nozzle adaptor
3. Hot-runner nozzle
4. Gate bushing
5. Extension plug
6. Ground connector
7. Thermoconnector

Figure 4.10: Hot-runner design 1.

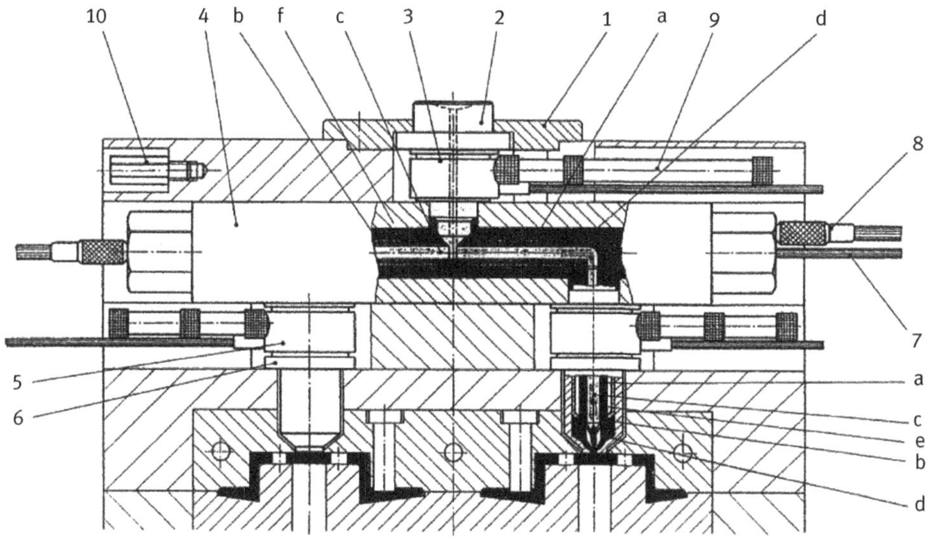

a) Heating tube
b) Insulation
c) Flowing material
d) Solidified material
e) Material tube

1. Locating ring
2. Machine nozzle adaptor
3. Connecting nozzle
4. Manifold
5. Hot-runner nozzle

6. Gate bushing
7. Thermocouple
8. Current supply cable
9. Extension plugs
10. Return connector

Figure 4.11: Hot-runner design 2.

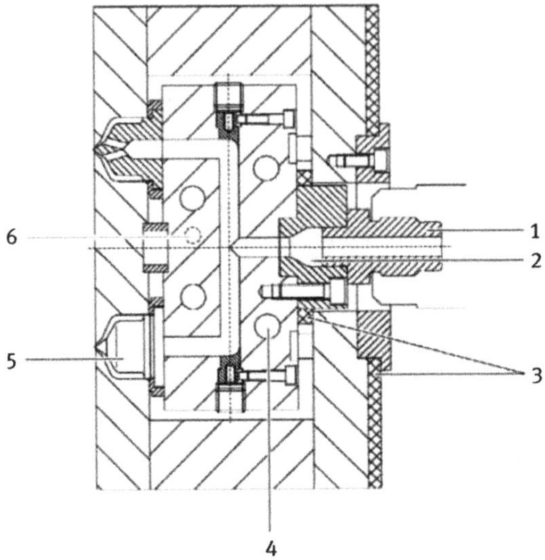

1. Injector cylinder
2. Nozzle
3. Insulation plates
4. Heater
5. Injection points
6. Thermocouple

Figure 4.12: Special hot-runner design (1): injection through 1, 2, 3 or 4 tips with a multitip nozzle.

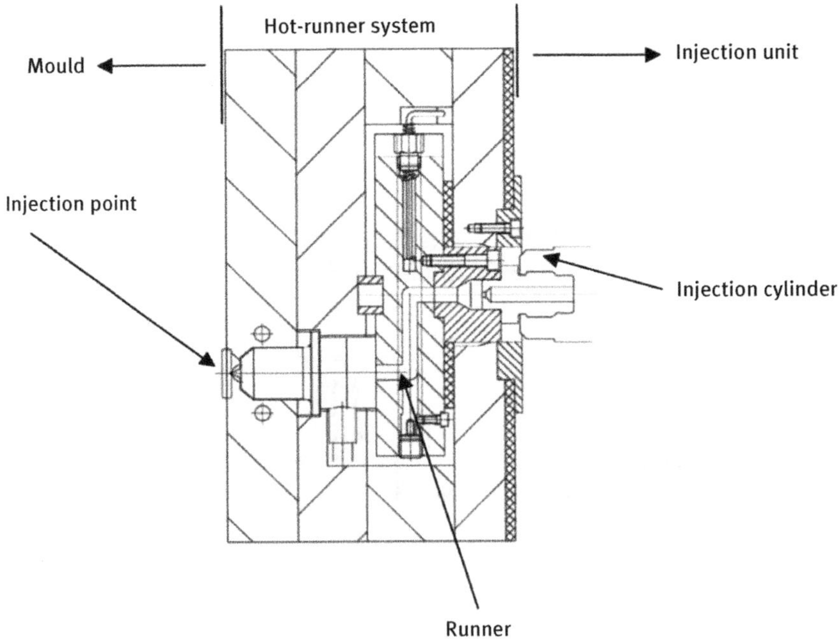

Figure 4.13: Special hot-runner design (2): edge gating of thick-walled parts, no touch-up of gate mark required.

4.3 Gate design

The gate is the connection between the runner system and the moulded part. The objective of the gate is to allow enough material flow for both mould filling and thermal shrinkage compensation. The moulding process and the properties of the final part are directly affected by the type of gate used, the location within the overall moulding and the size. There are a number of gate designs available, generally they fall into two basic types: large or offering only restricted flow.

The type and position of the gate is often dictated by the design of the component and the number of mouldings to be produced in each cycle. For guidance, the following sections provide information on different gating methods.

4.3.1 Sprue gate

This type of gate is the preferred gate and is normally used for single impression moulds, and is especially suitable when the component is cup shaped and involves a base. Compared with a side gate, the material flow is more direct experiencing minimal pressure loss and reduced shearing. However, there can be a zone of high stress concentration caused by the polymer at the gate area, as well as problems with material freeze-off which can lead to shrinking around the gate. There is also the need to remove the gate once the part is finished. This system may be extended to multiimpression moulds in conjunction with a hot-runner assembly. It is illustrated in Figure 4.14.

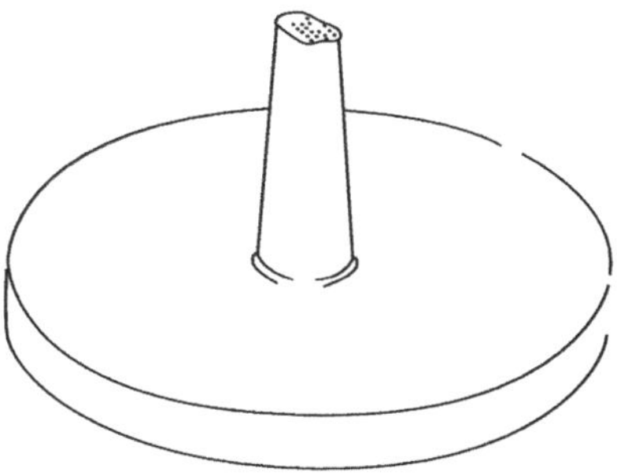

Figure 4.14: Sprue gate.

4.3.2 Restricted gate

This type of gating is used for multicavity tools and is shown in Figure 4.15. Finishing operations can often be eliminated as the small gate is broken off during the ejection of the moulding. The gate must not be too small otherwise the filling of the

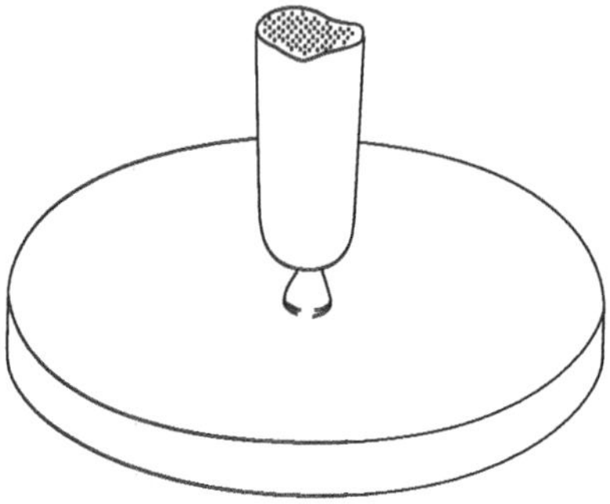

Figure 4.15: Restricted gate.

cavity is impaired. In addition, under the effect of high injection pressures, frictional heating of the material passing through the gate could lead to splash marking and burning on the finished component.

However, the gate must not be made too large otherwise it will not break off satisfactorily during ejection. As a guide, restricted gates should not be smaller in diameter than 0.6 mm or greater than 2 mm. It is also essential to have a generous runner system to prevent premature freezing of the melt.

To prevent any cracking around the gate during the ejection of the moulding (particularly where gates of 1.5–2.0 mm are being used), the gate should have a slight back taper so that it breaks off about 1.5 mm from the surface of the moulding.

Owing to the notch effect (the notch is recessed and therefore part thickness and strength are reduced), restricted gates should be located at a point in the moulding exposed to low mechanical stresses. Furthermore, where a clean finish is required, the pronounced orientation of the material in the gate area often hinders the removal of the gate mark by milling, due to small cracks occurring along the lines of orientation; hence care should be taken in the removal of any restricted gates.

4.3.3 Side or edge gate

This type of gating is shown in Figure 4.16 and is normally used for multi-impression moulds where components are relatively small and of a flat or shallow nature. The size of the gate is dependent upon the shape and thickness of the moulding. For thick sections, the gate thickness should be approximately 75% of the component thickness and as wide as the runner. With multicavity moulds, where the gates are arranged in series, it is necessary to balance the filling of the cavities. This is not always easy to predict at the design stage of the mould and it may be necessary to complete the balancing operations using trial runs. Generally, the gates furthest from the sprue are given the largest cross-section and those nearest the sprue the smallest.

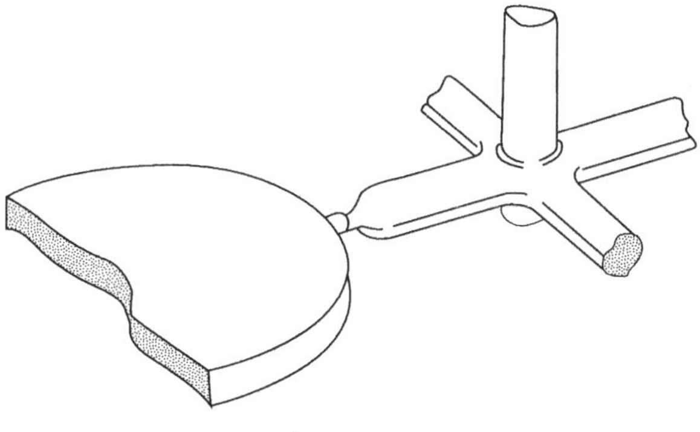

Figure 4.16: Side or edge gate.

4.3.4 Flash gate

This type of gate is shown in Figure 4.17 and can be used quite successfully for long flat components of thin section. It enables a large cavity to be filled quickly and consistently. The length and width of the article and the flow pattern required dictate the length of the gate. In some instances it is advantageous if the gate is the full length of the article, though usually a gate length which is about 50% of the longer side dimension is sufficient.

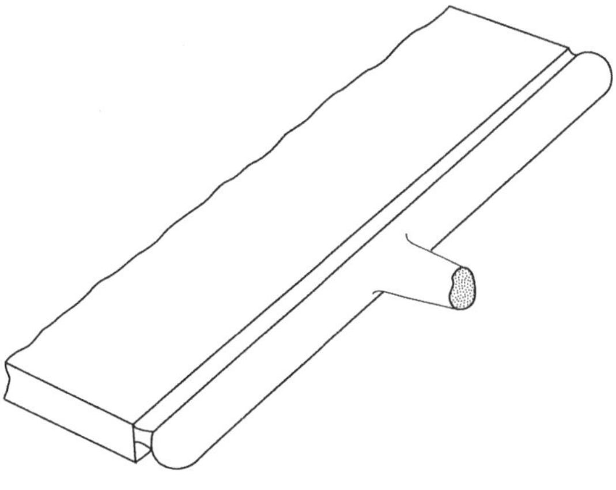

Figure 4.17: Flash gate.

4.3.5 Fan gate

For thick section mouldings, such as optical lenses, this type of gating is often used as it enables the runner to be made of an adequate size to help prevent the material from chilling off when it is injected slowly (as is necessary when making these components). It also allows sufficient follow-up pressure into the cavity during the cooling contraction phase. The design is shown in Figure 4.18.

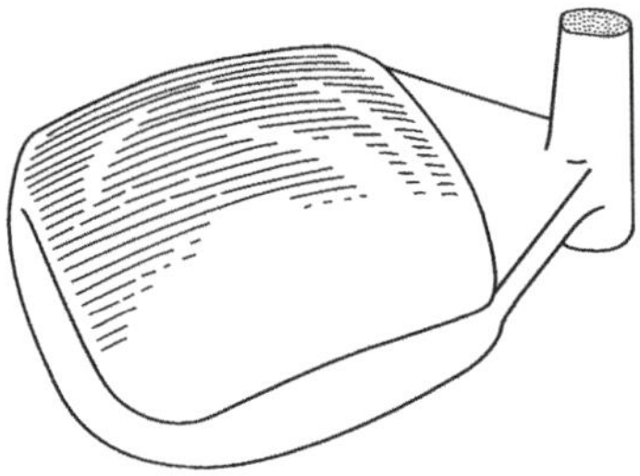

Figure 4.18: Fan gate.

4.3.6 Tab gate

This type of gating can be used as an alternative to side gating, especially in multi-impression moulds, to produce articles of a flat or shallow nature. It has certain advantages over normal side gates in that the design minimises the jetting of material into the mould cavity, which leads to weld lines and flow marks. It also creates turbulence which is an aid to dispersion when moulding dry-coloured material.

Tab gates (Figure 4.19) are normally used to produce elongated articles such as radio scales and rules. The tab in these instances is located towards one end so that the mould cavity is filled evenly down the greater part of its length. The longitudinal orientation of the material tends to strengthen the article and, because the gate is remote from the centre point of maximum stress, it avoids the risk of cracks developing at the gate area if the moulding is subsequently flexed.

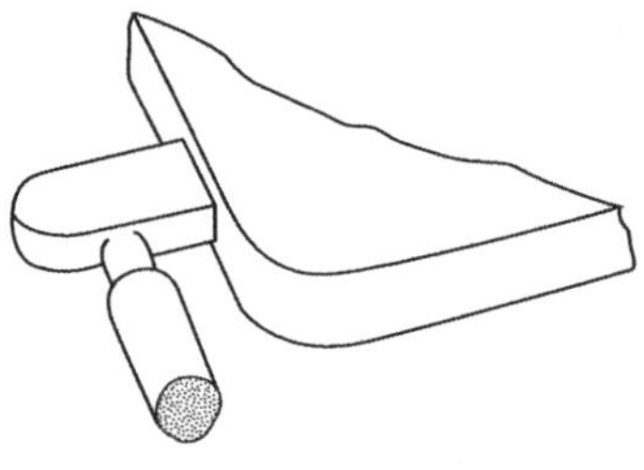

Figure 4.19: Tab gate.

4.3.7 Diaphragm gate

For single impression moulds which are produced with a central orifice, this type of gating can be used to obtain uniform radial mould filling. The diaphragm gate is removed by a subsequent machining operation– it is shown in Figure 4.20.

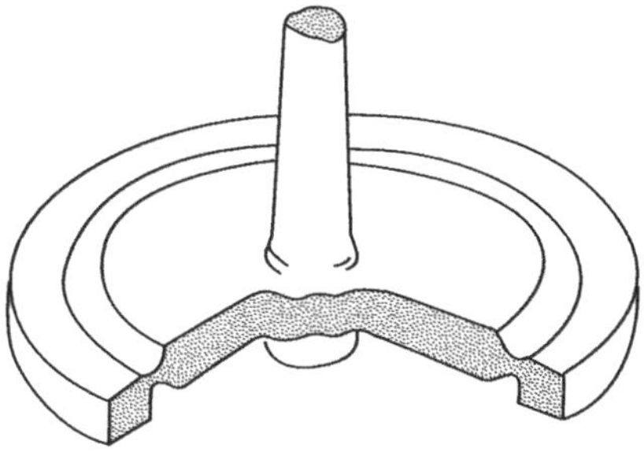

Figure 4.20: Diaphragm gate.

4.3.8 Spider gate

This is a variation of the diaphragm gate and is shown in Figure 4.21. It is normally used for moulding large diameter apertures and helps to reduce material wastage. A disadvantage is that the meeting of the separate flow streams creates weld lines and this factor needs to be considered at the component and mould design stages.

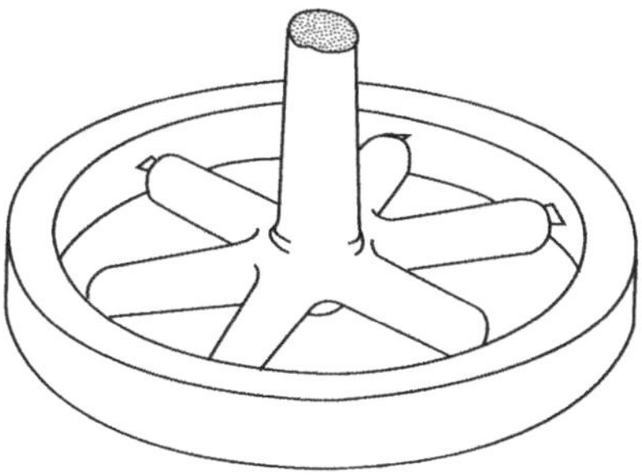

Figure 4.21: Spider gate.

4.3.9 Ring gate

For multi-impression moulds, which are used to produce tubular-type articles, the ring gate ensures consistent filling of the moulds. It also helps to ensure that the core pin is central to the cavity, whereas using an ordinary side gate the initial pressure would tend to displace the core pin and so cause the article to have an uneven wall section. Examples are shown in Figure 4.22.

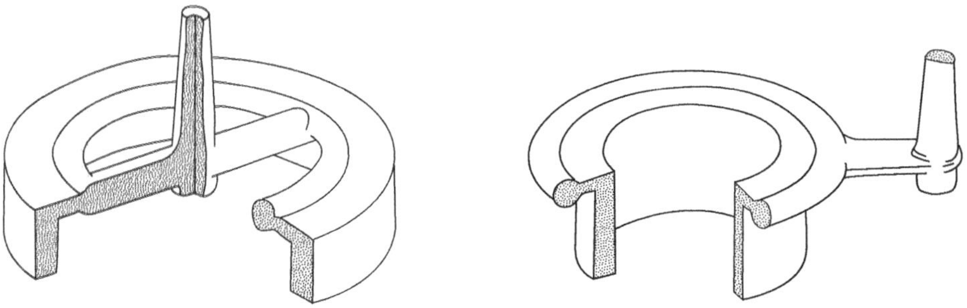

(a) (b)

Figure 4.22: Examples of ring gates.

4.3.10 Hot tip gating

The hot tip gate is the most commonly used gating method. Acceptable for crystalline and amorphous materials, this method relies on gate diameter and cooling at the gate area to optimise the application. The selection of gate diameter and proximity of the gate to the end product are therefore critical to final part quality.

Hot tip gating leaves a small mark on the moulded part surface. The less notch sensitive the thermoplastic or the larger the gate diameter, the more the vestige will protrude. Although the gate mark is only a few tenths of a millimetre high, it is common to recess the gate for larger diameters such that the vestige lies below the part surface. These are referred to as the 'cosmetic' or 'technical' gates, as shown in Figure 4.23, and appropriate details are shown below. Cosmetic gating must leave a sprue mark that is aesthetically acceptable, as opposed to a technical gate which is functional.

Gate diameters range from 0.6 to 1.8 mm and should be selected based on material and part considerations. With any gate diameter, the land should be no more than 0.1 mm. A sharp gate opening is essential to produce the required gate separation for cosmetically superior vestiges.

Gate and land tolerances are generally very tight. Adherence to a specific nominal gate diameter and land, although important, is in no way as critical as maintaining uniformity among all gate and lands.

For example, if one gate is 1.05 mm and all other gates in the system are 1.0 mm in diameter, the larger gate has a 10% greater cross-sectional flow area.

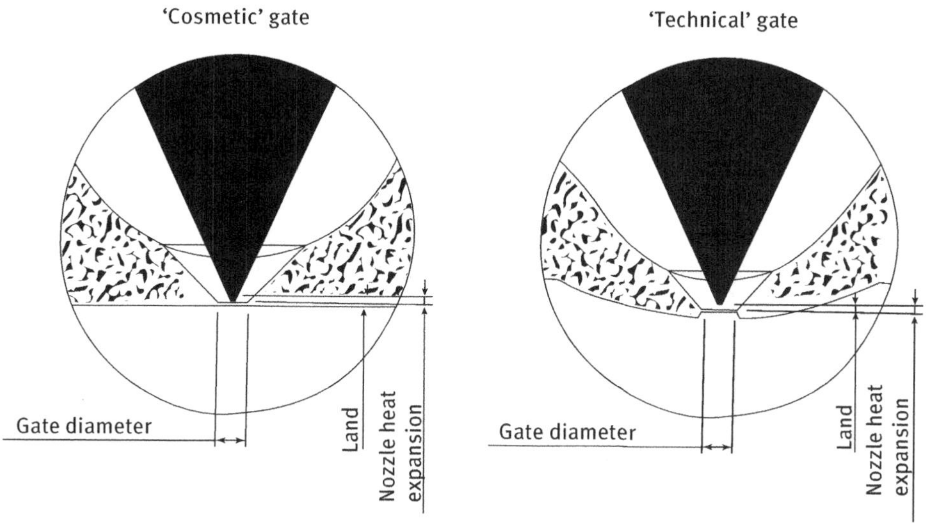

Figure 4.23: Hot tip gate cross-section.

Assuming a Newtonian laminar flow where flow is proportional to the diameter to the fourth power, the total flow in the larger gate is 22% greater; therefore, all gates must be opened to the slightly larger gate diameter to prevent unequal part filling.

4.4 Mould temperature control

With thermoplastics, the main purpose of the mould system is to minimise both the cycle time and thermal differences during mould part cooling. Mould cooling is therefore essential for both cost saving and quality control. Uniform cooling improves product quality by preventing differential shrinkage, high residual stress and mould release problems.

Therefore, when designing an injection mould tool, the size and layout of the cooling channel is an important consideration. Water circulation may be used for cooling or oil if higher temperatures are required. An oil-based system is suitable for temperatures up to 350 °C, whereas temperatures lower than 25 °C require a water/glycol (antifreeze) mixture system.

In order to achieve the uniform cooling of a moulding, the following points should be considered in conjunction with Figure 4.24.

Increases in D lead to a reduction in the heat transfer efficiency, and a large value of 'W' results in a non-uniform tool temperature.

The demoulding temperature requirements of various materials are given in Table 4.1.

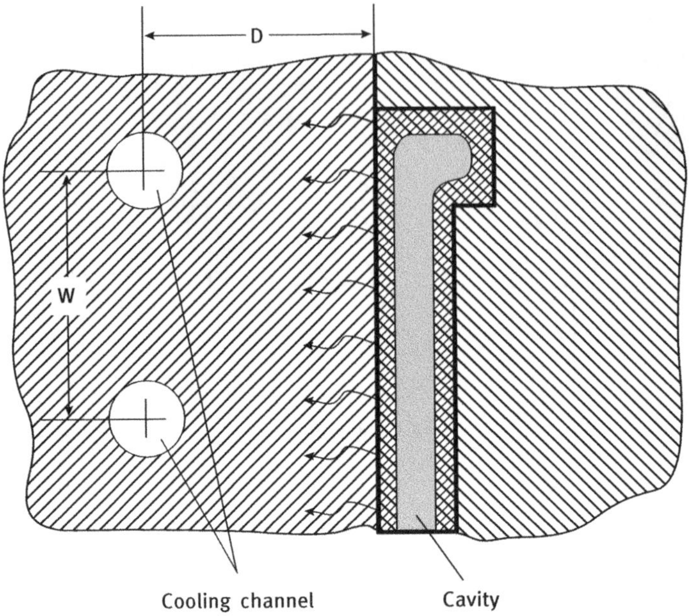

Figure 4.24: Layout of the cooling channel (D lies between d to 5 d and W lies between 2 d and 5 d where 'd' is diameter of the cooling channel, 'D' is the depth of the cooling lines from the moulding surface and 'W' is the width of the pitch).

Table 4.1: Recommended values for the demoulding temperature of different plastics.

Short designation according to DIN 7728	Demoulding temperature in °C		
	Lower temperature	Medium temperature	Upper temperature
Polystyrene	20–35	35–45	45–60
Styrene-butadiene	20–35	35–50	50–65
Styrene-acrylonitrile	35–50	50–70	70–85
Acrylonitrile-butadiene-styrene	35–55	55–75	75–90
Polyvinyl chloride (rigid)	45–65	65–80	80–100
Polyvinyl chloride (soft)	25–35	35–45	45–55
Cellulose acetate	35–50	50–65	65–80
Cellulose acetate butyrate	30–45	45–60	60–75
Cellulose propionate	30–40	40–55	55–70
Polymethyl methacrylate	50–70	70–90	90–110
Modified polyphenylene ether	65–80	80–95	95–110
Polycarbonate	60–85	85–110	110–130
Polyacrylate	120–140	140–160	160–185
Polysulfone	100–130	130–160	160–190

Table 4.1 (continued)

Short designation according to DIN 7728	Demoulding temperature in °C		
	Lower temperature	Medium temperature	Upper temperature
Polyether sulfone	130–145	145–165	165–185
Polyetherimide	135–150	150–170	170–190
Polyamide-imide	200–220	220–230	230–240
Polyethylene (soft)	30–40	40–50	50–65
Polyethylene (rigid)	40–50	50–60	60–75
Polypropylene	45–55	55–65	65–80
Polyamide 6	50–70	70–90	90–110
Polyamide 6.6	75–90	90–120	120–150
Polyamide 6.10	40–55	55–70	70–85
Polyamide 11	60–80	80–105	105–130
Polyamide 12	40–60	60–80	80–100
Polyamide amorphous	55–70	70–85	85–100
Polyoxymethylene	60–80	80–100	100–130
Polyethylene terephthalate	75–95	95–120	120–150
Polybutylene terephthalate	60–75	75–90	90–120
Polyphenylene sulfide	120–145	145–170	170–190
Tetrafluoroethylene hexafluoropropylene	160–180	180–200	200–220
Ethylene-tetrafluoroethylene	140–150	150–160	160–180
Polyarylether ketone	120–145	145–160	160–180
Liquid crystal polymer	60–100	100–140	140–180
Thermoplastic elastomer-ethylene type	25–35	35–50	50–65
Phenol formaldehyde	Mould temperature		
Urea formaldehyde	Mould temperature		
Melamine formaldehyde	Mould temperature		
Unsaturated polyester	Mould temperature		
Epoxide	Mould temperature		
Liquid silicone rubber	Mould temperature		

Because of the importance of the mould heating and cooling cycle to the economics of the process, considerable research has gone into speeding up this process. Rapid temperature cycling (also known as variothermal moulding) has been investigated using a number of approaches, and other mould heating methods instead of basic water or oil that have been investigated include methods such as using high pressure water, steam, induction heating (magnetic), infrared, cartridge heaters, passive heating to name but a few. Most of these methods are not well used (if at all) by industry with only high pressure water and induction heating being occasionally used. However, a low thermal mass mould coupled with a rapid heating and cooling cycle is extremely commercially important and thus of considerable interest to the industry.

Thermosets, unlike thermoplastics, require heating in the mould and not cooling. The demoulding temperature is therefore the temperature the mould is set at for crosslinking. More details on controlling cooling and heating systems can be found in Chapter 8.

4.5 Ejection systems

After a component has solidified and cooled down, it needs to be removed from the mould cavity. Ideally, this is achieved *via* gravity and the part falls to the floor as shown in Figure 4.25. However, some components with design features such as undercuts, adhesion or internal stresses may have to be removed from the mould manually or by robots.

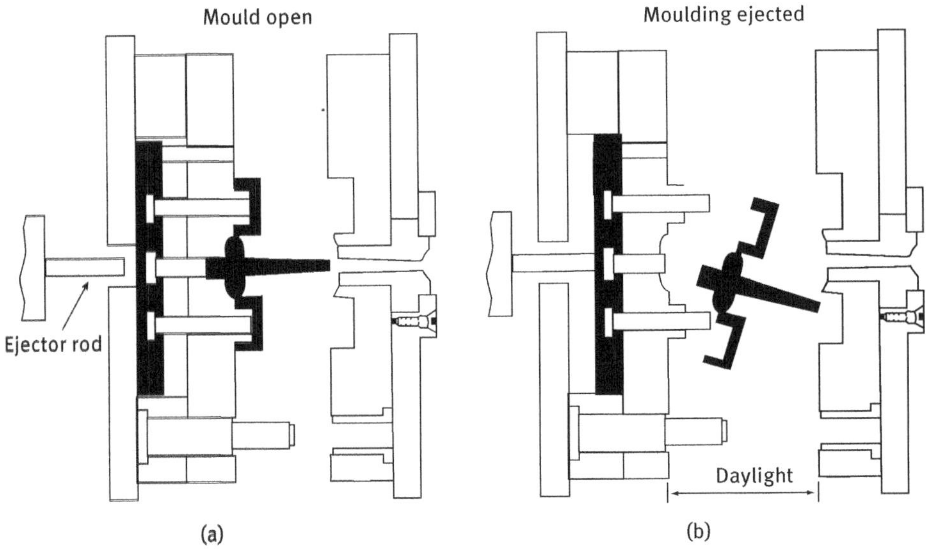

Figure 4.25: Moulding ejection.

The ejector rod in the resting position (Figure 4.25a) moves forward to a maximum preset position, which is set to push and release the moulding out of the cavity as shown in Figure 4.25b.

Ejection equipment is usually actuated mechanically by the opening stroke of the moulding machine. If this simple arrangement is insufficient, ejection can be performed pneumatically or hydraulically.

The ejector system is normally housed in the movable half of the mould. Mould opening causes the mechanically actuated ejector system to move towards the parting line and to eject the moulding. The result of this procedure is that the moulding stays on or in the movable mould half, which can be achieved by undercuts or by letting the moulding shrink onto a core. Taper and surface treatment should prevent too much adhesion. Methods of ejection for various components are shown in Table 4.2.

Table 4.2: Types of ejection systems.

Type	Ejection method	Components of operation	Applications
Standard system for small parts	During opening, the stroke is in the direction of demoulding. Ejection with pins, sleeves or stripper plate	Various: mechanical, hydraulic, pneumatic, manual, machine stop, lifting cylinder, cam, pivot, inclined plane, thrust plate	Mouldings without undercuts
Direction of ejection towards the movable side, stripping is used but usually for circular parts only	During opening, the stroke pulls in the direction of demoulding, ejection is *via* a stripper plate	Mechanical, hydraulic, pneumatic, stripper bolt, lifting cylinder, pin-link chain	Cup-like mouldings with internal gate
Demoulding at two parting lines for automatic operation including separation of the gate	During opening, the stroke thrusts in the direction of demoulding, ejection is *via* pins, sleeves or a stripper plate	Mechanical, stripper bolt	Mouldings with automatic gate separation
Demoulding of parts with local undercuts (slide mould)	During opening, stroke thrust is in the direction of demoulding, ejection is *via* pins, sleeves or a stripper plate after release of the undercut	Mechanical, cam pins, slide mechanism	Flat parts with external undercuts (threads)
Demoulding of large, full- side undercut (split-cavity mould)	During opening, stroke thrust is in the direction of demoulding, ejection is *via* pins	Mechanical, hydraulic, springs, links, pins, cams	Parts with external undercuts (ribs) or opening in side wall
Air ejectors usually provide support. Breaking is achieved mechanically	Thrust is in the direction of demoulding causing a first air shot followed by ejection with compressed air	Mechanical-pneumatic in stages	Cup-like, deep parts

4.6 Venting

Another design aspect of tooling is the need to provide vents for compressed air and gases to escape during moulding. Trapped air and gases can cause a variety of moulding defects, which are more fully described in Chapter 9, such as:

- Short shots (incomplete filling of the mould).
- Scorching or burning.
- Shrinkage (often seen as ripples or depressions in finished parts).
- In extreme cases volatile gases may cause etching on the mould surface.

Common venting methods are to incorporate parting-line vents, vent plugs and pins. More recent developments include the use of porous metals that allow gas to escape but not the polymer, and these materials often also allow for the venting area to be increased. Thermoset injection moulding must have a greater volume to allow the venting of vapour generated by the chemical crosslinking process.

4.7 Conclusion

This chapter has introduced the basics of tooling for injection moulding. It can hopefully now be appreciated how the machine, material and tool all play a part in the production of a successful injection moulding. From information here, and given in other chapters, on processes and machines, a more detailed picture can be built of the requirements of a mould tool.

Before injection moulding is selected (Chapter 1) – is there a better form of manufacturing this component? From Chapter 2 – is this the right machine for this tool? Will it physically fit; will the machine provide the required shot capacity and clamping force?

From Chapters 6 and 7 – is this the right material, how do properties such as shrinkage affect the design? Does the material flow sufficiently to fill the mould tool?

From Chapter 10 – does the part require insert or secondary operations?

From this chapter, it can be seen that as well as the type of tool, consideration is also needed when selecting the properties of the tooling material, such as rigidity and strength, heat transfer and thermal expansion. Depending on the moulded resins chosen, abrasion and corrosion may also need to be considered, as well as maintenance set-ups. The tool may require extra venting to be included, and the cooling channels need to be designed to optimise the heating and cooling cycle. The part will need to be ejected, including hot-runner removal if needed, and whether this is a multicavity process may also need to be considered.

Therefore, while this chapter serves as a brief introduction it should be remembered that the design of mould tooling is a specialist area in itself and the reader is referred to more specialist publications for more details.

Suggested further reading

1. G. Menges, W. Michaeli and P. Mohren in *How to Make Injection Molds*, 3rd Edition, Carl Hanser Verlag GmbH & Co. KG, Munich, Germany, 2001.
2. D.V. Rosato in *Plastics Engineered Product Design*, Elsevier Advanced Technology, Amsterdam, The Netherlands, 2003.

5 Process control systems

5.1 Introduction

The control system ensures repeatability during the moulding operation. It monitors both the hydraulic system and/or electric system as well as process parameters such as temperature, injection speed, screw retraction speed, and injection and back pressure. The ability to control the process has a direct impact on final part quality, part to part consistency and operational costs. The nature of the control system may vary from a simple relay switch to a complex microprocessor system with closed-loop control. Some of the components of the machine control system will now be introduced.

5.2 Explanation of the different concepts in control and regulation technology

5.2.1 Hydraulic or electric systems?

Machines can be driven by 100% hydraulic systems, 100% electrical systems or hybrid machines which combine some of the advantages of both systems. Whether either drive system is superior for all applications is a matter for fierce debate within the industry, but for specific applications one drive system may be advantageous over another. This section continues with an introduction to hydraulic systems, electric systems and hybrid drive systems.

5.2.1.1 Pump

The hydraulic pump generally draws hydraulic fluid from the supply reservoir and delivers it to the pump outlet. From here, it is conveyed through valves to the consumer and then returned. Figure 5.1 illustrates the symbol used to show a pump.

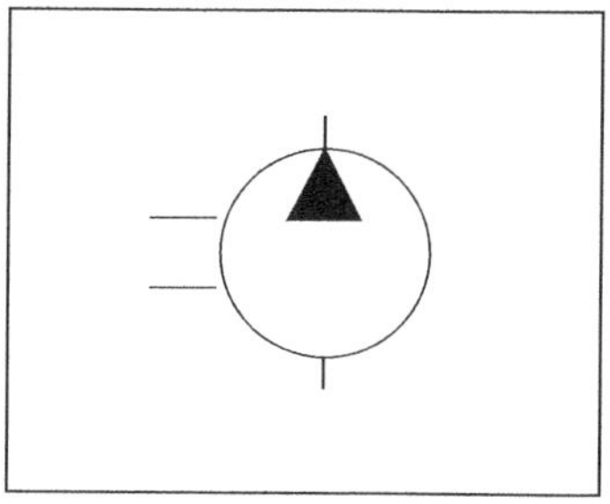

Figure 5.1: Symbol for a hydraulic pump with constant displacement volume.

https://doi.org/10.1515/9783110654813-005

5.2.2 Motor

Hydraulic motors transform the hydraulic energy supplied by the pumps back into mechanically utilised working force which has a rotary motion. Figure 5.2 illustrates the symbol used to represent a motor.

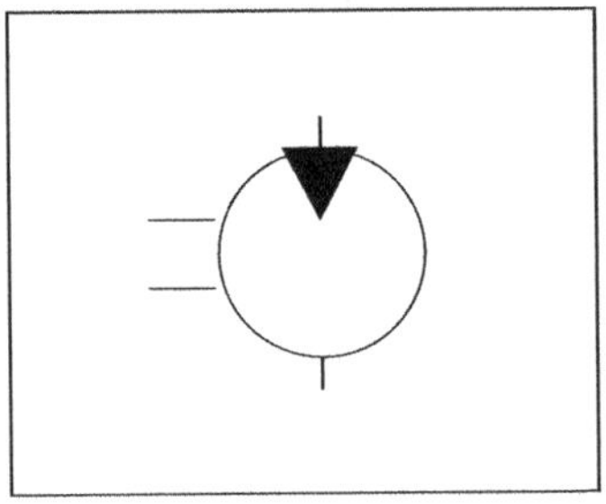

Figure 5.2: Symbol for a hydraulic motor with constant displacement volume.

5.2.3 Cylinder

The cylinder is charged with hydraulic fluid through valves in the base and the head, which results in motion being transferred through the piston surface of the working cylinder to the piston's connecting rod. Single- and double-acting cylinders are shown in Figures 5.3 and 5.4.

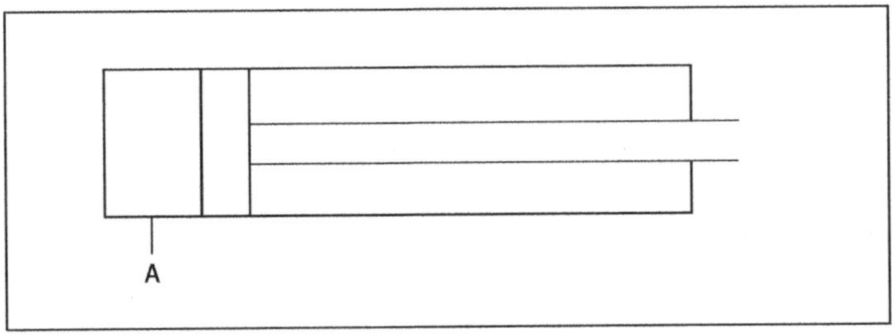

Figure 5.3: Single-acting cylinder with single-side connecting rod (force applied in only one direction). Refer to Table 5.1 for connection designation.

5.2.4 Directional valves

The function of directional valves is to close off different hydraulic lines from one another or to open them, and to continually create alternating line connections. In this manner, the effective direction of pressures and volume flows is regulated, and

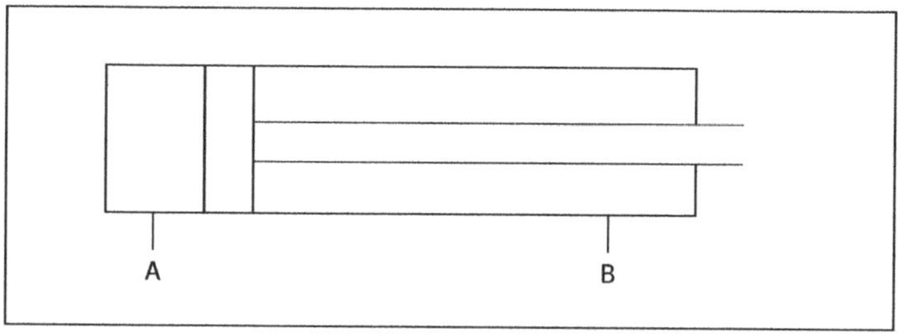

Figure 5.4: Double-acting cylinder with single-side connecting rod (force applied in both directions, uneven surfaces). Refer to Table 5.1 for connection designation.

Table 5.1: Connection designations.

P	Pressure connection
T	Tank connection
A, B	Working connections
L	Leakage return connection
X, Y	Control connections

the starting, stopping and the direction of motion of the consumer (cylinder or hydraulic) motor are thus controlled. A directional valve can have a number of connections and operational positions, which is of great significance and indicated by a prefix before each location. Every operational position is represented by a square, and arrows and dashes within the square identify the connections between the lines. The simplest form of a directional valve has two connections and two operational positions as shown in Figure 5.5 and 5.6. A more complex 4/3 directional valve is shown in Figure 5.7.

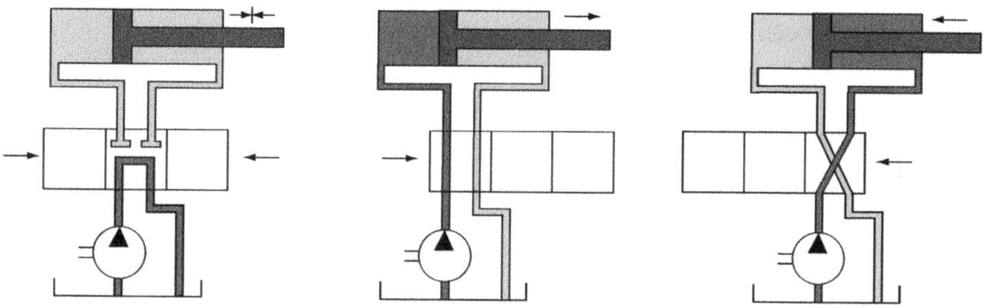

Figure 5.5: Directional valves.

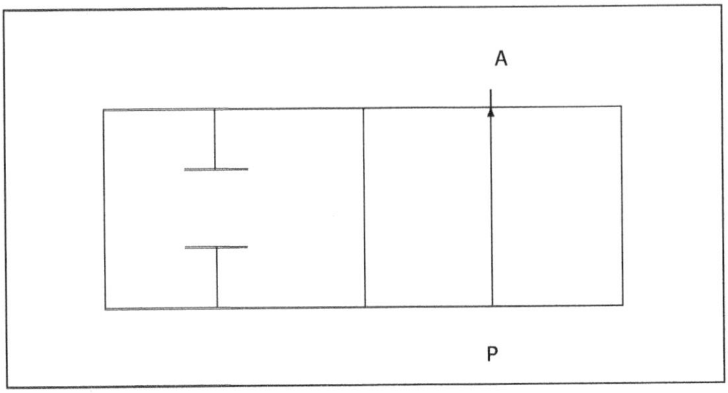

Figure 5.6: 2/2-port directional valve. Refer to Table 5.1 for connection designation.

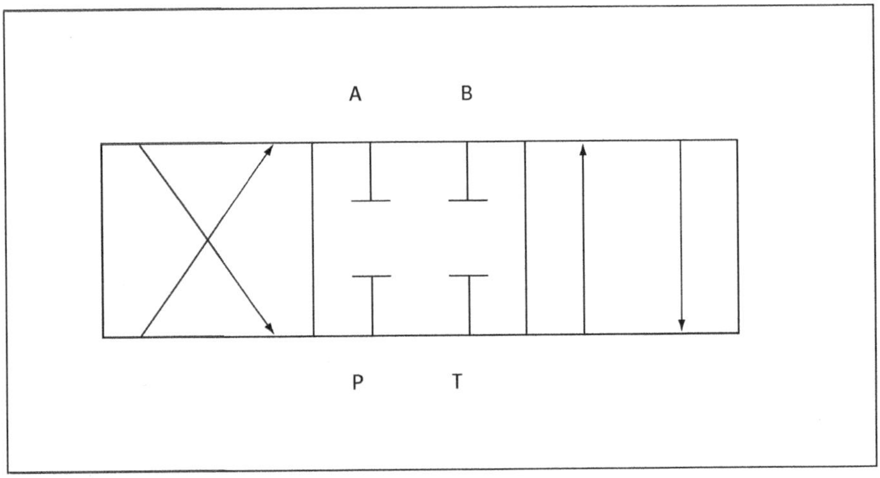

Figure 5.7: 4/3-port directional valve (4 is the number of connections, 3 is the number of operational positions). Refer to Table 5.1 for connection designation.

The outcome of the operational position is clear as soon as the total set of operating symbols has been situated into the fixed section of the service lines.

Larger directional values are hydraulically actuated by a pilot valve, which is controlled either electrically or pneumatically. A piloted directional valve is shown in Figure 5.8. The connection designations are shown in Table 5.1.

5.2.5 Pressure valves

Pressure control valves have the primary task of limiting pressure in the system and thus protecting individual components and lines from rupturing or overloading. The

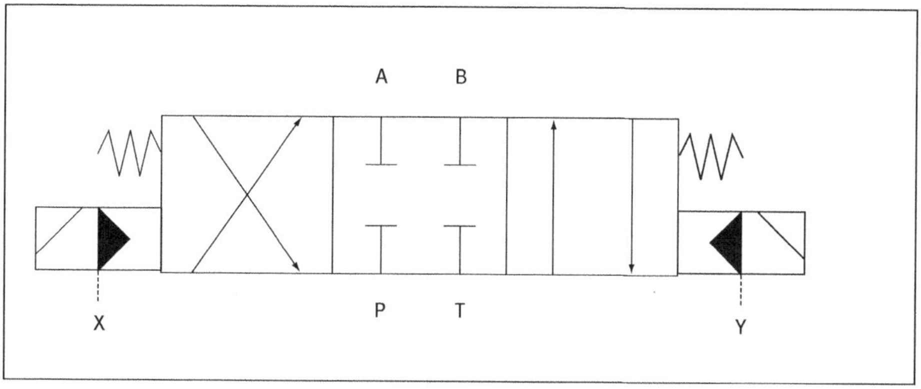

Figure 5.8: Piloted directional values. Refer to Table 5.1 for connection designation.

valve opens when a predetermined pressure is reached and conveys the pump's excess delivery flow back into the tank. A pressure control valve is shown in Figure 5.9. Pressure limiter valves limit the pressure in a specific circuit for a specific consumer. They close an open connection directly when the applied pressure exceeds a preset value. This is shown in Figure 5.9b.

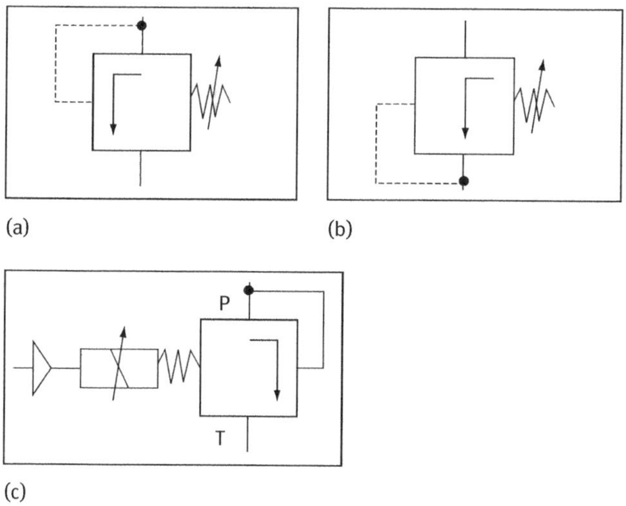

Figure 5.9: A pressure control valve (a), a pressure limiter valve (b) and a proportional pressure valve (c).

Proportional pressure valves convert an electrical input signal in the form of a voltage from 0…10 V, which is proportional to a value of hydraulic pressure. In principle, they are electrically controlled pressure valves in which the manual setting device is replaced by an electrical setting actuator, called the proportional solenoid. This is shown in Figure 5.9c.

5.2.6 Flow-regulator valves

The task of the flow-regulator valve is to regulate volume flow by changing the diameter of the valve governor, and thus to control the speeds of cylinders and hydraulic motors.

The two-port directional flow-regulator valve is installed when a constant speed is required for a consumer, regardless of load. Throughflow is independent of the pressure differential imposed upon it.

Proportional control valves or flow-regulator valves directly change the throughflow in proportion to an electrical input current (U = 0...10 V). The valves consist essentially of a sliding valve with precise control stages whose opening diameter is changed by a proportional solenoid. A proportional flow-regulator valve is shown in Figure 5.10.

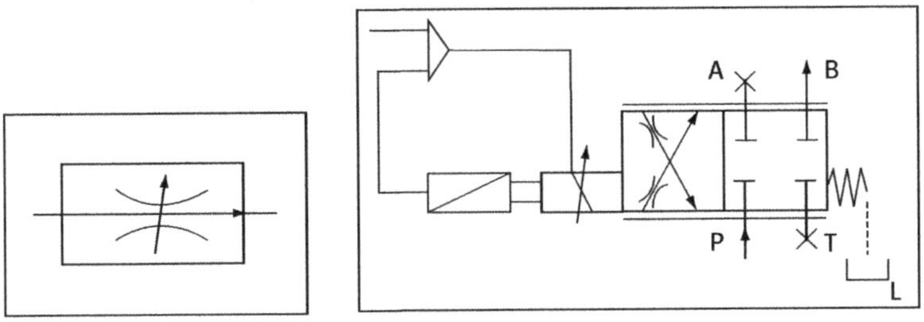

(a) (b)

Figure 5.10: A two-port directional flow-regulator valve (a) and a proportional flow-regulator valve (b).

5.2.7 Location and setting of proportional valves

The proportional valves for combined pressure and flow regulation are set directly on the radial piston pump. A further proportional pressure valve (mounted on the outer right-hand side of the hydraulic manifold) is responsible for the regulation of back pressure.

It is therefore possible to digitally set the pressure and volume with a position controlled proportional valve, which operates by:

- With proportional valves, a proportional offset of the magnetic armature is obtained and is dependent on the strength of the current applied to the magnets.
- The proportional magnet' is supplied with 24 V DC and is pulse-width modulated.

- The stroke of the magnetic armature is detected by an inductive displacement transducer and is transmitted to the electronic regulation on the proportional valve signal board where the difference between the command signal (input voltage) and the feedback signal (actual voltage proportional to the displacement of the transducer) is regulated.
- Position control of the magnetic armature is achieved.

5.2.7.1 Valve bodies

Depending on the type of valve, the valve body is designed as follows:
- As a throttle valve with the volume proportional valve.
- As a cylinder seat valve with the pressure proportional valve.

5.2.7.2 Function of the volume proportional valve

The volume proportional valve is located on the pump output channel and is double supplied from the pump *via* P and T to A and B, as shown in Figure 5.11.

The armature of the position controlled proportional magnet acts directly on the valve piston and pushes it directly against a centring spring. The position of the valve piston dictates the cross-section of the opening and consequently the flow. When the main switch is turned on, the proportional magnet is immediately supplied with DC voltage and the valve piston is set in the (P and T to A and B is closed) preadjusted zero position.

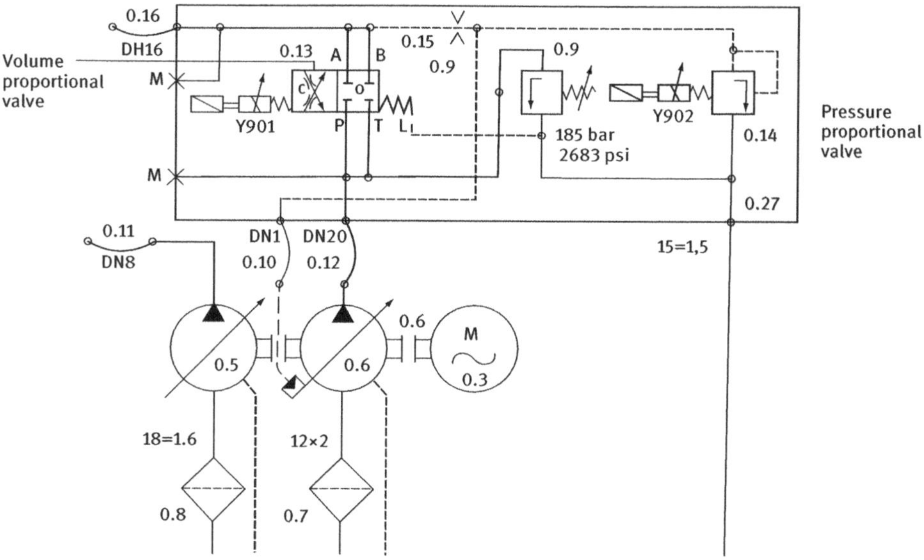

Figure 5.11: Volume and pressure proportional valve (an indication is given of what an actual system looks like).

Regulation of the position of the proportional amplifier, in conjunction with the displacement transducer, ensure that the magnet and valve piston go to and remain in the preadjusted position.

5.2.7.3 Function of the pressure proportional valve

Should the operating pressure of the pump rise above the preset value, this valve signals a reduction of the feed flow pressure to the pump regulator.

The flow supplied by the pump automatically reduces when the demand from the user is zero, i.e., the pump idles.

The 0.9 mm orifice plate acts as a flow control limit as the pressure proportional valve so demands (volume reduction because valve Y902 can only drain off 3 l/min).

When the main switch is turned on, the proportional magnet is immediately supplied with DC voltage and it pushes the valve spring to the preset zero position.

The position regulation of the proportional amplifier, in conjunction with the position transducer, ensure that the magnetic armature (pretensioned valve spring) goes to and remains in the preselected position.

A block diagram of a typical system is given in Figure 5.12.

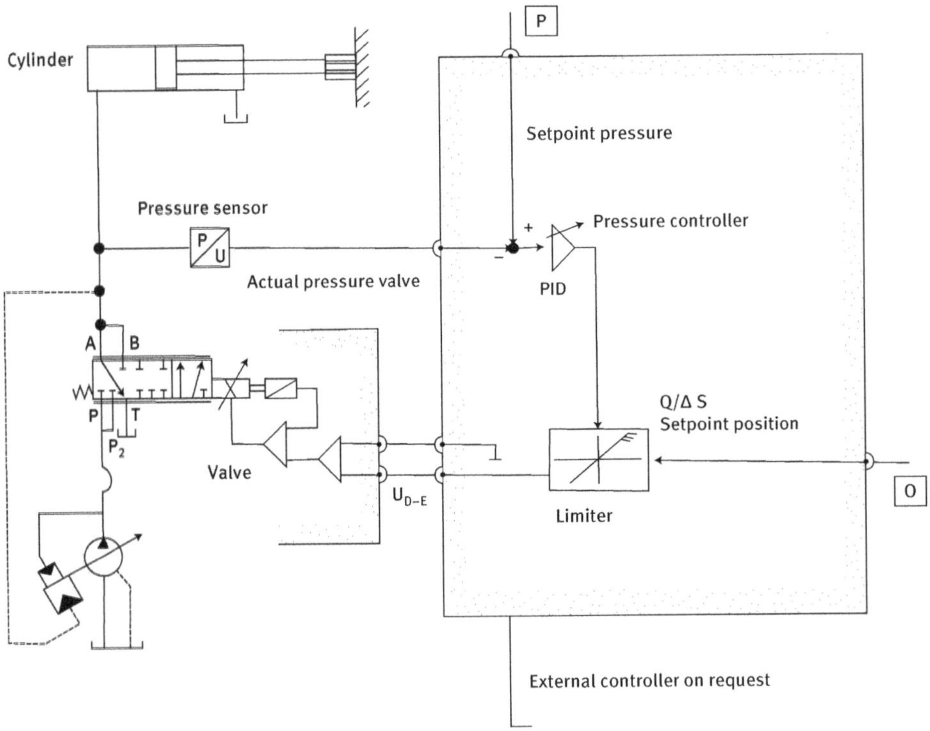

Figure 5.12: Hydraulic system block diagram.

5.2.8 Check valves

Non-return valves have the task of blocking the volume of flow in one direction and allowing free-flow in the opposite direction. In the closed direction, the valve should provide a completely leakproof seal and balls or cones are primarily used as sealing elements.

In order to hold a cylinder in position when external forces are being firmly applied, it is not enough just to block the working connections of the directional valve. Leakage in this operational position leads to a decrease in the load and in order to avoid this, unlocking non-return valves are used. With these valves, the closed position can be overridden by the control of the valve cone and hence throughflow, which was blocked in one direction, is thus permitted. Both types of non-return valves are illustrated in Figure 5.13.

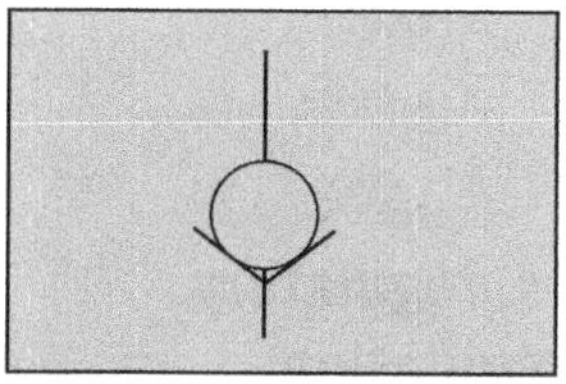

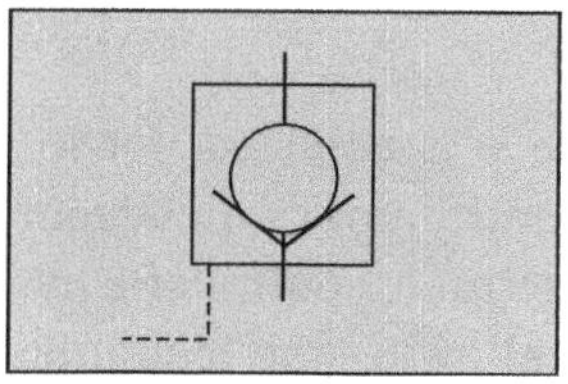

Figure 5.13: Symbols for a non-return valve (top) and unlocking non-return valve (bottom).

5.2.9 Receivers

Hydropneumatic receivers have the task of collecting and storing hydraulic energy, and then releasing it on demand. This type of receiver is used in conjunction with injection moulding machines (IJM) with very rapid injection (using an accumulator). Here, a high volume flow which can be partially accessed from the receiver is required periodically for brief intervals. The benefits of using a hydropneumatic receiver are the option to use relatively small pumps, drive motors and oil reservoirs.

The working principle of a receiver is that it is virtually impossible to compress the hydraulic fluid. If it is nevertheless to be stored under pressure, a gas is utilised, which in this case is nitrogen. The gas is compressed in a pressure reservoir by the hydraulic fluid and decompresses as needed through the release of fluid. In order to ensure that the gas does not mix with the hydraulic fluid, the pressure reservoir is divided into two chambers by an elastic separating wall (membrane). A hydropneumatic receiver is shown in Figure 5.14.

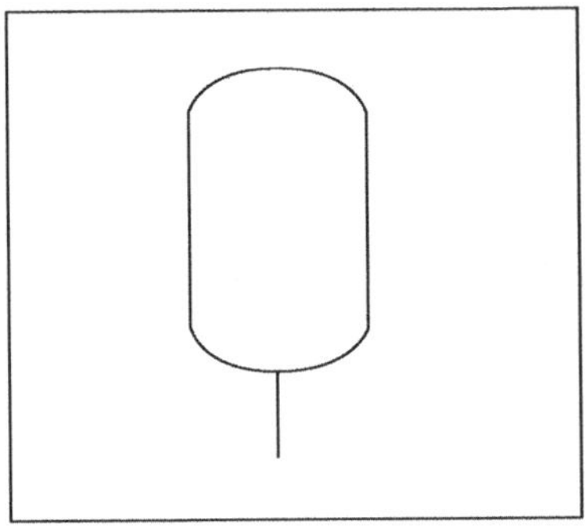

Figure 5.14: A hydropneumatic receiver.

5.3 General information regarding control units, regulators, position regulation and injection process regulation

5.3.1 Control unit

In the control unit, a desired valve – the nominal value – is adjusted, for instance, to change the injection speed in the SELOGICA control unit. Adjustment of this valve is transmitted as a command through the electronic control of a hydraulic valve, which opens to a specified position, thus allowing the hydraulic fluid to flow through at a desired volume rate per second, and so powering the stroke of the screw. The electronic controls and the control valve must be matched so that there is no throughflow of the fluid at a setting of 0% and for the 100% setting a maximum fluid volume is allowed to flow. The amount of fluid which actually flows per injection or the speed with which the screw is moved are not monitored; hence, temperature changes in the hydraulic fluid, viscosity variations in the plastic and other negative factors can create deviations between the adjusted or desired speed (the nominal value) and the true speed (the actual value).

Differences between the actual value and the nominal value are also possible if the characteristic curve of the control valve is not completely linear.

Finally, the most important disadvantage of the control unit should be mentioned: differences between the nominal value and the actual value are not independently recognised or adjusted.

5.3.2 Regulator

Using a regulator, the actual value is measured and compared by the regulator electronics to the adjustment value. When there are deviations of the actual value from the nominal value, the hydraulic valve is reset until the true speed (actual value) agrees with the desired speed (nominal value).

With the regulator, there is hence no relationship between a precisely fixed setting for the hydraulic valve and an adjustment value. The hydraulic valve is reset for as long or as often as is necessary for the measured actual speed to be equal to the desired speed.

For this reason, the regulator requires greater expenditure, from a technological standpoint, than a control unit. Test-data gauging devices for the measurement of the actual values are additionally required (speed and pressure gauges). In addition, faster hydraulic valves are necessary, compared with using a control unit, so that deviations can be corrected immediately.

5.3.3 Screw with position regulator

Position regulation of the screw is made possible by a double-action cylinder. In contrast to a single-action cylinder, this is referred to as a 'contained system'.

With a single-action system, pressure is applied to only one side of the injection piston. Using a position-sensing system and set time scale, a highly precise speed profile can be realised. However, in contrast to position regulation the piston can only be accelerated, it cannot slow down; the way this type of system works is comparable to an automobile without brakes. The defined speed cannot be maintained as quickly and precisely, because the only control options here are 'apply pressure' and 'withdraw pressure.'

With the 'contained system,' pressure can be applied to both sides of the injection piston. The actual pressure value of both piston chambers is measured by pressure gauges and the screw position is read by the position-sensing system. In the contained system, it is possible to define and regulate the position of the screw precisely. With this system, a precise injection speed or holding pressure profile may be followed, since there are now control commands for 'apply pressure injection side', 'withdraw pressure injection side', and also 'apply pressure retraction side' and 'withdraw pressure retraction side'.

A screw with position regulation, as shown in Figure 5.15, ensures a particularly high motive power and reproducibility of the injection process. Here, the hydraulic valve sits directly on the injection unit (for an injection governed by a screw which does not have position regulation, the valve sits on the control block of the injection side) and thus possesses significantly higher switching force from the valve to the cylinder through the short column of fluid. Furthermore, inaccurate control of the

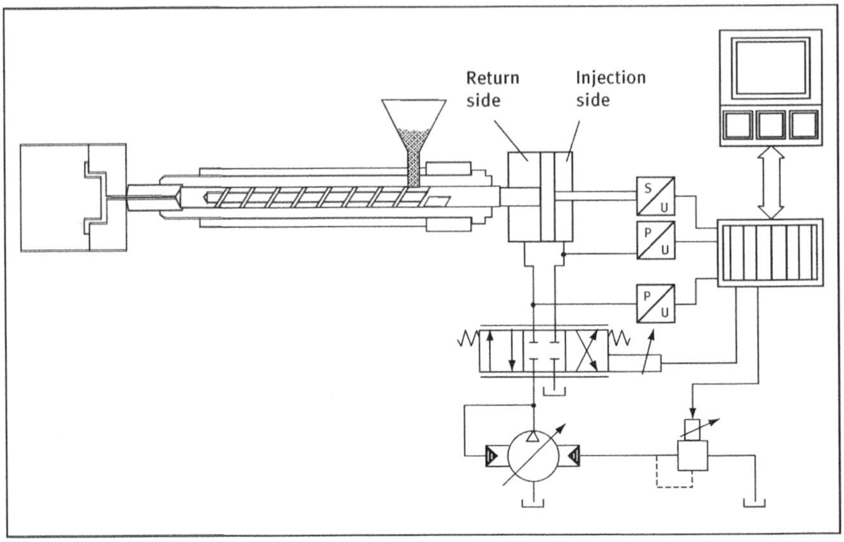

Figure 5.15: Screw with position regulation.

volume and pressure of the hydraulic fluid, such as losses which occur due to the friction and elasticity of the pressure lines, can be avoided. Two pressure sensors are positioned directly on the injection apparatus to determine the pressure differential between the injection side and the return side. Position regulation of the screw has enormous benefits during injection, especially for injection profiles, as well as in maintaining a position which has been achieved. Position regulation also balances outside forces such as gravity in vertical devices.

For an injection with a receiver, the receiver energy is proportioned through the regulator valve and as a result, position and pressure regulation are ensured even with high injection performance demands.

5.3.4 Injection process regulation

The properties on an injection moulded part depend upon the working material and the processing conditions. In the production of a series of parts, a certain deviation in quality features such as weight, dimensional consistency and surface characteristics must always be taken into consideration. These deviations vary from machine to machine and from material to material. Furthermore, external influences or negative factors have an effect on the quality of an injection moulded part. Examples of such negative factors where changes cannot be made are changes in the viscosity of the melt, temperature changes in the mould, viscosity changes of the hydraulic fluid and changes in the characteristics of the plastic.

These negative factors may arise from, for example, machine start-up after a long period of machine downtime, differences in the material properties of a new lot or a different colour, and environmental influences such as the ambient temperature at the time of processing.

The purpose of injection process regulation is to mitigate these negative factors, and thus attain even higher end-product reproducibility.

The internal pressure of the mould is the decisive factor for all quality features concerned with dimensions and weight. Constant maintenance of this pressure curve in every cycle guarantees uniformity of the quality of injection moulded parts. If the mould internal pressure curve is maintained at a constant, all of the negative factors mentioned above can be alleviated.

During injection moulding without injection process regulation, a specified pressure curve is established for injection and holding pressure, which can be maintained using a regulated machine; however, the mould internal pressure curve that arises can only be assumed. Pressure losses *via* the runner manifold as well as the mould-specific filling behaviour cannot be identified.

With the application of injection process regulation, the mould internal pressure is first measured and compared with a nominal value. If there is a deviation, a hydraulic valve that applies pressure to the injection cylinder is actuated. It is thus possible to follow the nominal value precisely and independently of negative factors. The switchover from injection to holding pressure also occurs as a function of internal pressure; thus, no pressure spikes occur as the switchover takes place when a specified threshold value is reached.

The following benefits are achieved through the application of injection process regulation:

- *Significant reduction in start-up cycles*: the required consistency in quality characteristics is achieved after just a few cycles.
- *Better reproducibility of the parts*: the deviation spread of the dimensions lies significantly below that of a non-regulated machine.
- *Cycle time reduction*: by visualising the internal pressure signal, the sealing point can be determined much more easily and accurately.
- *Restarts*: if the same internal pressure curve is applied at a restart, the resulting parts are exactly alike.
- *Improved quality of the parts through effective speed and pressure profiles*: internal pressure profiles without spikes enable the production of parts with low residual stresses. Switchover as a function of internal pressure prevents over-injection of the part, regardless of the selected dosage stroke.

The enormous significance of the mould internal pressure curve is characterised by the large number of parameters that can influence the properties of the curve. An example of such a curve is shown in Figure 5.16. Only the most important influencing factors are mentioned below:

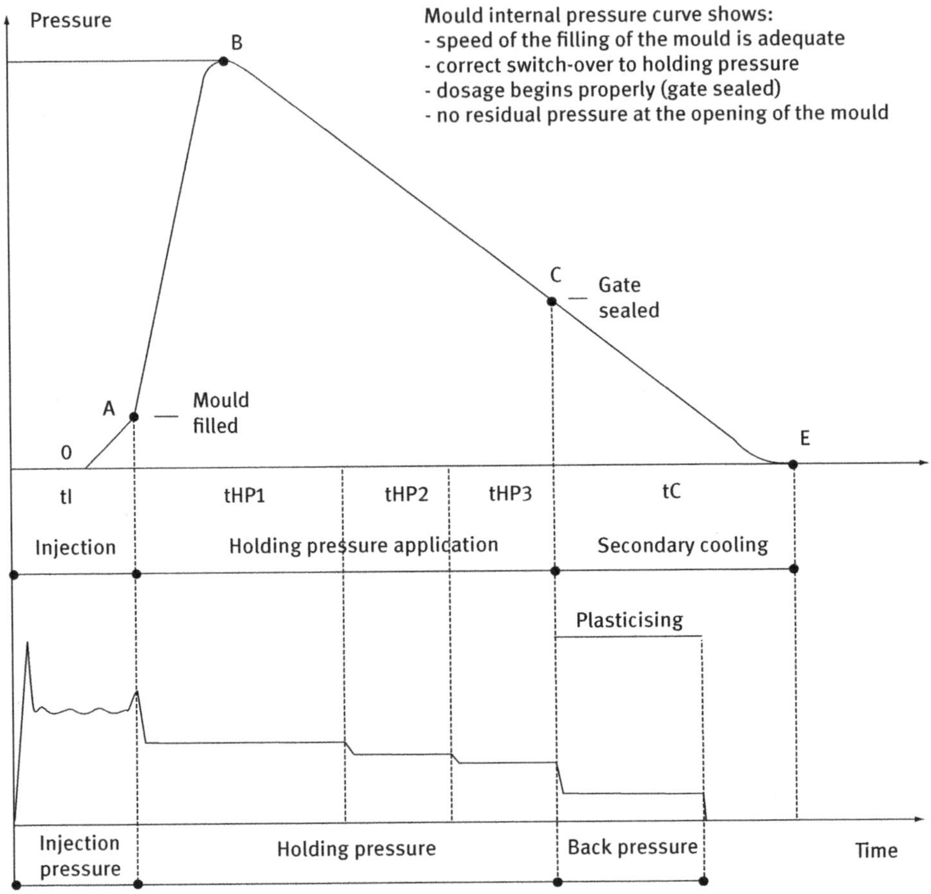

Figure 5.16: Ideal mould internal pressure and hydraulic pressure curve of an amorphous plastic.

- *Injection phase*: the injection speed, the flow resistance as a function of the type of plastic, the material temperature and the mould wall temperature.
- *Pressure holding phase*: the material temperature, the mould temperature, the level of and duration of the holding pressure.
- In relation to the maximum mould internal pressure: the injection speed, the material temperatures, the switchover point and the material flow.

The appearance of the internal pressure curve additionally contains important criteria which influence the following quality characteristics:

- *Injection phase*: the appearance, the surface characteristics, the orientation and the degree of crystallinity of the moulded part.
- *Pressure holding phase*: the formation of ridges, the weight, dimensions, shrinkage, shrink holes and sink marks, and the orientation.

5.4 Electrical systems

Machines can be all electrical injection moulders, or a combination of hydraulic and electric. In this section, the electrical systems will be the focus. A servo electric drive unit is shown in Figure 5.17.

Figure 5.17: Servo electric drive.

Components can be easily connected and removed *via* interfaces, such as can be seen on the injection unit in Figure 5.18 to produce an electrically driven injection process (Figure 5.19).

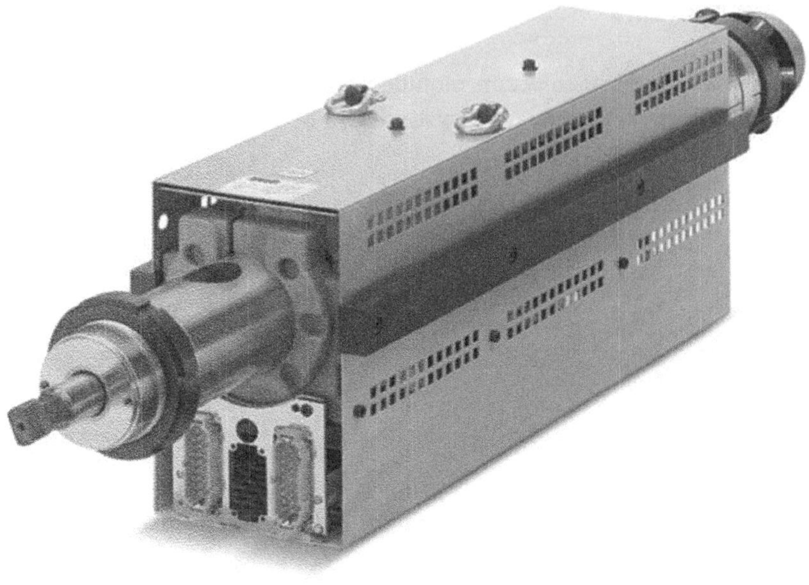

Figure 5.18: Electrical connection for coupling/decoupling.

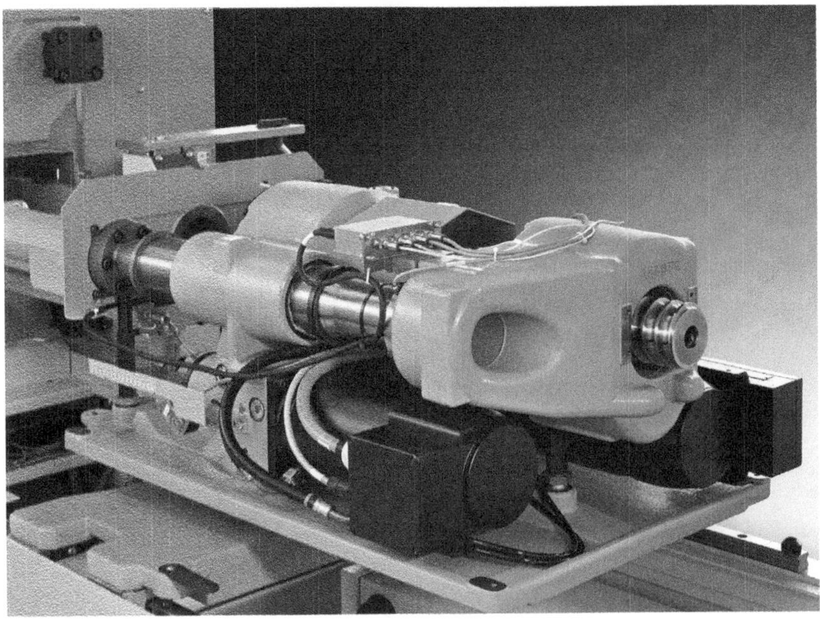

Figure 5.19: Electrically driven injection.

5.5 Hybrid machines

It has been stated that hydraulic or electric drive IJM can be advantageous in specific applications, but neither is proven to be advantageous to all applications. Therefore, an alternative approach is to combine both drive concepts in a hybrid machine, and they combine an electrical clamping unit, an electrical dosing and hydraulic injection. An example of such a hybrid drive system is shown in Figure 5.20.

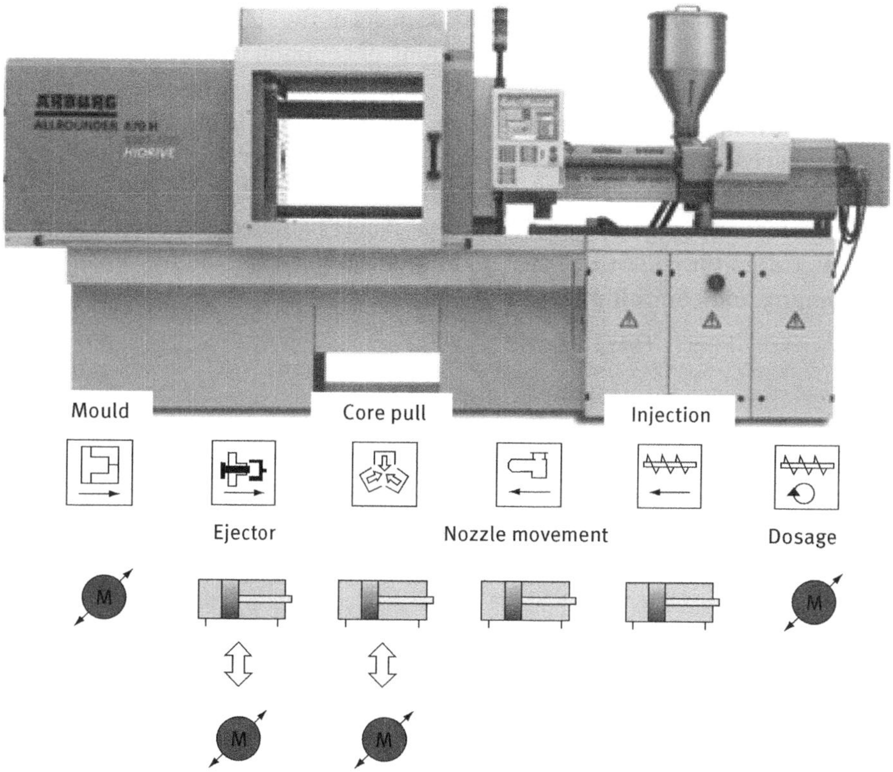

Figure 5.20: Hybrid drive concept.

This machine combines the hydraulics seen in Figure 5.3, with the electric drive shown in Figure 5.19. The movement of the moving platen of the mould tool, the injection unit heating and the screw movement are powered by electrical systems. The functions within the mould tool (ejection and core pulling functions in this case), as well as the backwards and forwards movement of the injection carriage, are hydraulically driven. In hybrid machines, this is a fairly usual spilt of hydraulic and electric functions. The injection unit consumes a considerable amount of energy and therefore it can make economic sense when electricity is cheap (such as in factories which can generate their own solar-powered electricity for instance), to power the injection unit in this way. However, there may be advantages in accuracy of mould movement positioning, as one specific example, to keep hydraulic mould movements in some cases.

5.6 The user interface

An injection machine operator will set all aspects of the IJM by inputting settings into a controller unit. The basic specifications of common units enable programming of injection moulding processes (Figure 5.21) and ideally other peripheral equipment within a production cell. This can include pickers, robots or mould heating devices, it is also possible to incorporate energy management systems, as well as link production to other higher order systems through the use of machine interface systems (Figure 5.22). Once such patented injection moulding system is called the SELOGICA system and is designed for the universal use of programming and controlling injection moulding systems. This enables just one

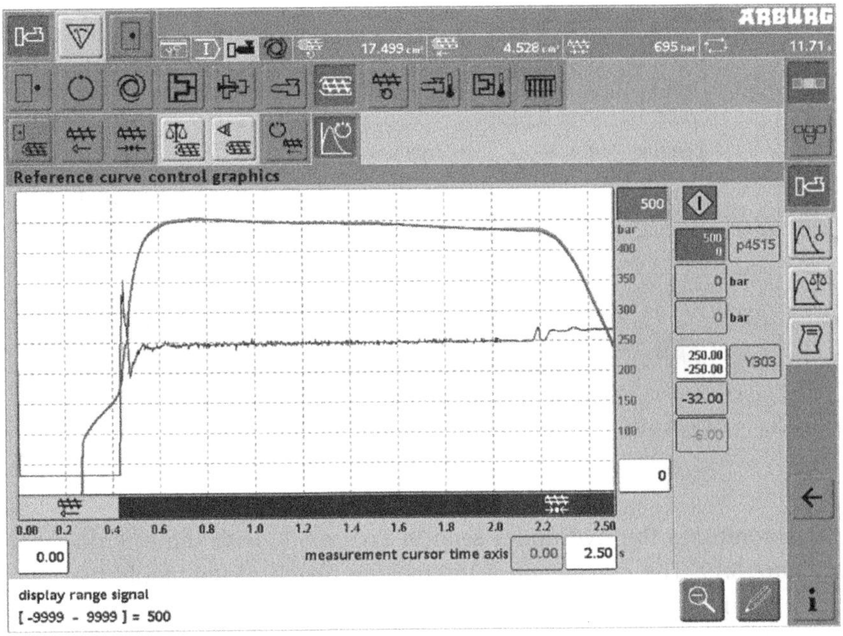

Figure 5.21: Process monitoring software.

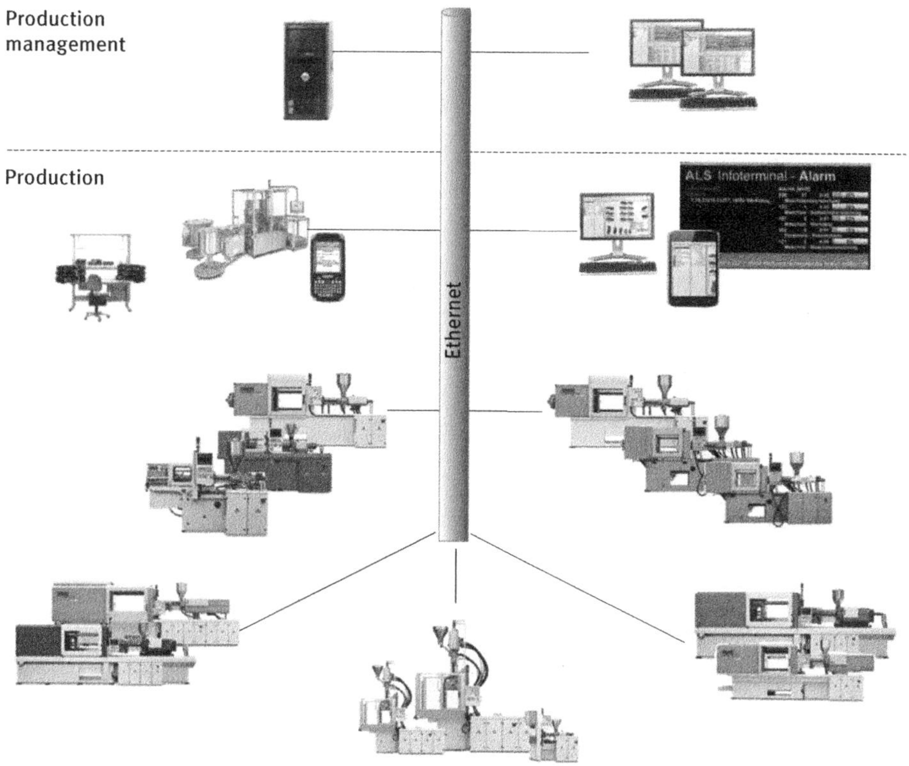

Figure 5.22: Integrated production control.

data set for production that can be stored *via* universal serial bus (USB) and accessed *via* a touchscreen. Access to programming and parameter changes can be granted or limited to various levels of the software by individual user access cards which are used to log-on and log-off the software and allow traceability. In this manner, a basic machine operator will have limited access to make system changes compared with those requiring access to more advanced machine settings or programming system interface.

Mould removal and installation can also be fully automated by control systems, with users not required to input data once the mould install subroutines have been initiated.

Individual axes such as mould height and daylight are referenced and zeroed automatically.

Extensive process optimisation is possible and both in-process monitoring and documentation can be saved, output or transmitted to other systems, such as remote maintenance systems, as required.

Process monitoring systems can maintain several reference curves for production monitoring, more complex systems may also offer user-specific freely programmable inputs and outputs.

A control system will supply functions such as:
- Automatic start-up/shutdown sequences.
- Cycle time diagrams.
- Maintenance displays and reminders.

6 Amorphous and semi-crystalline thermoplastics

6.1 Introduction

Thermoplastics can be subdivided into two distinct classes based upon differences in molecular structure. These differences can have a significant effect on the behaviour of the material during processing as well as impacting on the performance of moulding parts in service.

Materials such as polystyrene (PS), polycarbonate (PC), acrylics [polymethyl methacrylate (PMMA)], acrylonitrile-butadiene-styrene (ABS) and polyvinyl chloride (PVC) are said to be amorphous thermoplastics. This signifies that in the solid state their molecular structure is random and disordered, the long chain molecules that make up the structure are all entangled; rather like solidified spaghetti.

Materials such as most of the nylons [polyamides (PA)], polyacetal [polyoxymethylene (POM)], polypropylene (PP), polyethylene (PE) and thermoplastic polyesters [polyethylene terephthalate (PET)] have a much more ordered structure in the solid state, with a considerable proportion of the long chain molecules closely packed in regular alignment; these materials are known as semi-crystalline thermoplastics. It should be noted however, that at sufficiently high temperature (this is when the material is in its melt state) the molecular structure of both semi-crystalline and amorphous materials is amorphous. Table 6.1 classifies some common materials into these two groups.

Most amorphous thermoplastics are transparent in their natural, unpigmented form, although ABS, for example, is an exception, whereas most semi-crystalline thermoplastics in their solid unpigmented form are translucent or an opaque white colour. It is interesting to observe (for example, during the purging stage in injection moulding machines) that fully molten natural PP or acetal are initially transparent, but as the melt cools they cloud over becoming translucent in the case of PP, and opaque white in the case of acetal. This clouding is due to the molecular structure of the material gradually rearranging itself from the tangled amorphous state in the melt to the more ordered semi-crystalline state in the solid state.

The main differences in behaviour between amorphous and semi-crystalline materials observed during injection moulding are:

- *Melting and solidification*: amorphous thermoplastics exhibit a progressive softening over a wide temperature span, whereas semi-crystalline materials rapidly change from the solid melt condition over a quite narrow temperature band.
- Conversely, when amorphous materials are cooled they slowly solidify over a wide range of temperature, as opposed to semi-crystalline plastics, which change from melt to solid over a narrow range of temperature.

https://doi.org/10.1515/9783110654813-006

Table 6.1: Classifying plastics.

Amorphous	Semi-crystalline
Acrylic	Acetal
PVC	Nylon
SAN	Polyester
PS	PE
PC	PP
ABS	PTFE

PTFE: Polytetrafluoroethylene
SAN: Styrene-acrylonitrile

- *Shrinkage*: amorphous thermoplastics display very low shrinkage when they solidify, typically between 0.5 and 1%. Semi-crystalline materials shrink considerably more, usually between 1.5 and 5% depending upon the particular material.

The higher shrinkage of semi-crystalline materials is due to the repeat units along the molecular chains being in a formation that they can pack very closely together in an ordered manner. By use of appropriate moulding conditions it is possible to vary the extent of the crystalline areas, for example, when semi-crystalline thermoplastics are moulded in hot moulds, cooling rates are slow allowing more time for the molecular chains to disentangle themselves and take up their crystalline formation. This results in a greater proportion of the material being in its crystalline state (higher crystallinity) giving a product with superior mechanical strength and dimensional stability, but with relatively high shrinkage. If the same material is moulded in a cold mould, the more rapid cooling inhibits the formation of crystalline areas. The resulting lower level of crystallinity will give the product inferior mechanical properties and a lower shrinkage, which is accompanied by a tendency for dimensional stability and distortion during later service due to after-shrinkage.

Section 6.2 describes the properties of some typical amorphous materials; semi-crystalline materials can be found in Section 6.3.

6.2 Amorphous plastics

This section details the properties and applications for the following 11 amorphous materials:
- Standard PS.
- Styrene copolymers [styrene-butadiene (SB), SAN, ABS)].

- Modified polyphenylene oxide (PPO) (e.g., Noryl).
- PVC.
- Cellulose acetates (CA) [cellulose acetate butyrate (CAB) and so on].
- PMMA.
- PC.
- Polyacrylates (PAR) [polyacrylic ester (PAE), aromatic polyester (APE), polyester carbonate (PEC)].
- Polysulfones (PSU) [polyether sulfone (PES)].
- Polyetherimide (PEI).
- Polyamide imide (PAI).

6.2.1 Overview and common properties

Amorphous plastics are basically transparent (exceptions are: styrene copolymers containing butadiene, such as SB and ABS). Amorphous plastics have lower shrinkage values than semi-crystalline compounds, hence with amorphous compounds higher levels of accuracy can be obtained at lower cost than with semi-crystalline materials.

Amorphous plastics are subject to strong elastic deformation between the melting point and freezing point. They must therefore be processed with a holding pressure low enough to avoid overloading in the moulds, which results in ejection difficulties and internal stresses. Hence, there is a high risk with amorphous plastics, as opposed to semi-crystalline plastics, that the injection moulded components will have induced regions of higher stress. If a higher injection pressure is required for filling the mould, two pressure stages (holding pressure control) must be used during processing. The general properties of selected materials are given in Table 6.2.

6.2.2 Standard polystyrene homopolymer

Typical characteristics and applications:
- Compound plastic.
- Unplasticised and brittle.
- Transparent.
- Very low dielectric losses.
- Economy price mass article (housings, packs).
- Used for toys, office equipment.

Processing data is given in Table 6.3.

Table 6.2: Amorphous plastics: examples and properties.

Section	Abbreviation as per DIN 7728	Tensile modulus of elasticity[2] (N/mm^2)	Tensile impact strength[3] (Nmm/mm^2)	Maximum operation temperature in air without load		Plasticising (freezing) temperature approx. (°C)	Processing temperature (°C)	Flow behaviour of material[6]
				Permanent[4] (°C)	Temporary[5] (°C)			
6.2.2	PS	3,200	1.5–3	50–70	60–80	90	160–250	I
6.2.3	SB	1,800–2,500	5–20	50–70	60–80	90	160–250	I
6.2.4	SAN	3,600	2–3	85	95	100	180–260	M
6.2.5	ABS	1,900–2,700	7–25	75–85	85–100	105	180–260	MS
6.2.6	PPO mod.[1]	2,500	8–15	80	150	–	280–320	S
6.2.7	PVC hard	1,000–3,500	2–5 or 50[7]	55–65	70–80	90	160–180	S
6.2.8	PVC soft	–	3 WB[8]	50–55	55–65	–	150–170	I M
6.2.9	CA	2,200	2–40	50–70	80–90	100	160–230	I
6.2.10	CAB, CP	1,000–1,600	8–15	60–115	80–120	125	160–230	I
6.2.11	PMMA	2,700–3,200	1.5–3	65–90	85–100	105	200–250	M
6.2.12	PC[1]	2,100–2,400	20–35	135	145	150	250–300	S
6.2.13	PSU[1]	2,600–2,750	3–4	150	200	–	320–400	S
6.2.14	PES[1]	2,450	4–6	200	260	–	320–400	S

1) These compounds also exist in reinforced forms (with fibreglass, etc.) which will exhibit differing properties from those shown.
2) The ranges given are indicative of different grades and different processing conditions.
3) 1 Nmm/mm^2 = 1 KJ/m^2 = 1 $kpcm/cm^2$.
4) Months to years.
5) Up to a few hours.
6) I = easily flowing, M = average flow characteristics, S = flows with difficulty.
7) Impact-resistant types.
8) WB = without break.

Table 6.3: Processing data for PS.

Predrying	Generally not required; only if high surface quality is desired: forced-air drying cabinet, 2–3 h at 70–80 °C. High surface moisture content leads to the formation of waviness.
Compound temperature	Inside the cylinder: 160–280 °C depending on type. Excessively low compound temperature causes internal stresses. If compound temperature is too high, a sweet smell, a yellow colour and the formation of waviness on components indicate the onset of decomposition.
Mould temperatures	20–50 °C.
Flow behaviour	Good to very good.
Holding pressure	Must be sufficiently low to avoid overloading, especially because PS is susceptible to stress cracking, so it is usually advisable to inject with two pressure stages (holding pressure control). Excessive holding pressure causes a tendency to rupture during ejection, as well as high internal stresses.
Processing shrinkage	0.2–0.6%. Practically no after-shrinkage. If lifting tapers are too slight on external contours, difficulties during ejection are possible.
Peculiarities	PS is suitable for the antechamber through the injection process.

6.2.3 Styrene copolymers

There are also a number of important commercial materials based on mixtures of PS monomers with other components such as:
- SB.
- SAN.
- ABS.

Typical characteristics and applications:
- SB:
  - Somewhat more resistant to impact, but less hard and rigid than PS.
  - Not transparent (opaque) due to butadiene.
  - More susceptible to ageing than PS due to butadiene.
  - Housings and components for record players, electrical equipment, household equipment, refrigerator parts, camping equipment and shockproof packs.

- SAN:
  - More rigid and harder than PS, impact resistance is higher than PS, but lower than SB.
  - Can be transparent.
  - Technical components as per SB, especially when transparency is required: light covers, scales and precision equipment housing.
- ABS:
  - Significantly higher impact resistance than PS and SB.
  - Not transparent due to butadiene.
  - More susceptible to ageing than PS due to butadiene.
  - Galvanisable types are generally relatively problem-free.
  - Technical components, as per SB however, with higher susceptibility to stress.

Processing data is given in Table 6.4.

Table 6.4: Processing data for styrene copolymers.

Predrying	SB: as for PS, usually not required. SAN: predrying to eliminate moisture (leads to the formation of waviness in components). ABS: surface moisture advisable: forced-air drying cabinet, 2–3 h at 70–80 °C.
Compound temperature inside the cylinder	SB: 160–250 °C. SAN: 200–260 °C. ABS: 180–240–(260) °C (onset of decomposition indicated by yellowing and the formation of waviness).
Mould temperatures	SB: 50–70 °C (80 °C if high surface gloss is desired). SAN: 40–80 °C. ABS: 50–85 °C (higher values for high surface gloss).
Flow behaviour	SB: somewhat worse than PS. SAN: less than PS and SB. ABS: comparable to SAN.
Holding pressure	Sensitive to overloading, as with PS, for all three.
Processing shrinkage	SB: up to 7%. SAN: 0.5–0.6%. ABS: 0.4–0.74%.

6.2.4 Modified polyphenylene oxide

The properties detailed here represent an approximate 50/50 polyblend (mixture) of PPO and PS or polyacrylonitrile:

- Rigid and impact-resistant, similar to ABS, however slightly less tendency to creep and higher temperature resistance.

- Generally self-extinguishing.
- Not transparent.
- High value mechanical and electrical components, similar to ABS only with susceptibility to stress.
- Often replaces metallic materials.

Processing data is given in Table 6.5.

Table 6.5: Processing data for modified PPO.

Predrying	In general, only if high demand on surface quality is required, forced-air drying cabinet, approximately 2 h at about 85–100°C. Drying too abruptly can lead to material feeding difficulties.
Compound temperature inside the cylinder	260–300 °C depending on type. Higher temperatures only for short dwell times. Overheating the melt reduces mechanical properties, and in particular, causes brittleness.
Mould temperatures	80–150 °C depending on type. Higher temperatures lead to better flow behaviour, fewer internal stresses, higher surface finish at higher temperatures.
Flow behaviour	Lower than for ABS, rapid filling of mould is required, as solidification begins as early as 240 °C; the moulding system should be designed accordingly.
Injection pressure	1,000–1,400 bar, in order to fill the mould faster.
Processing shrinkage	0.5–0.9%, practically no after-shrinkage.
Peculiarities	Only fibreglass-reinforced types should be used for inserts.

6.2.5 Polyvinyl chloride

PVC is a widely used and durable commodity plastic which can be tailored through the use of various additives for a wide range of applications from chemically resistant bottles to PVC windows to floor coverings:
- Rigid PVC = PVC without plasticiser.
- Soft PVC = PVC with plasticiser.

Typical characteristics and applications:
- Compound plastic similar to PS, however, it is more often processed into half-finished products by extrusion moulding, as there is less play in the processing temperatures with injection moulding and thermal decomposition releases corrosive hydrogen chloride (HCl).

- Components which must be flame-resistant and self-extinguishing, or requirements for chemical strength (or when there are requirements for salt solutions, many acids, alkalis, petrol, mineral oils, fats and alcohol).
- Unplasticised PVC: hard and rigid like PS, slightly less brittle.
- Plasticised PVC: rigidity depends on plasticiser type and content, often brittle at low temperatures.
- Even unplasticised PVC cannot be used at above 60–70 °C.
- Slight tendency to stress crack formation.
- Can be transparent.

Processing data for PVC is given in Table 6.6.

Table 6.6: Processing data for PVC.

Predrying	Generally not required.
Compound temperature inside the cylinder	Unplasticised PVC: 160–180 °C (200 °C) – decomposition can be recognised by brown or black colouration of moulded components and pungent smell of HCl. Plasticised PVC: 150–170 °C – depends on type and addition of stabilisers. Short temperature loading, i.e., short dwell time in the cylinder reduces the danger of decomposition.
Mould temperatures	30–70 °C, higher values advisable to improve flow behaviour.
Flow behaviour	High melt viscosity for unplasticised PVC. Injection speed not too high to avoid overheating through excessive shear rate sprue runners and gates as wide and short as possible. Avoid pin-point gates and sharp turns.
Injection pressure	From 400 bar (plasticised PVC) and 1,000 bar (unplasticised PVC) up to 1,500 bar.
Screw circumferential speed, back pressure	To reduce overheating due to friction, work with a lower screw speed (circumferential speed maximum 0.08–0.1 m/s) and low back pressure (40–80 bar).
Processing shrinkage	Unplasticised PVC: 0.2–0.5%. Plasticised PVC: 1–2.5%. Practically no after-shrinkage with unplasticised PVC.
Peculiarities	Use only suitable material grades for injection moulding. Only screw injection moulding machines are suitable for PVC injection moulding.

Table 6.6 (continued)

	Only in exceptional cases can a few types of unplasticised PVC be processed on plunger-type machines. To keep the dwell times in the cylinder short, the shot volume should be not less than 20% of the maximum cylinder volume. Dead spots in the compound flow are to be avoided so work only with open nozzles and also if possible without a non-return valve. Even during normal processing, decomposition phenomena cannot be entirely avoided. Since the HCl released is very corrosive, cylinder fittings and moulds must be protected against corrosion (mould hard-chrome plated or made from corrosion-resistant steel).

6.2.6 Cellulose materials

Cellulose is a component found naturally in the cells of all green plants and algae and is a bioplastic:
- CA.
- CAB.
- Cellulose acetopropionate.
- CP.

Typical characteristics and applications:
- Brilliant and scratch-resistant surfaces due to self-polishing effect.
- High viscosity.
- Rigidity depends on type and quantity of plasticiser (always present).
- High water absorption.
- Low tendency to becoming dusty.
- Can be transparent.
- Handling elements of all kinds, which must have a high surface finish, such as control keyboards, tool handles, filling container housings, and so on, and especially those with metal inserts.

Processing data is given in Table 6.7.

6.2.7 Polymethyl methacrylate

Polymethyl methacrylate (PMMA) is also commonly known as acrylic and the commonly used trade name Perspex. It is a transparent plastic that can be used to replace glass in some applications. Copolymers include:

Table 6.7: Processing data for cellulose materials.

Predrying	If water absorption is above 0.2%: forced-air drying cabinet, 3 h at approximately 80 °C. Excessively high drying temperatures lead to migration of the plasticiser (in some circumstances leading to corrosion).
Compound temperature inside the cylinder	180–230 °C depending on type. Higher compound temperatures increase strength. When running in, start at lower limiting values, in view of the danger of decomposition and separating off of the plasticiser.
Mould temperatures	40–70 °C.
Flow behaviour	Good: aim for the highest possible injection speed.
Injection pressure	800–1,200 bar.
Processing shrinkage	0.2–0.5–0.7%. Lower values with lower wall thickness. Practically no after-shrinkage, but dimensions change due to the absorption of moisture.

- Acrylonitrile methyl methacrylate has a higher chemical strength than PMMA.
- Methacrylate-butadiene-styrene has a higher impact strength than PMMA, without any significant impairment of its transparency.

Typical characteristics and applications:
- High transparency and high resistance to ageing.
- Brittle, hard and rigid.
- Breaks without plastic deformation, but does not splinter.
- High static charge.
- High-class optical components (lenses, prisms), weather-resistant elements such as light covers, reflectors, lightweight windows, advertising panels and visual education models.

Processing data is given in Table 6.8.

6.2.8 Polycarbonate

Typical characteristics and applications of PC:
- Hard and rigid.
- Extremely impact-resistant from -100 °C up to +135 °C (for short periods).
- Low creep tendency.

Table 6.8: Processing data for PMMA.

Predrying	Efficient predrying essential: forced-air dying cabinet, 2–3 h (without air circulation, 4–6 h) at 70–110 °C, depending on type. Moisture leads to the formation of blisters and waviness of components.
Compound temperature inside the cylinder	200–250 °C. To avoid sink marks on very thick-walled components, it may be necessary to work at 170 °C. This gives a very high melt viscosity which requires a high screw torque (risk of overload).
Mould temperatures	50–90 °C. High temperature improves the flow behaviour and reduces the internal stresses.
Flow behaviour	Average to high melt viscosity, depending on type. Filling speed should not be too high, excessively strong shearing action leads to overheating of the compound and causes the formation of waviness. Large runner and gate cross-sections are especially necessary where high demands are made on optical quality.
Injection pressure	400–1,500 bar (2,000). Select sufficiently low holding pressure because of overload risk. Injection with two pressure stages may be necessary.
Screw circumferential speed and back pressure	Excessively high screw speeds lead to overheating through frictional heat.
Peculiarities	Where the component has to fulfil high optical requirements, care should be taken to ensure that the screw, cylinder and so on, are as clean as possible, the granules being free from dust (risk of attracting dust through electrostatic charge). Subsequent post cure of the component is advisable, if it can come into contact with media which may generate stress cracks, it should be glued or lacquered. 5 °C below the temperature at which deformation begins to occur (approximately 60–90 ° C, depending on type), cool slowly.

- Low water absorption (maximum 0.5%) and low heat expansion, giving components with stable dimensions.
- Susceptible to stress cracks with specific chemicals; susceptible to long-term effects of hot water and steam.
- Can be transparent (often with light yellow or blue tinge).

- Very impact-resistant (transparent) construction elements, housings and protective covers, especially for electrical units and installations, spools, plug boards, precision components, and eating utensils.

Processing data for PC is given in Table 6.9.

Table 6.9: Processing data for PC.

Predrying	Absolutely essential. Forced-air drying cabinet, 8–14 h at 120–130 °C. Important to keep dry until fed into the screw. Injection using prewarmed granules improves the surface quality, and using heated granule hoppers or dry conveyor units is recommended. The moisture absorption of PC is relatively low (maximum 0.5%), but absorption takes place very rapidly, however, for processing it may only reach 0.02% at most, any moisture in excess of this leads to a reduction in the mould filling capacity and to the formation of wavi-ness on the component.
Compound temperature inside the cylinder	250–330 °C. Thermal damage (discolouration) begins above 340 °C. Results in a reduction of impact resistance and so on.
Mould temperatures	80–120 °C. High temperatures to improve flow behaviour and reduce risk of internal stresses.
Flow behaviour	Melt very viscous. Mould filling more strongly influenced by rise in temperature than by increase in pressure. Select injection speed as high as possible. Sprue runners and gates with suitably large cross-sections and as short as possible.
Processing shrinkage	0.7–0.8 % in the direction of flow and almost the same transversely (low tendency to orientation); practically no after-shrinkage.
Injection pressure	800–1,600 bar
Peculiarities	If production is interrupted, lower cylinder temperatures to 160–170 °C only, as cool PC adheres strongly to the cylinder wall. Cylinder, screws and nozzles must usually be thoroughly cleaned if the injection moulding compound is changed; otherwise, dark particles may become detached from the heat-damaged surface layer upon restarting and influence the component quality.

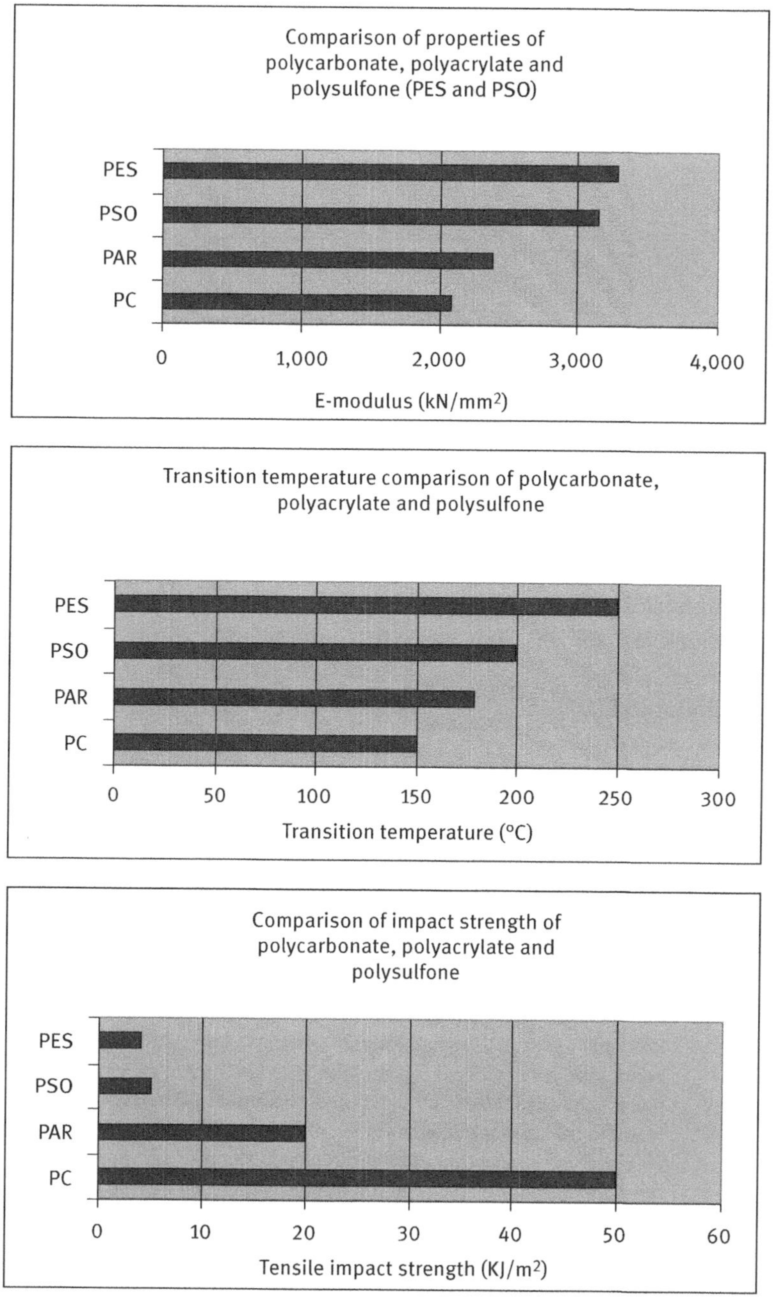

Figure 6.1: Typical properties of PAR.

6.2.9 Polyacrylates

Although they are plastics, polyacrylates are more generally applied as the polymeric components within paints, surface coatings and adhesives:
- PAR.
- PAE.
- APE.
- PEC.

The construction and properties of PAR are similar to those of PC. The processing temperature range is between PC and PSU.

Typical properties are similar to PC, but:
- Higher rigidity.
- Higher processing temperature.
- Lower viscosity.
- Electrical properties are similar to PC, durability is similar to PC (chemicals, ambient influences). Typical values for modulus, transition temperature and tensile impact strength are shown in Figure 6.1.
- Typical applications as for PC, but also for components used at higher temperatures.
- Identification: similar to PC.

Typical processing data is given in Table 6.10.

Table 6.10: Processing data for PAR.

Processing	Similar to PC, but with higher material temperatures (330–370 °C), higher melt viscosity.
Predrying	Absolutely necessary: circulating air dryer 5–8 h at 110–120 °C, dehumidifier for 3–5 h at 120–140 °C. PAR only absorbs little humidity (maximum 0.5%), but it absorbs it very quickly; for processing, the humidity should not exceed 0.02%; more humidity leads to a reduction of the mould fill behaviour and leads to part surface waviness. To avoid the absorption of humidity inside the feed hopper of the injection moulding machine, the hopper should be heated or a dry feeder used.
Compound temperature inside the cylinder	(320)–330–370 °C. Longer dwelling at more than 350 °C (e.g., in the case of faults) should be avoided. Thermal damaging begins from 370–380 °C, perceivable by discolouration. Consequence: reduction of the impact resistance and so on.

Table 6.10 (continued)

Mould temperatures	40–100 °C. Upper range temperatures improve the flow behaviour of the mass and reduce the formation of internal stresses. Higher temperatures can make demoulding more difficult.
Flow behaviour	The melt is more viscous than PC, similar to PSU. The viscosity of the melt does not diminish with increasing shear speed. The cross-sections of the sprue channels and the feed orifices should have corresponding cross-sections. The diameters of the feed orifices should not be smaller than 1.2 mm.
Injection speed	Should not be too high to keep the internal stress as small as possible.
Injection pressure	1,000–1,600 bar.
Holding pressure	600–1,300 bar to keep internal stress as small as possible.
Maximum internal pressure	350–650 bar.
Back pressure	80–120 bar.
Shrinkage	Processing shrinkage: 0.7–0.9 %, independent of the flow direction. After-shrinkage: almost 0.
Peculiarities	Reduce the cylinder temperatures to 160–170 °C when production is interrupted as frozen PAR adheres strongly to the cylinder wall. When you change the material type it is usually necessary to clean the cylinder, screw and nozzle carefully; otherwise dark particles of thermal damaged material from the wall layer will reduce the quality. With the processing of transparent PAR it is recommend reserving one special nozzle exclusively for this purpose. When you change the material to PAR you should inject PC preliminarily, as PAR has to be processed with high temperatures. For cleaning, PC is also recommended.

6.2.10 Polysulfones

Polysulfones are high-performance engineering thermoplastics able to withstand high service temperatures:
- PSU
- PES

Typical characteristics and applications:
- Hard, rigid and at the same time extremely impact-resistant.

- Similar to PC, however, it has low creep tendency over an even greater temperature range.
- (−100 to +170 °C).
- Stress crack formation can occur in some media.
- May be transparent, but with yellowish-brown colouring.
- Construction elements subjected to high mechanical, heat and electrical stress, especially if transparency is required.

Processing data for PSU is given in Table 6.11.

Table 6.11: Processing data for PSU.

Predrying	Required: forced-air drying cabinet, 120–140 °C for 4–6 h, vacuum cabinet: 120–140 °C for 2.5–3 h.
Compound temperature inside the cylinder	330–400 °C. No processing difficulties worth mentioning, in spite of this high processing temperature.
Mould temperatures	100–160 °C (180 °C).
Injection pressure	700–1,400 bar.
Processing shrinkage	0.7–0.8%.
After-shrinkage	Practically no after-shrinkage.
Peculiarities	Do not use silicone-based mould release agents with PSU. Postcure of the component is recommended to reduce the internal stresses and improve the mechanical properties: up to 5 min in oil or glycerine, up to 5 h in air at approximately 165 °C.

6.2.11 Polyetherimide

Polyetherimide (PEI) is an amorphous engineering high-performance transparent thermoplastic material, e.g., Ultem (Sabic).

Typical applications take advantage of the material's ability to retain properties (physical and mechanical) at high service temperatures. These materials find use in aerospace and automotive applications among others and properties include:
- Very high solidity and rigidity, also not reinforced, up to +210 °C.
- Low creep tendency.
- High endurance under completely reversed stress.
- Low thermal expansion coefficient.
- Very low dielectric losses, high dielectric rigidity, very resistant against chemicals, hydrolysis, ambient influences and radiation.

- High non-flammability, low development of smoke.
- For parts which are put under a lot of mechanical, thermal and electrical stress, e.g., bearings, ball bearing guides, cogs, gear pans, carburretor housings, valve covers, connector pans, flow solderable flat connector blocks, integrated circuit housings and parts of microwave ovens.
- Natural condition: transparent amber.

Processing data is given in Table 6.12.

Table 6.12: Processing data for PEI.

Processing	Very high flow capability. Frozen melt tends to adhere strongly to the cylinder.
Predrying	Preliminary drying absolutely necessary. Circulating air dryer 4–6 h at 150 °C. Dehumidifier 3–5 h at 150 °C. For processing, the humidity should be lower than 0.05% (absorption of humidity up to 0.25% within 24 h).
Compound temperature inside the cylinder	340–425 °C. In many cases a temperature of 360 °C is optimal (Ultem).
Mould temperatures	60–175 °C. In many cases a mould temperature of 95 °C is optimal. A high mould temperature improves the mould fill and the rigidity of the joint line. It reduces internal stresses (can stand higher temperatures and has higher resistance to chemicals).
Flow behaviour	Very high flow capability. Minimal wall thickness of 0.25 mm possible. Frozen melt tends to adhere strongly to the cylinder.
Injection speed	Select midrange values.
Injection pressure	750–1,500 bar.
Holding pressure	400–750 bar.
Maximum internal pressure	350–650 bar.
Back pressure	40–80 bar.
Processing shrinkage	0.7–0.5–0.1% reinforced with fibreglass.
Peculiarities	Be careful that the melt does not solidify in the cylinder (extremely adhesive). Reduce the cylinder temperature to 180–200 °C when production is interrupted. Clean the cylinder carefully after production has finished.

6.2.12 Polyamide imide

Thermoplastic polyamide imide (PAI) resins offer a very high level of shear resistance more commonly seen in thermoset materials, e.g., Torlon (Solvay).

Typical properties:

- Very high strength and rigidity between −200 and +260 °C.
- High impact resistance.
- Low creep tendency.
- High endurance under completely reversed stress.
- Very resistant against abrasion.
- Low thermal expansion coefficient (9–30 × 10^{-8} °C).
- Very low dielectric losses.
- Low stress cracking susceptibility.
- Very resistant against chemicals.
- Very stable against ultraviolet (UV) and is UV-resistant against radiation of high energy.
- Not resistant against hot water.
- Low exhalation under high vacuum.
- High non-flammability.
- High resistance against oxidation.
- Typical applications: for items which are put under a lot of mechanical, thermal and electrical stress, up to +260 °C and under abrasion stress, such as cams, bearings, sliding rings, vanes for hydro and pneumatic motors, connector parts and spark suppression covers.
- Natural condition: brown.

Typical processing data is given in Table 6.13.

6.3 Semi-crystalline plastics

This section details the properties and applications for the following 11 material types:

- PE.
- PP.
- PA.
- Polyacetals (POM).
- Polyterephthalates [PET, polybutylene terephthalate (PBT)].
- Polyphenylene sulfide (PPS).
- Polyfluorolefins [polyfluoroalkoxyalkane, tetrafluoroethylene hexafluoropropylene (FEP), ethylene-tetrafluoroethylene (ETFE), polyvinylidene fluoride].
- Polyacrylic acid (PAA).
- Polyphthalamide (PPA).

- Polyaryl ether ketones (PAEK) [polyether ether ketone (PEEK), polyether ether ketone ketone (PEEKK), polyether ketone (PEK)].
- Liquid crystal polymers (LCP).

Table 6.13: Processing data for PAI.

Predrying	Necessary: l6 h at 150 °C or 8 h with 180 °C.
Compound temperature inside the cylinder	340–360 °C.
Mould temperatures	Approximately 230 °C.
Flow behaviour	Very low. Very high melt viscosity after injection, heat treatment is recommended.
Injection speed	As high as possible. It is recommended to use an accumulator.
Injection pressure	750–1,550 bar.
Holding pressure	550–1,050 bar.
Maximum internal pressure	450–750 bar.
Back pressure	40–80 bar. Some types need a special screw with a compression ratio of 1:1.
Peculiarities	To reach optimal properties heat treatment is necessary; up to 3 mm wall thickness: 48 h (2 days), for thicker walls: 3–4 days, for some types 7–17 days.

6.3.1 Overview and common properties

Semi-crystalline plastics are naturally of a milky opaque colour (thus non-transparent) as a result of light diffusion at the crystallite borders. Semi-crystalline thermoplastics as a whole have higher shrinkage values

than amorphous plastics between solidifying and transition temperatures. The risk of overloading is therefore considerably less, hence ejection difficulties due to this reason scarcely occur in semi-crystalline components. Furthermore, increased shrinkage is equalised by favourable antifriction behaviour to give a positive effect. An overview of the general properties is shown in Table 6.14.

Table 6.14: Semi-crystalline plastics: examples and properties.

Section	Abbr. as per DIN 7728	Module of elasticity[3] (N/mm²)	Tensile impact strength as per DIN 53453[4] (Nmm/mm²)	Max. operation temp. in air without load		Plasticising (freezing) temp. approx. (°C)	Processing temp. (°C)	Flow behaviour of material[7]
				Perm.[5] (°C)	Temp.[6] (°C)			
6.3.2	PE soft	200–500	18–NB	60–75	80–90	105–115	160–240	I M
	PE rigid	700–1,400	4–NB	70–80	90–120	125–140	190–280	M S
6.3.3	PP[l]	1,100–1,300	4–18	100–110	130–140	158–168	200–270	M S
6.3.4	PA6.6[1, 2]	2,000–2,900	5–NB	80–120	170–200	250–265	260–300	–
	PA6[1,2]	1,400–3,200	3–NB	80–100	140–180	215–225	230–260	I
	PA6.10[1,2]	2,000	7–10	80–100	150	210–225	220–260	I
	PA11[1]	1,000	30–40	70–80	140–150	180–190	200–250	I
	PA12[1]	1,600–1,700	10–20	70–80	140–150	175–185	190–250	I M
	PA amorphous[1]	2,000	13	80–100	130–140	150–160	260–300	M
6.3.5	POM[1]	2,800–3,200	5–10	90–110	110–140	165–175	200–220	I
6.3.6	PET[1]	3,100	3–6	100	200	255–258	260–280	I
	PBT[1]	2,000	3–6	100	165	220–225	230–270	I
6.3.7	PPS[1]	3,400	3–8	200	300	280–288	315–360	I

6.3.8	FEP	350–650	NB	205	250	285–295	340–360	S
	ETFE	1,100	NB	150	220	270	315–365	S
6.3.9	PAA[1]	11,300–17,700	7.5–27.5	115–145	–230	235–240	250–290	I
6.3.10	PPA[1]	6,300–14,500[9]	–	to 185	–280	310	320–350	I
6.3.11	PAEK[1]	4,000	NB	250	–300	335–370	350–420	I
6.3.12	LCP[1]	10,000–35,000[8]	20–70	120–240	–	270–380	300–450	I M

1) These compounds also exist in reinforced forms (with fibreglass, asbestos fibres and so on) Then some of the values given in the table are replaced by others.
2) The mechanised values given are valid for the conditioned state, i.e., with a sufficiently high water content.
3) The ranges given are conditioned by differences in types and processing conditions; NB = without break.
4) 1 $Nmm/mm^2 = 1\ KJ/m^2 = 1\ kpcm/cm^2$.
5) Months to years.
6) Up to a few hours.
7) I = easily flowing, m = average flow characteristics, 5 = flows with difficulty.
8) With optimal orientation.
9) Filled or reinforced material.

6.3.2 Polyethylene

PE can be purchased in a variety of densities and properties which are dependent upon the formation of the polymer chains. The most common materials encountered in injection moulding are low-density polyethylene (LDPE) and high-density polyethylene (HDPE).

The density range for LDPE is 0.91–0.94 g/cm^3 and the low density is due to strongly branched molecules; whereas for HDPE, the density range is 0.94–0.96 g/cm^3 and the high density is due to molecules with few branches.

The operating and processing properties of PE are strongly dependent on the density (molecule shape) and the degree of polymerisation [molecule length or molecular weight (MW)].

As the density and the degree of polymerisation increase and melt-flow index falls, the changes observed include:

- Increases in, for example, hardness, rigidity, upper operating temperature (from 80 to 105 °C, 120 °C for brief periods).
- Decreases in, for example, brittle temperature, tendency to stress crack formation.
- Becomes more difficult to process by injection moulding because flowability is reduced. Typical characteristics and applications:
- Flexible (LDPE) to tough and hard (HDPE).
- Cold-resistant down to −50 °C and below.
- Practically indestructible however, becomes brittle under the effect of UV (can be stabilised against UV).
- No water absorption.
- Very low dielectric losses.
- High chemical strength (against nearly all common solvents).
- Gas permeability is higher than for many other plastics, though steam permeability is very low.
- Very transparent (semi-crystalline).
- Drain lugs, covers, containers, bottle crates, household articles and toys. It is not used for precision components because of high shrinkage and after-shrinkage.

Processing data is given in Table 6.15.

6.3.3 Polypropylene

The way the atoms are arranged in the polymer molecule has a great influence on the properties of a polymer component; this order is named the tacticity and with PP there are two different arrangements commonly available, i.e., isotactic and atactic. The degree of tacticity determines the level of crystallisation possible. Atactic polymers are those with side groups placed in a random order. Atactic PP with

Table 6.15: Processing data for PE.

Predrying	Generally not required (no water absorption). Short predrying period only needed if surface moisture could have built up after long storage in an unsealed container.
Compound temperature	160–280 °C depending on flow behaviour. At lower temperatures, dull points appear on the component.
Mould temperatures	20–70 °C. Higher surface gloss and higher degree of crystallisation at higher temperatures thus, producing less post shrinkage. At higher temperatures above 70 °C, ejection difficulties are due to the compound adhering to the mould.
Flowability	Good to medium, depending on density and MW.
Shrinkage	Processing shrinkage 1.5–3.5% for easy flowing types 2–4% for difficult flowing types. Postshrinkage unavoidable, as even at 70 °C mould temperature the cooling speed are still so great that the crystallisation in the mould cannot be completed.
Injection pressure, holding pressure time	500 bar (LDPE)–1,200 bar (HDPE). If holding pressure time ends before the gate freezes, this leads to formation of sink marks or cavities, because of the high shrinkage.
Special features	PE is suitable for a prechamber through the injection process. Gluing, printing, lacquering and other processes are only possible after pretreatment (because of high chemical strength).

non-uniformly arranged methyl groups has a density of 0.94 g/cm^3. Isotactic polymers are those whose side groups are all on the same side. Isotactic PP has a density of 0.90 g/cm^3. There are also copolymers with ethylene, to increase impact resistance below 0 °C.

Typical characteristics and applications:
- Higher rigidity and hardness than PE (usually no longer possible to slit with a finger nail).
- Impact resistance falls off sharply.
- Upper operating temperature of 110 °C and possibly higher in stabilised types.
- Good electrical properties, similar to PE.
- No tendency to stress crack formation.
- Lower chemical strength than PE.
- Colourless, more translucent (less opaque) than PE.
- Fan wheels, heating ducts in vehicles, internal components for washing machines and dishwashers, transport cases and containers with film hinges.

Processing data for PP is given in Table 6.16.

Table 6.16: Processing data for PP.

Predrying	Usually not required, as with PE.
Compound temperatures	(170 °C) –200–270 °C– (300 °C).
Mould temperatures	50–100 °C. Upper temperatures give better surface gloss, dull points at lower temperatures.
Flow behaviour	Flowability is better than PE, especially in a lower processing temperature range. Shut-off nozzles are usually required if working without material decompression.
Shrinkage	Processing shrinkage in the direction of flow 1.3–2%, crosswise to direction of flow 0.8–1.8%.
Injection pressures, holding pressure time	1,000–1,500 bar. Select a sufficiently long holding pressure time (similar to PE).
Special features	PP is suitable for a prechamber through the injection process. The plasticising efficiency of the machine is lower, due to the lower density (only approximately 70% compared with PS).

6.3.4 Polyamides

The PA types with numerals as part of their name are semi-crystalline. They are distinguished by the number of carbon (C) atoms of the monomers (building blocks) from which the thread-shaped molecules have been formed thus, for example, PA6 is made from a monomer with 6 C atoms. PA6.10 is made from two different kinds of monomer, one of which has 6 C atoms and the other has 10 C atoms.

The properties of PA are vitally affected by the degree of crystallinity and the water content. Water absorption leads to dimensional changes.

As the number of C atoms increases (higher index numbers), the water absorption capability falls rapidly (increased dimensional stability). Strength and rigidity show a slight tendency to fall and the price increases (PA12 is twice as expensive as PA6).

The most important types of PA are shown in Table 6.17.

Typical characteristics and applications:

- High rigidity and toughness, good friction and wear properties.
- PA6.6 has the greatest strength and wear resistance.
- As water absorption increases, the rigidity decreases and toughness increases.

- Cold-resistant down to approximately –40 °C. Upper operating temperature of 80–120 °C, depending on type, can withstand 140–210 °C for brief periods (cf. melt temperature).
- Partly self-extinguishing.
- High dielectric losses [not suitable for high frequency (HF) use].
- Low tendency to become dusty (water absorption reduces surface resistance) but good creep resistance.
- Resistant to petrol, oil and numerous solvents, and weather-resistant (use stable types for thin-walled components).
- Low tendency to stress crack formation.
- Only amorphous types can be transparent.
- Construction elements subjected to friction stress and wear stress, such as slide elements, bearings and similar coupling elements operating on a ball or roller bearing principle, tread rollers, cams and so on, fan wheels, fittings, housings for electrical tools, electric motors, pumps and so on (usually with fibreglass reinforcement).
- PA6 and PA6.6 are not suitable for precision components on account of high water absorption (low dimensional stability).

Table 6.17: Common types of PA.

Type	Density	Water absorption
PA6	1.12–1.15 g/cm^3	2.8–3.2%
PA6.6	1.12–1.15 g/cm^3	2.5–2.7%
PA6.10	1.06–1.08 g/cm^3	1.2–1.4%
PA11	1.04 g/cm^3	0.8–0.9%
PA12	1.01–1.02 g/cm^3	0.7–0.8%
PA amorphous types	1.06–1.08 g/cm^3	0.3–1.1%

Processing data for nylons is shown in Table 6.18.

6.3.5 Polyacetals

Homopolymers or copolymers which are made from formaldehyde, e.g., Delrin (DuPont). Typical characteristics and applications:
- High levels of hardness and rigidity.
- Good elastic properties.
- Low creep tendency.

- Good friction and wear properties (slightly lower wear resistance than PA6.6, but lower friction coefficient).
- Cold-resistant down to −50 °C, upper operating temperature 100 °C, can withstand 150 °C for brief periods.
- Low moisture absorption.
- Favourable dielectric behaviour.
- Resistant to petrol, oil and numerous solvents, weak acids and copolymers are also resistant to alkaline solutions.
- Practically no tendency to stress crack formation in air.
- Strong natural colour: white and opaque.
- Applications: precision components, parts subjected to friction stress and wear stress (bearings, gears and so on), similar to PA, however with better dimensional stability. All kinds of parts with high permanent loadings and are frequently used to replace metals. This material is also used in components with flexible parts, especially for tear-off closures.

Table 6.18: Processing data for PA.

Predrying	Absolutely necessary, vacuum drying cabinet for a few hours at 80 °C. Exception: granules packed in tin canisters in the same condition as when delivered. Moisture causes surface waviness and impairment of mechanical properties.
Compound temperatures	10–40 °C (50 °C) above melt temperature: PA6: 215 °C. PA6.6: 250 °C. PA6.10: 200 °C. PA11: 187 °C. PA12: 177 °C. If material feed difficulties arise, select cylinder temperatures rising toward throat side. Dwell time in cylinder to be as short as possible. Melt tends to oxidise if air enters (yellowing). Semi-crystalline PA have an excellent melt temperature. So take care that the gate does not freeze too early, so that the holding pressure can be effective for as long as possible (gate dimensions, compound temperature, mould temperature).
Mould temperature	Generally 50–90 °C. If higher crystallinity is desired: 100–120 °C. The higher the degree of crystallisation (crystallinity) the lower the after-shrinkage due to after-crystallisation. The degree of crystallisation (if low) can also be increased by a subsequent reheating treatment. If crystallinity is uniform and rising, the wear resistance, in particular also rises.

Table 6.18 (continued)

Flow behaviour	Very good, melt very thin. Shut-off nozzles are required if it is not possible to work with compound decompression. Fluidity causes the formation of scratches on moulds not closing well.
Shrinkage	Processing shrinkage 0.2–2.5 %, depending on type and component. Post shrinkage: decreases as processing shrinkage increases. Injection pressure: 450–1,550 bar. Holding pressure: 350–1,050 bar.
Special features	Because of the high heat of fusion, usually only approximately 60 % of melting efficiency is achieved in comparison with PS. The typical properties of PA such as, for example, great toughness, only come into play if the water absorption is sufficient. Injection fresh (dry) components means that the moisture content to be expected under operation conditions is deliberately introduced by storage in water or by atmospheric humidity. This is also important for checking the operating dimensions.

Typical processing data is given in Table 6.19.

Table 6.19: Processing data for polyacetals.

Predrying	Generally not required, only necessary for very moist material, very high humidity (associated) reduces heat resistance during injection moulding. On microcomponents, to increase flowability. Treatment in vacuum drying cabinet recommended to remove volatile constituents.
Compound temperature	Homopolymer: 215 ± 5 °C. Copolymer: 205 ± 5 °C. Copolymers are somewhat less susceptible to overheating than homopolymers. Dwell time in the cylinder should be as brief as possible, decomposition starts at temperatures above 230 °C. If possible, work with open nozzles.
Mould temperatures	50–120 °C. Upper limits on precision components, to keep after-shrinkage very low.
Flow behaviour	Very good.
Shrinkage	Processing shrinkage 1–3.5%, depending on wall thickness (upper values for large wall thickness) and processing conditions. Post shrinkage decreases as processing shrinkage increases. After-shrinkage can be forestalled by reheating.

Table 6.19 (continued)

Injection pressures	800–1,700 bar. Holding pressure must be sufficiently high and of sufficient duration. No pressure jumps should occur during the holding pressure phase, so as not to interfere with the uniform crystallisation cycle. For precision components, gates of approximately 2/3 of the maximum wall thickness are required (so that the holding pressure can operate as long as possible).

6.3.6 Polyterephthalates (linear polyesters)

PET is the most common of the polyterephthalates, being used extensively in carbonated drinks bottles and water bottles due to its transparency. In contrast, PBT is opaque in colour and finds its main commercial use as a result of its electrical insulating properties. Both materials have lower moisture absorption properties than Nylon 6 (PA6) and 66 (PA66) in service:g

- PET.
- PBT = identical to polytetramethylene terephthalate.
- PET has a very low crystallisation rate. Certain types remain amorphous (transparent), if cooling is rapid (low mould temperatures).

Typical characteristics and applications:

- High levels of hardness and rigidity.
- Very low creep tendency.
- Friction and wear properties better for PET than for POM at correspondingly high levels of crystallinity, about the same wear resistance as PA6.6 but a lower friction coefficient.
- Cold resistance: PET cold-resistant down to –40 °C, PBT down to –60 °C. Upper operating temperature for (crystalline) PET is 100 °C, higher for short periods; for PBT, 110 and 170 °C for short periods. Very low water absorption and favourable dielectric behaviour.
- Resistant to oil, petrol, dilute acids and alkalis, salt solutions, not resistant to hot water and steam (hydrolysis).
- No tendency to stress crack formation in air.
- With amorphous PET, cloudiness can arise at high temperatures owing to crystallisation.
- Applications include components subject to friction stress and wear stress and components under high permanent load, similar to POM. High crystallinity required for semi-crystalline precision components, with PET obtainable only through high mould temperatures.

Processing data is given in Table 6.20.

Table 6.20: Processing data for PET.

Predrying	Moistened granules 3–4 h at 100–120 °C. Moisture decreases thermal stability during processing.
Compound temperature	PET: 260–280 °C. PBT: 230–270 °C. Only brief dwell times at upper temperatures, because of the risk of heat damage.
Mould temperatures	PET: 30–60 °C (amorphous) up to 140 °C (semi-crystalline). PBT: 30–60 °C. For precision components made from PET: 140 °C. With semi-crystalline PET, excessively low temperatures lead to non-uniform crystallisation (mottling on non-pigmented compounds).
Flow behaviour	For PBT favourable, similar to PA, for PET not as good.
Shrinkage	Processing shrinkage 1–2%. Post-shrinkage with semi-crystalline PET only becomes negligibly small at a mould temperature of 140 °C.
Injection pressures	1,000–1,700 bar. Secondary pressure should be sufficiently high and of sufficient duration. Gate cross-sections should be made sufficiently large, so that sealing does not occur too early (see POM).

6.3.7 Polyphenylene sulfide – typical characteristics and applications

A rigid high-performance engineering thermoplastic with the combined properties of good electrical insulation and high heat resistance. It is therefore of interest as a metal replacement material where these properties can combine, for example, in electrical applications, e.g., Ryton®, Solvay:
- Very high strength and rigidity even at relatively high temperatures (up to 230 °C).
- Very low tendency to creep, high wear resistance and low deformation capability.
- Very low moisture absorption.
- High electrical resistance, very low dielectric losses.
- Resistant to solvents up to 200 °C and to many acids and alkalis, with good resistance to hydrolysis.
- Flame resistance.
- Non-transparent (semi-crystalline).

For components subjected to mechanical, thermal, electrical and chemical stress. Processing properties are given in Table 6.21.

Table 6.21: Processing data for PPS.

Predrying	Neither necessary nor advisable.
Compound temperature	315–360 °C. Because of the high temperature, the cylinder and nozzle should first be thoroughly cleaned to remove traces of other compounds.
Mould temperature	20–200 °C. Low mould temperatures give higher impact-resistance values and less processing shrinkage (but, at the same time, they produce higher post shrinkage at relatively high operating temperatures). With mould temperatures above 130 °C, components have better surface finish values and higher dimensional stability at high operating temperatures (lower after-shrinkage as a result of lower after-crystallisation).
Shrinkage	Processing shrinkage 0.5–1.5%. Higher value for higher mould temperatures and large wall thicknesses.
Injection pressures	Injection pressure: 750–1,500 bar. Holding pressure: 350–750 bar.
Special features	PPS can be made to form nets by special (relatively expensive) heat treatment.

6.3.8 Polyfluorolefins

The polyfluorolefin family provides speciality plastics for applications needing low friction and high chemical resistance:

- PTFE density = 2.1–2.2 g/cm^3 has such a high viscosity above the crystallite melting point (327 °C) that injection moulding is not possible.
- FEP copolymer density = 2.1–2.2 g/cm^3.
- Both PTFE and FEP are known by the trade name Teflon™.
- ETFE copolymer has the trade name Tefzel™ and density = 1.7 g/cm^3. Typical characteristics and applications:
- PTFE:
  - Flexible to tough and hard (depending on the crystallinity).
  - Relatively high creep tendency.
  - Very low friction coefficient (same at rest and in motion, so no slip-stick effect), however, wear resistance is relatively low (sliding friction partner must have a very smooth-surface roughness of 0.1 m).

  - Operating range: (−270 °C) −200 to +260 °C (+280 °C) non-flammable, no water absorption.
  - Very high electrical resistance, even at high atmospheric humidity, very low dielectric losses and high creep resistance.
  - Very high chemical resistance to almost all corrosive media, no stress crack formation.
  - Non-transparent (semi-crystalline).
- FEP/ETFE:
  - Properties similar to PTFE but rigidity is significantly higher, whereas anti-friction properties and electrical insulating properties are not quite as good.
  - Application ranges: FEP from −100 to +205 °C; ETFE from −100 to +155–180 °C, flame-resistant and self-extinguishing. The chemical resistance of ETFE is not quite as good.
  - Low stressed sliding mountings, seals, rod-packing glands, piston rings, electrical components (resistant to cold and heat) including HF technology, high resistance components in chemical equipment construction, for pumps and laboratory equipment and similar applications.

Processing data for FEP and ETFE is given in Table 6.22.

Table 6.22: Processing data for FEP and ETFE.

Predrying	Not required.
Compound temperatures	FEP: 340–370 °C. ETFE: 315–365 °C.
Mould temperatures	FEP: 150–180 °C. ETFE: 80–120 °C.
Flowability	Work with a low injection rate only, because of high susceptibility to shear breaks. For the same reason, sharp bends and abrupt cross-section alterations in the mould are to be avoided.
Shrinkage	ETFE (at 3–5 mm wall thickness). 1.5 to 2% in the direction of flow. 3.5 to 4.5% in the transverse direction. Reheating treatment for preform is recommended, to forestall possible after-shrinkage.
Special features during processing	If the melt comes into contact with iron (or steel), fluorine is split off (corrosive); the plasticising unit and mould must therefore be protected against corrosion. Good suction (ventilation) in working areas is vital.

6.3.9 Polyacrylic acid and polyarylamide

Polyarylamide is a semi-crystalline engineering thermoplastic. In combination with fillers, a variety of high-performance engineering properties can be achieved, e.g., IXEF® polyacrylamide.

Typical properties and applications:

- Excellent mechanical properties.
- Apparent limit of elasticity at 255 N/mm^2 with 2% breaking elongation.
- Modulus of elasticity up to 20.000 N/mm^2; modulus of bending up to 17.500 N/mm^2.
- Stiffness 10 times higher than non-reinforced PA.
- Low creep tendency.
- High resistance to oscillation (higher than other PA–glass reinforced parts).
- High continuous usage temperature.
- Crystalline melt temperature 235–240 °C.
- Low thermal expansion coefficient (11–21 × 10^{-6}/C), i.e., within the same range as metals, low water absorption and low absorption and expulsion speed.
- Natural condition: non-transparent (partly crystalline and filled) light colours possible.
- Typical applications: high requirement mechanical parts used in machine engineering, automotive industry, precision mechanics, electrical industry and so on.

Processing data is given in Table 6.23.

Table 6.23: Processing data for PAA.

Processing	Easy to inject; fibreglass content also with 50–60 °C due to high flow capability. Injected parts are practically torsion free due to low shrinkage and small differences in shrinkage as well as minute pressure differences in the mould as a consequence of the low viscosity. You have to consider that the joint lines are weakened mechanically (as with all reinforced compounds), especially when they are far from the sprue (solidity of the joint line approximately 80 N/mm^2).
Predrying	3–5 h with 100 °C with warm air, 12 h with maximum 80 °C or in a vacuum cabinet for 5 h with maximum 120 °C. The remaining humidity has to be less than 0.3 %.
Compound temperature inside the cylinder	250–290 °C. Formulations for flame protection, maximium temperature of 270 °C. With a longer dwell time above 300 °C, thermal decomposition starts.

Table 6.23 (continued)

Mould temperatures	120–150 °C. The effects of lower mould temperatures are insufficient, resulting in inhomogeneous crystallisation and after-crystallisation with torsion tendency. Mould temperatures in the upper range improve the surface quality.
Flow behaviour	Very high flow path/wall thickness ratio at least 200:1 with 2 mm wall thickness; we recommend using a shut-off nozzle, but you can also work with an open nozzle.
Injection speed	High, the filling time should be as short as possible.
Injection pressure	1,000–1,500 bar.
Holding pressure	350–800 bar. Higher dimensional stability in the upper range.
Maximum internal pressure	300–700 bar.
Back pressure	40–80 bar.
Screw circumferential speed	Up to 10 m/min. Higher back pressures and higher circumferential speeds can lead to a damaging of the fibreglass.
Processing shrinkage	0.2–0.6%.
After-shrinkage	Low. The higher the mould temperature, and therefore the processing shrinkage, the lower the after-shrinkage.
Peculiarities	If you have to interrupt production for more than one hour you should empty the cylinder and clean it with a cleaning material.

For example, Amodel® (Solvay) is only available in reinforced fibreglass and/or mineral filled varieties. These materials tend to be used where PA cannot withstand elevated service temperatures, humidity or some chemicals.

Typical properties:

- Excellent mechanical properties.
- Modulus of elasticity at 225 N/mm^2 with 2% breaking elongation.
- Flexural strength up to 305 N/mm^2.
- Modulus of bending up to 14.500 N/mm^2.
- High tenacity.
- Low creep tendency.
- High resistance to oscillation.
- High continuous usage temperature, up to 165 °C for 20,000 h.
- Crystalline melt temperature is 310 °C.
- Low thermal expansion coefficient.

- Low absorption of humidity; practically no influence on the mechanical properties.
- High tracking resistance.
- Very resistant against chemicals.
- High non-flammability.
- Natural condition: non-transparent (partly crystalline and filled).
- Typical applications: high requirement mechanical parts used in the car, aerospace, chemical, sanitary industries and so on.

Processing:
Easy to inject, the melt is not aggressive. Injected parts almost do not contort due to small differences in shrinkage as well as low absorption of humidity. Contrary to PC and other materials, predrying is not critical. Further processing data is given in Table 6.24.

Table 6.24: Processing data for PPA.

Predrying	Not necessary when the material is processed directly from the rolls; otherwise 16 h at 80 °C.
Compound temperature inside the cylinder	320–350 °C.
Mould temperatures	135–165 °C.
Flow behaviour	High, similar to PA.
Injection speed	Middle to high.
Injection pressure Holding pressure	700–1,500 bar. 350–800 bar.
Maximum internal pressure	300–700 bar.
Back pressure	40–80 bar.
Processing shrinkage	0.2–1.1% according to the type.

6.3.10 Polyacryl and ether ketones

Polyacryl ketones are composed of acryl ether groups and acryl ketone groups. PEK are composed of ether and ketone groups. Generally, they are partly crystalline thermoplastics with high mechanical, thermal and chemical strength. PEEK is the material that is most commonly seen in bulk processing from this group, where high specification engineering material is required.
Examples:
- PEK.
- PEEK.

- Polyether ketone ketone.
- PEEKK.
- Polyether ketone ether ketone ketone.
- PAEK.

A high share of ketone groups increases the processing temperature due to higher transition temperatures and higher crystallisation melt temperatures. The ether groups have positive effects on the processability of the material.

Typical properties:

- Very high strength and rigidity, almost constant up to 140 °C.
- Processable up to 250 °C, in the short run up to 300 °C.
- Very low creep tendency.
- High endurance under completely reversed stress.
- High abrasion resistance up to 250 °C and favourable sliding behaviour.
- Low thermal expansion coefficient.
- Low absorption of water.
- Up to 220 °C, constantly good electrical properties. Especially small dielectric losses.
- Very resistant against chemicals, (except against acetones). Resistant against hydrolysis up to 280 °C.
- Very resistant against stresses.
- Very resistant against radiation of high energy, low resistance to UV-radiation when not coloured.
- High non-flammability.
- Least development of smoke among all thermoplastics.
- Natural condition: non-transparent (partly crystalline).
- Typical applications: for parts which are put under a lot of mechanical, thermal and electrical stress, e.g., bearing, ball bearing guides, cogs, valves, seals (car and aerospace industry). Parts for hot water flow meters, pump turbines, connector parts and printed circuit boards.

Processing data is given in Table 6.25.

6.3.11 Liquid crystalline polymers

LCP are block copolymerisates – their filamentary molecules consist of rod-shaped hard segments (mesogenes), which are connected to each other by short and more flexible segments. Therefore, an ordered structure is already present in the liquid state caused by the parallel order of the little rods (semi-crystallisation). Perhaps the most well-known example is that of the aramid Kevlar, which is used for bullet-proof vests.

Table 6.25: Processing data for PAEK.

Predrying	3 h with 150 °C.
Compound temperature inside the cylinder	350–420 °C. 440 °C should not be exceeded.
Mould temperatures	150–180 °C.
Flow behaviour	Flows easily.
Injection speed	Middle to high.
Injection pressure	800–1,500 bar.
Holding pressure	450–800 bar.
Maximum internal pressure	400–700 bar.
Back pressure	60–90 bar.
Processing shrinkage	1% non-reinforced, 0.1–0.4% reinforced.

The above-mentioned LCP, called thermotropic LCP (as they show a liquid crystalline order in the melt), are suitable for injection. Lyotropic LCP are liquid crystalline in solvents and are used to produce fibres (e.g., Aramid fibres). There are also side chain LCP with little rods which are connected at the side with the main chain (which are flexible in the melt).

During injection, the rods orientate into the flowing direction, therefore, they have exceptional mechanical and thermal properties. This so-called self-reinforcing effect only appears along the orientation direction; hence, in this direction, LCP exhibit a fibrous structure that is reminiscent of wood. Most LCP have aromatic (circle-shaped) structures (special APE).

There are different types with especially developed properties:

- The material has very high strength, rigidity, resistance to abrasion and high thermal resistance, but with weaknesses in impact resistance and processability (high temperatures, high viscosity of the melt).
- Highest impact resistance and good processability but with the surrender of extreme strength and rigidity.
- Very resistant against chemicals.
- Best processability with high strength and rigidity but with the surrender of the highest thermal resistance.

To consider: along the orientation direction (privileged direction) the LCP do much better than conventional thermoset plastics; however, vertically to the orientation direction they are in the range of technical thermoset plastics.

Typical properties:
- Very high strength and rigidity in the orientation direction:
  - Strength = 150–300 N/mm^2.
  - Rigidity = 10,000–35,000 N/mm^2.
- The processing temperature is generally above 200 °C.
- Very little creep tendency.
- Danger of splitting parallel to the orientation direction, high to modest impact resistance (depends on the type).
- High endurance under completely reversed stress.
- Very small thermal expansion coefficient (values as found with steel or smaller), adjustable *via* processing conditions; therefore, excellent dimensional stability.
- Little dielectric losses.
- Very resistant against chemicals, no tendency to cold crack formation, very resistant against ambient influences, very resistant against radiation of high energy and high non-flammability.
- Little development of smoke.
- Natural condition: in general, non-transparent (already in the melt) as they are semi-crystalline, certain types can be transparent.
- Typical applications: for parts which are put under a lot of mechanical, thermal, electrical and chemical stress, possibly to replace metallic materials, ceramic or thermoset plastics, for the car and aerospace industry, precision mechanics (electronics, optical industry), electrotechnical industry, also for motor parts, gearing, safety systems and so on.

Processing: predrying is necessary. The flow behaviour is modest to excellent, depending on the type. Due to the molecule orientation, the joint lines represent weak points. The cycle periods are shorter than with conventional partly crystalline thermoplastics. Further processing data can be found in Table 6.26.

Table 6.26: Processing data for LCP.

Predrying	2–4 h with 150 °C. Permitted remaining humidity = 0.015%.
Compound temperature inside the cylinder	280–450 °C, depending on the type.
Mould temperatures	30–160 °C, depending on the material type.
Flow behaviour	Flows easily.
Injection speed	High.

Table 6.26 (continued)

Injection pressure	400–1,500 bar.
Holding pressure	350–1,000 bar.
Maximum internal pressure	300–800 bar.
Back pressure	40–60 bar.
Screw circumferential speed	0.15 m/s.
Processing shrinkage	−0.1 to +0.6%. Lower values in orientation direction, upper values vertical to it.
After-shrinkage	Practically nil.
Peculiarities	Parts behave very anisotropically, e.g., among others that they are very solid in orientation direction; therefore, the position and form of the sprue are of critical importance.

6.4 Conclusion

This chapter has detailed common properties for a number of amorphous and semi-crystalline materials. It has also highlighted any major processing issues attached to using these materials for injection moulding. All the necessary information to successfully mould these materials has been detailed in the current chapter and Chapters 8 and 9. Chapter 7 will consider other types of injection mouldable materials, for example, thermosets, elastomers and liquid silicone rubber.

7 Processing thermoset, liquid silicone rubber and ceramic materials

This chapter is split into three sections covering processing and materials focusing on thermosets, elastomers [specifically liquid silicone rubber (LSR)], and briefly, ceramic and metal powder injection mouldable materials.

The first section will give details of the injection moulding of thermosets and applications of the most common thermoset materials.

7.1 Thermosets

7.1.1 Overview and common properties

Commercial thermosets can provide perfect application solutions where high-performance long-term properties are required. Properties such as high mechanical strength, dimensional stability, temperature resistance, non-flammability, resistance to abrasion and chemical resistance are of particular interest for replacing metal materials, and can provide lighter weight commercially viable (price/performance) alternatives to traditional metal products, as well as being a dedicated market sector in their own right.

Thermoset plastics are very different to thermoplastics and require dedicated processing routes and machinery, due to the different heating cycle requirements, chemical reactions that occur and resultant gas emissions that require tool venting.

Thermoset plastics are chemically setting synthetic resins which set in the hot mould *via* a crosslinking (chemical) process, as opposed to thermoplastic freezing in a sufficiently cold mould (physical process). Synthetic resins tend to shrink during curing which can cause distortion and void formation, since plastic deformation is practically impossible. The various resin types are shown in Table 7.1.

Thermoset plastics are almost solely processed in combination with fillers and reinforcing materials, in order to extend and reinforce synthetic resin and, in particular, to minimise shrinkage. Examples of common fillers and reinforcing materials are shown in Table 7.2.

Traditionally, thermoset plastics are processed by compression moulding or for small volumes by hand layup techniques. However today, injection moulding is one of the most economical ways of mass-processing thermosets. If it is to be cost-effective, the technology used must be appropriate to the shape and properties of the material and all of the processing components. Thermosets can be supplied in a variety of forms, from granulates to rods to doughy polyesters (dough moulding compound/bulk moulding compound/sheet moulding compound). For a more basic introduction to the processing options for structural composites the reader is referred to [1].

https://doi.org/10.1515/9783110654813-007

Table 7.1: Resin types.

PF	Phenol formaldehyde resins	Phenoplasts	Polycondensate, i.e., crosslinking *via* the elimination of H_2O
UF MF	Urea formaldehyde resin Melamine formaldehyde resin	Aminoplasts	
UP DAP EP	Unsaturated polyester resins Diallyl phthalate Epoxide (epoxi-, ethoxyl) resins	Crosslinking *via* polymerisation or addition, without the formation of fission products	

Table 7.2: Filling/reinforcing compounds.

	Form	Powder flour	Fibre	Pellets	Chips	Mats webs	Weave
Inorganic	Stone	X					
	Asbestos		X	X	X	X	
	Glass		X	X		X	X
Organic	Wood	X					
	Cellulose		X		X	X	
	Textiles		X		X		X
	Carbon		X			X	X

7.1.2 Materials for screw injection moulding machines

The decisive factor for processing thermosets is the very structure of the material.

The right material must meet the following requirements:

- Free-flowing.
- Fast plastification is possible.
- Extended stay in cylinder is possible.
- Fast and high-quality setting in the mould.

Processing thermoset materials requires dedicated processing routes and machinery due to the chemical reactions and gas emissions and requirement for precise temperature control throughout the process, cylinder modules able to withstand highly abrasive wear and specially designed screw geometries that suit the materials being selected.

Therefore, not all materials are suitable for injection moulding; a range of thermosets from soft to medium is recommended for processing on injection moulding machines (IJM). A temperature guide for continuous processing is shown in Table 7.3.

Table 7.3: Temperatures for the continuous processing of thermosets.

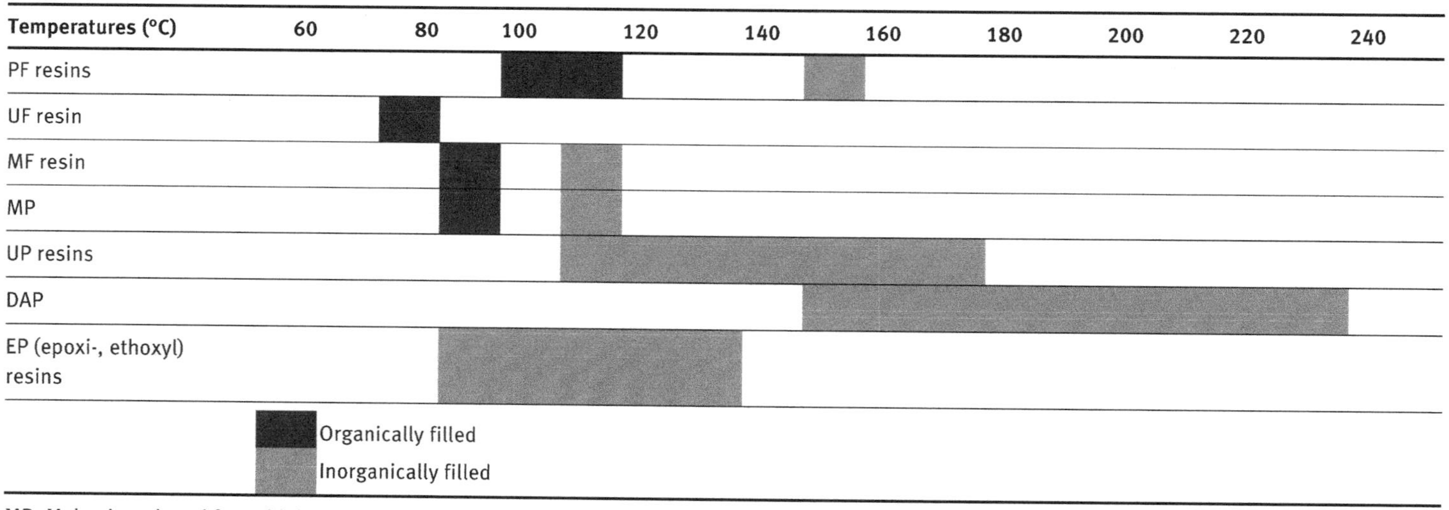

MP: Melamine-phenol formaldehyde

7.1.3 Processing procedures for thermoset plastics

7.1.3.1 Compression moulding

During the classic compression moulding process, a blank, prepared in shape and weight, is put into the opened mould and pressed under high pressure as shown in Figure 7.1.

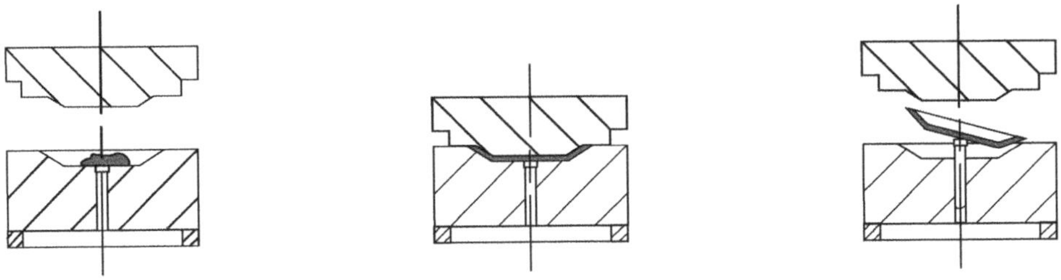

Figure 7.1: Compression moulding.

The disadvantages of this production method compared with injection moulding are:
- Long cycle times.
- High consumption of material.
- Considerable additional treatment is necessary.
- Accuracy of multiple parts are difficult to obtain.
- Only semi-automatic operation is possible.

7.1.3.2 Injection moulding

The more economical process is often the injection moulding of thermosets on screw IJM, where the material is plasticised with a special screw. The optimum preparation of the melt includes the use of a plasticising unit with a corresponding feed unit and screw for the material-specific dosage. This is because dosing and plasticising for a new shot takes place while the material is still forming crosslinks in the mould. The whole process can run automatically but is generally a much slower injection cycle than that used in thermoplastic injection moulding.

Advantages of processing thermosets on IJM:
- Considerably shorter cycle times compared with alternative processes.
- The material is heated and plasticised homogeneously.
- High-quality, high-precision parts.
- Material saved by injection.

- Parts with minimum flash, because of injection into a closed mould.
- Less wear on the mould.

7.1.4 Screw injection machines for thermosets

For injection machines to be equipped for processing thermosets, the following equipment is required:
- Thermoset cylinder assembly.
- Mould blow unit.
- Dosage delay mechanism.
- Venting control mechanism.
- Special control equipment for cleaning or brushing the mould.

Screws and cylinders for thermosets are treated with, for example, Arbid. The screws have a special geometry, a suitable tip and no backflow prevention valve; other versions with backflow prevention valves are also available.

There will be wear on the screw and cylinder, especially when processing materials with abrasive fillers, such as asbestos, rock meal and glass fibre, primarily due to the high injection pressure. The backflow of material will put additional strain on the screw and the front part of the cylinder. Gentle processing is required to prevent premature crosslinking reactions occurring in the material and to ensure smooth flowpaths where material cannot accumulate. Hence, particular attention has been paid to minimising return flows during injection and the holding pressure phase.

To increase the wear resistance and durability, the following measures can be taken:
- Use of hard alloy (bimetallic) bushings for the cylinder.
- Use of a screw with a backflow prevention valve.
- Use of a hard alloy screw.
- Use of a screw with a double thread in the metering section.

The interplay play between the screw and nozzle section is important to prevent excess material, and constant shot weights are achieved by a mechanical antireversing lock on the screw.

For cylinder heating, often two temperature control devices are needed and precise temperature control is required for optimum moulding. Inside the cylinder a temperature between 50 and 80 °C is generally (but is material specific) needed in the feed zone and between 75 and 115 °C in the metering zone. Traditionally, a temperature control device operated with water is used for the feed zone and an oil-operated device is used for the metering zone.

7.1.5 Mould

There is an increased importance of adequate venting during thermoset injection moulding and a fundamental reverse in heating requirement in that heating (for curing completion) and not just cooling (for solidification) is provided by the tool. Therefore, it should be noted that thermoplastic mould designs are not suitable for direct transfer for use in the moulding of thermoset materials. There are also design limitations on certain aspects, such as minimum wall thicknesses, that can be achieved with many materials and therefore the reader is referred to material-specific requirements as well specialist thermoset mould makers for mould design.

The mould is heated electrically and it is recommended that control of both mould halves is separate. The heating power for small moulds should be between 1.600 and 2.200 Watts per mould half, varying according to the mould size (e.g., 215 × 300 mm). The mould temperatures required vary according to the thermoset material but standard values range from 150 to 240 °C. The mould should have exchangeable inserts and insulating platens of 10 mm thickness.

7.1.6 Guide values for cylinder and mould temperature settings

Always observe the specific instructions of the material manufacturers.

When adjusting cylinder temperatures please note:

- The melt temperature (T_m) is not only influenced by the cylinder temperature, but also by the frictional heat created by the plasticising screw.
- The higher the screw speed and the higher the back pressure, the higher the T_m will be (at the same cylinder temperature).
- The upper temperature limits should be selected only when the melt stays in the cylinder for a very short time (high cycle speed).

A guide to the effects of the various stages of injection moulding thermosets is given in Figure 7.2.

As a result of the injection operation, the heat of friction is increased through the nozzle and cross-sections, while the material is compressed at holding pressure. The residual heat of reaction, which is necessary for crosslinking of the injected part, is conveyed *via* the mould-tempering unit. Since the material is cured while heat is supplied, temperature management during the plastication and injection operations is of great significance for the cost-effective production of high-quality injected parts. To achieve this, special temperature control units are installed. The injection cylinder is divided into several temperature zones in order to optimally adapt the heating equilibrium and shot volume to one another. This principle is illustrated in Figure 7.2. The heat introduced into the melt during plastication can therefore be maintained within strict limits.

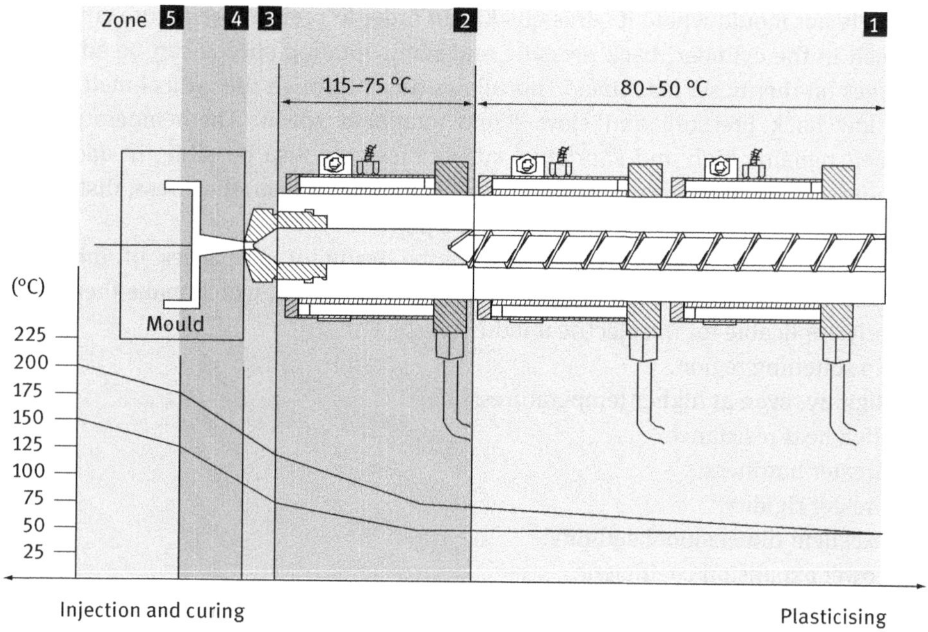

Zone 1	Preheating zone	Effect of screw geometry, rotational speed, back pressure and cylinder temperature control
Zone 2	Plasticising zone	
Zone 3	Friction dependent upon injection capacity (injection pressure and injection speed) and nozzle bore	
Zone 4	Friction caused by sprue and gate cross-section	
Zone 5	Effect of mould heating and curing	

Figure 7.2: Thermoset injection moulding.

In the next phase, the friction of the nozzle significantly increases the T_m. As a result, the viscosity of the material is reduced, the melt becomes thin and can be introduced into the cavity in the optimum manner. Machines with high injection capacities (injection pressure and speed) are required to handle this.

The mould heating and curing time influence the last phase of heat supply. Only a small amount of heat is fed to the melt in order to form the part and to complete the forming process.

In a typical processing sequence the crosslinking of the prepared material is delayed in the relatively cool injection cylinder for a period until the melt reaches the relatively hot mould where it cures quickly. In order to prevent premature curing of the melt in the cylinder, back pressure and screw rotation speeds can be adjusted in stages on thermoset machines. This allows travel through the critical melt range with low back pressure and slow screw rotational speed. The temperature of the melt remains high and shortened curing times are also possible; in addition, the more uniform temperature structure has positive effects on the stress, distortion and shrinkage characteristics of the moulded part.

When compared with thermoplastics, the resultant properties of injection moulded thermosets have several unique and clear benefits, which make them particularly applicable for the electrical industry:
- No softening region.
- Rigidity, even at higher temperatures.
- High heat resistance.
- Greater hardness.
- Greater rigidity.
- Excellent dimensional stability.
- Lower expansion coefficient.
- Lower cold flow.
- More favourable combustion behaviour.
- Excellent price/performance ratio.

7.1.7 The injection unit for granulated thermosets

The screw in the injection unit is not compressed and is characterised by its relatively low length/diameter (L/D) ratio of 15:1 (compare this with 20:1 for thermoplastics), which is a result of shifting the feed zone forward; in addition, a surface-cured screw is used as standard. A fully cured version is available for use with extremely abrasive materials. The bimetal construction of the cylinder enables an extremely long service life. The heating control system can be adjusted across three or four independent temperature zones. The open, cured nozzle is screwed directly into the nozzle socket and can therefore be tempered precisely.

7.1.8 The injection unit for processing moist polyester

The screw diameter in this case is greater, since lower pressures and a shorter L/D ratio of around 12:1 are sufficient for processing purposes. In comparison with units for granulated thermosets, there are also differences in the following areas.

The geometry of the screw is specifically designed for moist polyester melts and has a plain check valve to reduce fibre destruction. This was illustrated in Chapter 3 (Figure 3.6).

The feed aperture is optimised for the non-destructive introduction of the material.

7.1.9 Nozzle temperature (if Nozzle Heating is Applied)

The nozzle temperature should be adjusted to 5–15 °C above the temperature suggested for the metering zone of the plasticising cylinder and suggested values are given in Table 7.4.

Table 7.4: Temperature guide for thermosets.

Type of material	Temperatures (guide values) (°C)		
	Mould	Plasticising cylinder	
		Metering zone	Feed zone
PF	145–175	75–90	45–60
UF	140–165	80–95	45–55
MF	150–180	75–100	50–60
MP	150–180	70–100	45–55
UP	140–165	80–105	40–65
DAP	150–190	85–110	50–60
EP	155–200	85–115	50–60
LSR	180–250	5–20	5–20

7.1.10 Common thermoset materials: Properties and applications

This section will detail properties and processing data for the following five classes of thermoset materials:
- PF.
- UF resin.
- MF resin.
- UP and DAP.
- EP compounds.

7.1.10.1 Phenol formaldehyde compounds

Including grades such as:
- FS 11–with inorganic (e.g., mineral) fillers.
- FS 31–with wood (e.g., wood flour) fillers.
- FS 51–with paper (e.g., paper fabric) fillers.
- FS 71–with fabric (e.g., cotton) fillers.

N.B.: The higher the first digit in the designation is, the higher the percentage of fillers in the resin. The specific properties depend largely on the type and percentage of the filling material in the resin. ^pTypical properties and applications:

- Hard and stiff.
- Good heat resistance, even in long-term applications (up to 150 °C for resins with inorganic fillers).
- Strong water absorption of resins with organic fillers.
- Low rate of creep.
- No light colours.
- Often physiologically problematic (contact with foodstuffs is forbidden).
- Applications: mass products for the electrotechnical industry, such as housings, sockets for switches and relays, connectors and so on.

Processing data for PF compounds is given in Table 7.5.

Table 7.5: Processing data for PF.

T_m inside the plasticising cylinder	45 to 90 °C	
Back pressure	10 to 15% of injection pressure	
Mould temperatures	145 to 175 °C	
Flow	Depending on the material, medium to low flow	
Shrinkage	Immediate mould shrinkage	0.5 to 0.9% in flow direction 0.7 to 1.1% normal to flow direction
	Post shrinkage	Up to 0.4%
Injection pressure	Injection	800 to 1,500 bar
	Holding pressure	40 to 60% of injection pressure
	Processing practically without material cushion	
Storage	Limited storage (6–24 months), material forms crosslinks	

7.1.10.2 Urea formaldehyde compounds

This includes UF compounds such as FS 131–with organic fillers (e.g., cellulose). Typical properties and applications:

- Hard and stiff.
- Maximum temperature stability for long-term use.
- Application at 80 °C.

- Stress cracking due to strong shrinkage.
- Ageing at high temperatures.
- Low rate of creep.
- Light colours possible.
- Often physiologically problematic (contact with foodstuffs is forbidden).
- Applications: light coloured mass-produced items for the electrotechnical industry, especially installation materials.

Processing data is given in Table 7.6.

Table 7.6: Processing data for UF.

T_m inside the plasticising cylinder	45 to 90 °C	
Back pressure	10 to 15% of injection pressure	
Mould temperatures	140 to 165 °C	
Flow	Depending on the material, medium to low flow	
Immediate mould shrinkage	0.2 to 0.6% in flow direction	
	0.6 to 1.3% normal to flow direction	
Post shrinkage	Up to 1.7% depending on the filler	
Injection pressure	Injection	800 to 1,500 bar
	Holding pressure	40 to 60% of injection pressure
	Processing practically without material cushion	
Storage	Limited storage (generally no longer than 6 months), material forms crosslinks	

7.1.10.3 Melamine formaldehyde compounds

MF resins with organic and/or inorganic fillers, e.g., wood flour, cotton wool and/or rock meal, asbestos:
- FS 150–MF resins.
- FS 180–MP resins.

Typical properties and applications:
- Hard, stiff and a low rate of creep.
- Strong heat resistance, even in long-term applications (up to 130/150 °C for resins with inorganic fillers).
- MF tends toward stress cracking, due to strong shrinkage at high temperatures.
- Higher creep resistance compared with PF and UF.

- MF hardly flammable, self-extinguishing.
- Light colours possible (especially with MF).
- Some MF types are physiologically harmless (contact with foodstuffs is allowed).
- Applications: light coloured parts for the electrotechnical industry (housings, sockets for switches and relays, connectors and so on) even for higher creep resistance requirements, household appliances and tableware. Processing data is shown in Table 7.7.

Table 7.7: Processing data for MF and MP.

T_m inside the plasticising cylinder	55 to 100 °C
Flow	Depending on the material, medium to low flow
Injection pressure	800 to 2,000 bar
Back pressure	10 to 15% of injection pressure
Holding pressure	40 to 60% of injection pressure
	Processing practically without material cushion
Mould temperatures	150 to 180 °C
Immediate mould shrinkage	0.2 to 0.6% in flow direction
	0.6 to 1.3% normal to flow direction
Post shrinkage	Up to 1.7% depending on the filler
Storage	Limited storage (generally not longer than 6 months), material forms crosslinks

7.1.10.4 Unsaturated polyester

UP with mostly inorganic fillers (glass fibre, minerals):
- UP.
- DAP and special UP (allyl resin).

Typical properties and applications:
- Hard and stiff.
- Strong heat resistance, even in long-term applications (up to 180 °C according to type, DAP up to 230 °C).
- Even at low temperatures, relatively insensitive to impact.
- Low water absorption.
- Weathering resistance dependent on the coupling agent between the resin and glass fibre.

- Very high creep resistance.
- Light colours possible.
- There are some physiologically harmless types (contact with foodstuffs for these are allowed).
- Applications: electrotechnical and precision mechanical items for high mechanical, thermal and electrical requirements (coil cores, connectors, sockets, housings, covers and so on).
- Due to the very low shrinkage, varying material concentrations and sharp edges are less critical than with other materials.

Processing conditions for UP and DAP are given in Table 7.8.

Table 7.8: Processing guide for UP and DAP.

T_m inside the plasticising cylinder	UP (generally): 65 to 105 °C	
	DAP: 55 to 110 °C	
Back pressure	10 to 15% of injection pressure	
Mould temperatures	UP (generally): 140 to 165 °C	
	DAP: 150 to 190 °C	
Flow	Depending on the material, medium to low flow	
Shrinkage	Immediate mould shrinkage	Up to 0.3%
	Post shrinkage	Practically zero
Injection pressure	Injection	800 to 2,500 bar
	Holding pressure	40 to 60% of injection pressure
	Processing practically without material cushion	
Storage	Limited storage (generally weeks to a few months). Material forms crosslinks. In addition to the free-flowing materials there are some pasty materials, mostly filled with long glass fibre. These can also be further processed in combination with an INJESTER-type feeding unit.	

7.1.10.5 Epoxide compounds

EP compounds only with inorganic fillers (glass fibre, minerals):

Typical properties and applications:

- Hard and stiff.
- Tougher than UP.
- Strong heat resistance, even in long-term applications (up to 130 °C, special types even higher).

- Even at low temperatures, relatively insensitive to impact.
- Very low water absorption (less than for UP).
- High weathering resistance, since there is no coupling agent necessary between the resin and glass fibre (EP resins are excellent adhesives).
- High creep resistance.
- Light colours may be possible but in these cases items do not have durability to light exposure.
- EP resins are physiologically harmless once they have set (are crosslinked). However, avoid contact with the raw material as skin rashes may occur.
- Applications: electrotechnical and precision mechanical parts for high mechanical thermal and electrical requirements (coil cores, connectors, sockets, housings, covers and so on).
- Due to the very low shrinkage, varying material concentrations and sharp edges are less critical than with other materials.
- Generally, high weathering resistance allows outdoor use.

Guide values for processing by injection moulding are given in Table 7.9.

Table 7.9: Processing data for EP.

Drying prior to injection	When you have stored EP resins at low temperatures to counteract crosslinking, dry and heat the material in a closed container prior to injection.	
Melt temperature inside the plasticising cylinder	55 to 115 °C	
Back pressure	10 to 15% of injection pressure	
Mould temperatures	155 to 200 °C	
Flow	Depending on the material, medium to low flow	
Shrinkage	Immediate mould shrinkage	0.1 to 0.5%
	Post shrinkage	Practically zero
Injection pressure	Injection	800 to 2,500 bar
	Holding pressure	40 to 60% of injection pressure
	Processing practically without material cushion.	
Storage	Limited storage (generally weeks to a few months), material forms crosslinks.	

7.2 Elastomer injection moulding compounds

This section will discuss issues relating to the injection moulding of rubbers, and specifically, the moulding of LSR, which is of large commercial interest. Firstly however, a quick introduction to rubber materials is given.

There are two types of rubber, natural rubber (NR) and synthetic rubber.

NR has for a long time been the only raw material of the rubber industry. It originates from the sap of the rubber plant from the rubber plantations in Asia (rubber milk = latex). The thickened material is rolled, dried and pressed.

Synthetic rubber is a product of organic chemistry. Its basic material is raw oil, especially heavy petrol that is produced when cracking raw oil in the refinery. It is polymerised by means of various chemical combinations and additions, which results in synthetic latex milk. As with NR a solid material remains, which can be extracted. The raw material for LSR is quartzite or quartz powder

Both rubber types, natural as well as synthetic rubber can be processed using moulding machines.

7.2.2 Processing procedures for elastomeric materials

7.2.2.1 Compression moulding

Using this procedure, a blank, prepared in shape and weight, is put into the opened mould and pressed under high pressure (Figure 7.1).

The disadvantages of this production method are similar to those identified with thermoset compression moulding:

- Long cycle times.
- High material consumption.
- Considerable additional treatment is necessary.
- No parts of accurate dimensions are obtained.
- Only a semi-automatic operation is possible.

7.2.2.2 Injection moulding

In this method, the injection moulding screw plasticises the prepared rubber cords by homogeneously heating up the elastomeric material.

Advantages of heating the rubber:

- Shorter vulcanisation times, 12–16 s/mm wall thickness.
- A higher accuracy of the parts.
- A better quality of the parts.
- Any additional treatment is not necessary.
- 10–20% material is saved.
- Flash-free parts or parts with minimum flash are obtained.

- Less wear on the mould.
- Automatic production.

7.2.2.3 Injection moulding machines for rubber compounds

Standard machines can be prepared for the processing of elastomeric materials. The machines must, however, have the following equipment:
- Cylinder for elastomeric material.
- Mould blow unit.
- Dosage delay mechanism.
- Special control equipment for cleaning and brushing devices in the mould (to remove bloom).

N.B.: Special 10 D cylinders with compressionless screws and a big feed bag for particularly sensitive rubber materials are available for many injection units.

7.2.2.4 Peripheral device

A temperature control unit is needed to heat the plasticising cylinder. Depending on the rubber material, a temperature between 50 and a maximum of 170 °C is needed inside the plasticising cylinder. In most cases, such a temperature control unit works using oil.

7.2.2.5 Mould

Mould heating is achieved by means of electrical heating and it is recommended that both mould halves are controlled separately. The heating power for each mould half should be between 1.200 and 1.600 Watts. The mould temperatures vary according to the elastomeric material but standard values range from 110 to 210 °C. In view of the high mould temperatures, insulating platens are required for the mould which have a thickness of 10 mm.

7.2.3 Silicone elastomers – liquid silicone rubber systems

The increasing expense of machines, staff and especially raw materials forced manufacturers of plastic materials and custom moulders to look for new solutions; hence, liquid silicone rubber was introduced by the chemical industry in 1978. Its basic material is quartzite or quartz powder (SiO_2), making it independent of raw oil (in contrast with synthetic rubber).

When you crack quartz powder (SiO_2), silicium remains. By chemical joining and adding, silicium is built into polymeric liquid silicon rubbers consisting of two components, A and B. These are mixed in a 1:1 ratio shortly before being processed into formed components, and are then vulcanised into silicone rubber in the mould. Preforms made from LSR, in spite of the high material prices, are often considerably more cost-effective than those made from conventional types of rubber because the processing costs are considerably lower.

Among the variety of plastics and elastomers, the silicones occupy a special position. Their significant difference from other plastics lies in their chemistry. Most plastics are essentially bridged over carbon–carbon bonds; in contrast to this, the silicones are distinguished by alternating silicon and oxygen atoms.

Since there are very many different possibilities for varying organic groups on the chains within the molecule, there is an extensive range of products on the silicone marketplace with widely differing properties. The primary products which are produced on a silicone base are silicone elastomers, silicone oils, which are used as antifoaming agents or parting agents, chemical specialities such as cosmetics or polishes, silicone resins, textile products and paper coatings.

7.2.3.1 Classification

Silicone rubbers may be classified by the crosslinking method, and viscosity and vulcanisation temperature. Differentiation can be made between hot and cold [room temperature vulcanisation (RSV)] rubber types. Within each of these two groups are found single-component and two-component systems, grouped by viscosity range. The viscosity of the rubber (fluid mouldable, pasty, plastic-firm) determines the processing method and influences the characteristics of the vulcanised material. Figure 7.3 depicts the classification of the silicones.

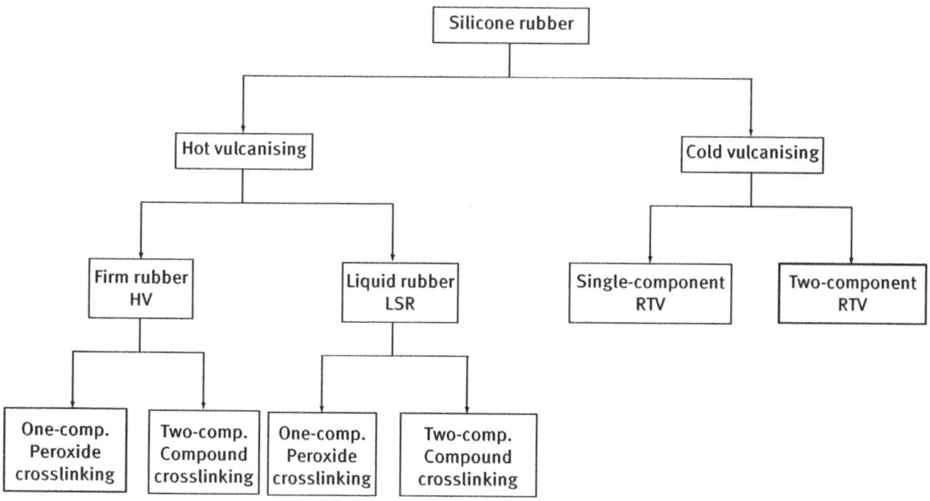

Figure 7.3: Classification of the silicones (RTV: room temperature vulcanization and HV: hot vulcanisation).

Silicone rubber, HV, liquid rubber LSR-2K compound crosslinking, are shown in Figure 7.3 and are discussed in the following sections on LSR processing. All other LSR types require different machine equipment and process conditions, which are beyond the scope of this book.

7.2.3.2 Crosslinking

Compound crosslinking is based on the compounding of SiH to double bonds and this reaction takes place relatively quickly. A catalyst must be present for the initiation of the reaction. Platinum complexes, which are added in very low concentrations to the rubber, are used as catalysts. The crosslinking reaction begins at room temperature (RT) and the reaction speed increases rapidly with increasing temperature. Figure 7.4 depicts a schematic view of compound crosslinking.

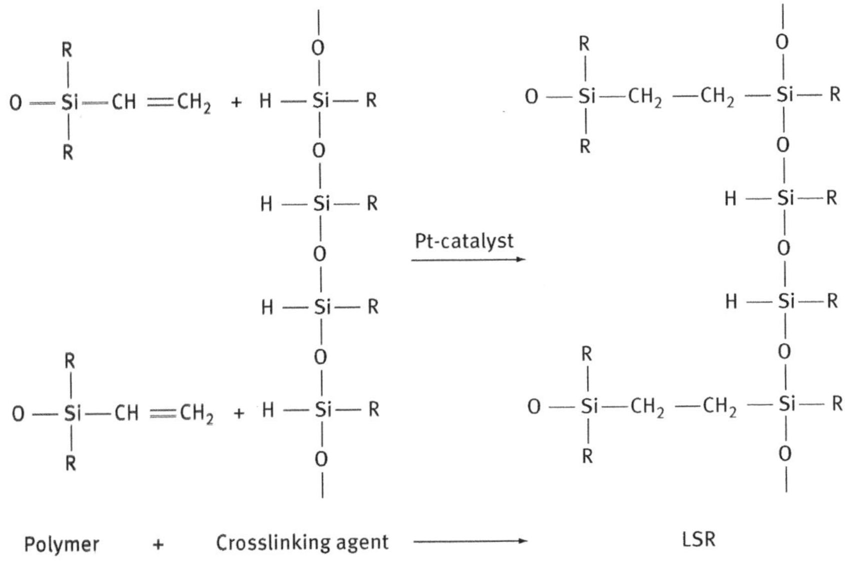

Figure 7.4: Compound crosslinking.

Rubber systems with compound crosslinking are based on two-components, since a separation into two components is necessary for storage. One component must contain the catalyst, the other the crosslinking agent.

7.2.3.3 Processing of two-component liquid silicone rubber

Because of the particular material properties of LSR, in comparison to other materials which are capable of being injected, a special processing technology is required.

This technology is known by its abbreviation, LIM, which stands for liquid injection moulding.

Injection moulding of LSR is possible with an LSR cylinder assembly; however, a multicomponent dosage unit is required. A multicomponent dosage unit mixes equal amounts of components A and B and feeds the resulting compound into the injection cylinder of the IJM. A typical processing system will include:

- Multicomponent dosage unit with containers.
- Component 'A': crosslinking catalyst.
- Component 'B': components to be crosslinked.
- Component 'C': colouring paste.
- Mixer (not necessary for some machines if using a mixing screw).
- Feeding device for the injection unit.
- Clamping unit with mould.

7.2.3.4 Advantages of processing liquid silicone rubber materials

There are a number of advantages in moulding LSR materials:

- Low injection pressure processing on small, low-cost IJM with low clamping forces is possible.
- The number of mould cavities can be increased.
- Easy ejection of moulded parts.
- High-precision parts can be produced.
- Few burrs on parts.
- Considerably shorter cycle times in comparison with conventional rubber materials.
- Vulcanising times of 5–6 s per mm wall thickness with a mould temperature of 200 °C.
- Temperature range for the end product: −60 to +300 °C.

7.2.3.5 Injection process

You can process LSR with modern IJM, however, you need special LSR equipment for the machine. Today, injection moulding of LSR is solely performed with screw-IJM, in which the screw plasticises the prepared compound. During the plasticising stage, the material is kept at a temperature of 5 to 25 °C, which is followed by the low pressure injection of material by the screw into the mould.

7.2.3.6 Configuration of injection moulding machines and processing data

For injection moulding of LSR materials, the following equipment is required:

- Special LSR plasticising cylinder with cooling jackets.
- Hydraulically actuated needle-type shut-off nozzle with cooling jacket.

- A special pressure relief valve for the material feed line from the dosage and mixer units.
- A mould blow unit.
- If necessary, a special control circuit for cleaning or spraying the mould.

N.B.: to keep the cylinder temperature at 5 to 25 °C, the cooling jackets will be connected directly to a cooling water distributor. To reach extremely low temperatures, an external cooling device can be used. For injection, use a single-stage non-compression mixing screw with backflow prevention valve.

General processing data for LSR is given in Table 7.10.

Table 7.10: Processing data for LSR.

Cylinder temperature	5–25 °C. Higher cylinder temperatures carry the risk of transient scorch on the cylinder wall. Moulded-on particles ripped into the mould can impair the preform's properties.
Screw speed, back pressure	Average screw speed (100–200 rpm). Back pressure = 0. Because of the low viscosity of LSR, the back pressure considerably reduces the dosing rate or completely prevents dosing (screw return stroke).
Flowability	Very good, as LSR systems have low viscosity. Therefore, the injection pressure and mould clamping forces required are also very low.
Injection pressures	100–200 bar.
Mould temperatures	180–250 °C. Higher mould temperatures accelerate vulcanisation, e.g., a 10 g component gives: 50 s at 150° C, 5 s at 200 °C.
Shrinkage	Processing shrinkage and post shrinkage are practically zero.
Special features	Processing of LSR systems is possible on normal screw injection moulding machines with special LSR cylinder fittings (hydraulic needle shut-off nozzles, screws with non-return valves, but without compression, cylinder temperature controlled by fluids). The two LSR components are dosed in a special apparatus in a 1:1 ratio and mixed together and then fed into the drawing-in aperture of the injection unit. At temperatures under 20 °C, the 'shelf life' of the mixed components is very long (Dow Corning gives at least 30 h), i.e., when work is interrupted overnight, the dosing and mixing equipment and the cylinder do not need to be cleaned.

7.2.3.7 Specific properties

The use of LSR for the production of elastic parts using injection moulding technology is gaining ever-increasing significance in the processing of elastomers. The low viscosity in comparison to solid silicone rubber and other elastomers is characteristic of LSR and other elastomers, and typical characteristics are:

- High heat resistance up to constant temperatures of 180 °C.
- Good low-temperature resistance and cold flexibility to −50 °C.
- Good tensile strength and tear strength.
- High resistance to weathering and ageing.
- High electrical insulating properties as well as resistance to shock.
- Constant mechanical and electrical properties within a wide temperature range.
- High resistance to alcohol, polar solvents and weak acids.
- Anti-adhesive and hydrophobic characteristics.
- Excellent physiological properties, i.e., can be used in foodstuffs and is compatible with human tissue.

The producers of the LSR raw material provide a number of differing types of material in order to cover as broad a spectrum of applications as possible. The following typical commercial products are currently available:

- Standard types with a Shore hardness of 20 to 70 Shore A.
- Electrical conducting types.
- Oil-sweating, self-lubricating types.
- Flame-retardant products.
- Types for medical technology, as they are physiologically harmless.
- Oil-resistant types for the automotive industry.
- Fluorosilicone rubbers.

Based on the wide number of excellent properties and the fact that LSR can be dyed or tinted without problems, there are application areas in nearly all branches of industry. A few examples here are the electrical and automotive industries, machine construction, medical technology and the food industry. The applications range from keyboard covers, seals, plug connectors and medical apparatus to nipples for baby bottles.

7.2.3.8 Crosslinking reaction

LSR vulcanise in heat (170 to 210 °C) following compound crosslinking. In this technology, the crosslinking agent and the polymer material react under the influence of a catalyst. In contrast to peroxide crosslinking, no fission products occur here.

In order to begin the reaction at a desired point in time, the monomers are held in separate containers, A and B. The A component contains the catalyst, and the B component contains the crosslinking agent. The compound crosslinking begins immediately after mixing the two components, even at RT; however, the reaction at that point proceeds very slowly and unevenly, so that the vat time (the processing time after the mixing of the two components) amounts to three days at RT. Figure 7.5 depicts the vat time of a typical LSR as a function of temperature. Observe the rapid decrease in processing time with increasing temperature.

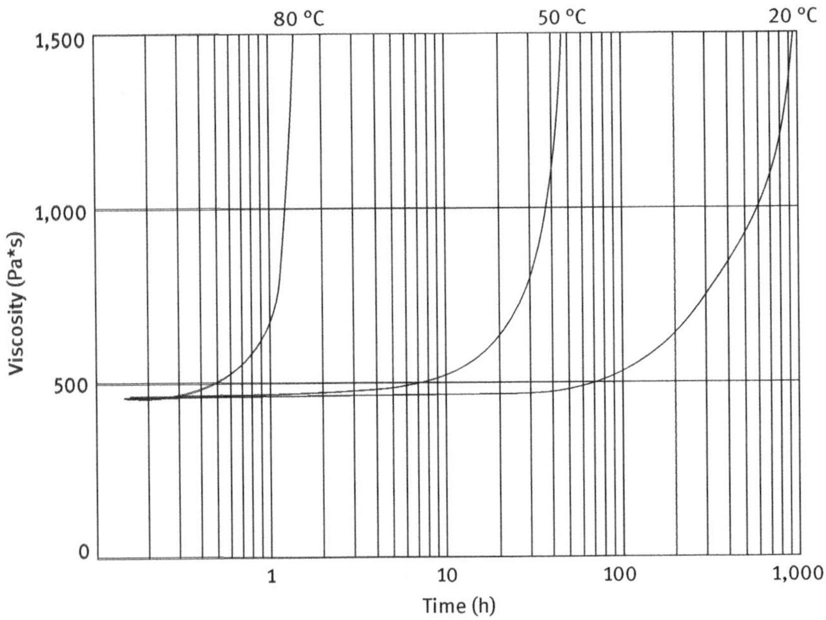

Figure 7.5: Vat time of LSR as a function of temperature.

From a temperature level of approximately -20 °C and below, there is no discernible reaction. This characteristic temperature dependence requires increased monitoring during longer periods of production stoppage. Hot machine parts which come into contact with the mixed material for longer periods can become coated with material which has vulcanised.

7.2.3.9 Structural viscosity

The consistency of the non-vulcanised two-component LSR is flow-capable and pasty. Although the pure polymer material has the characteristics of a Newton fluid in its flow behaviour, that is, the viscosity is independent of the shear rate of the process, a mixture of polymer materials with extenders, especially silicic acid, displays a structural viscous property. Structural viscosity means that the viscosity of the mixture decreases significantly with increasing shear velocity. This property, manifested to a greater or lesser extent, may be observed in every thermoplastic and has consequences which relate to the processing method because the material is subjected to widely different shear stresses:

- Low shearing (= 1 to 10 s^{-1}) of the individual components during conveying and dosing.
- Medium shearing (= 50 to 500 s^{-1}) during mixing and homogenising.
- High shearing (= 500 to 50,000 s^{-1}) during the injection process phase and filling the mould.

Figure 7.6 depicts the viscosity characteristics curve as a function of shear rate and temperature for silicone rubber components.

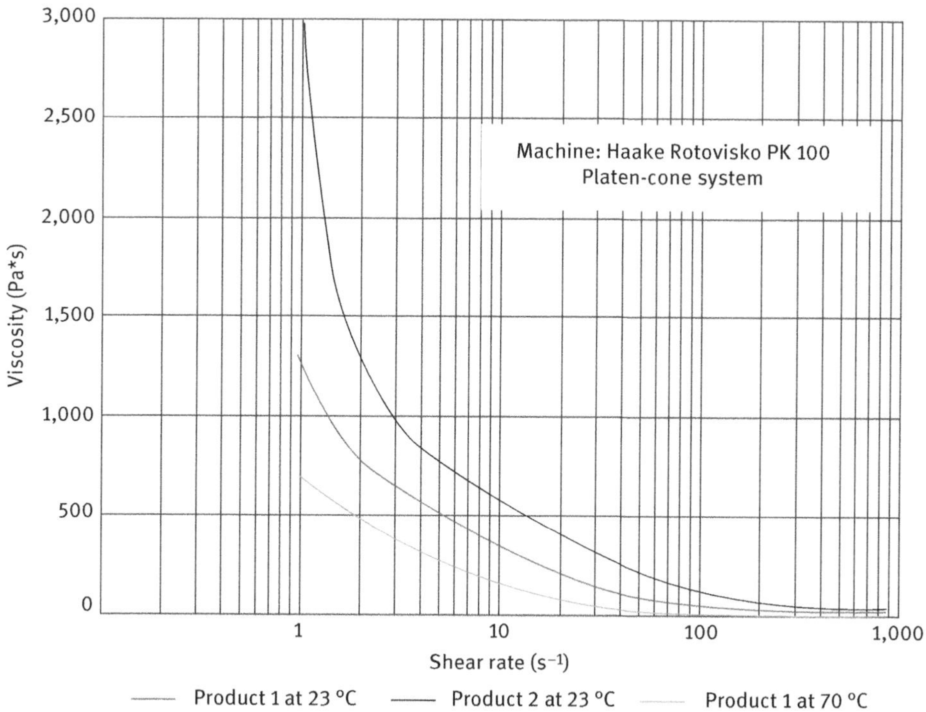

Figure 7.6: LSR viscosity characteristics curve.

The sharp decrease in viscosity with increasing shear stress is very clear, as is the shape of the viscosity curve with increasing temperature. Since the material experiences a very strong shear stress just at the injection process phase, the viscosity decreases dramatically at that point. The result of this is that the entire machine technology, and especially the mould, must be matched to a very ‘thin’ material.

7.2.3.10 Vulcanisation

In general, the vulcanisation of standard LSR materials increases strongly from approximately 110 °C, which can therefore be established as the starting point of the compound crosslinking during processing. For this reason, mould temperatures during LSR processing lie between roughly 170 to 210 °C. Vulcanising periods with optimal heat economy and suitable mould geometry in a temperature range of 180 °C, are approximately 3 to 5 s/mm of wall thickness. This means a crosslinking speed, which is 3 to 4 times faster than with other rubber systems, and hence a correspondingly higher number of moulding cycles.

7.2.3.11 Internal mould pressure

As with any HV material, LSR also undergo an increase in volume due to the heating in the mould and related vulcanisation. Internal mould pressures of 100 to 150 bar are the result; however, specific injection pressures of 1,000 bar and more are required for filling the mould. The necessary injection pressure is highly dependent upon the sprue system. In order to produce parts that are as consistent as possible, this data must be applied during the calculation of the clamping force. In addition, because of the low viscosity of the material which is introduced into the mould, all of the tolerances of the processing machine and the mould must be kept especially low in order to avoid over-injection.

7.2.3.12 Shrinkage

LSR parts undergo shrinkage during processing, which is a result of the significantly different expansion coefficients of the mould steel and the polymer material. In addition to the material-specific data, the degree of shrinkage is highly dependent upon the mould temperature, the generated pressure level and the direction of injection.

7.2.3.13 Tempering

Compound crosslinking silicone rubbers do not form fission products during crosslinking. However, despite the high degree of complete chemical reaction, a 100 % reaction cannot be achieved, which means that volatile polymer components remain behind in the finished part. For applications in the food industry or in medical areas, liquid as well as solid silicone rubber products must be tempered in accordance with regulations, for example, the German Bundesinstitut für Gesundheitlichen Verbraucherschutz Und Veterinärmedizin (BgVV) and the Food and Drug Administration (FDA) in the USA.

This tempering takes place in ventilated kilns with a fresh air-supply at approximately 200 °C over a period of 2 to 4 h.

7.2.3.14 Preparation of liquid silicone rubber

LSR is produced, prepared, mixed and filled ready-for-use by the raw goods manufacturer. The crosslinking agent required for vulcanisation is added to one portion of the material and the initiating catalyst is added to the other portion.

As a result, two components are produced, A and B, which are matched in such a way that the ready-for-use LSR systems may be mixed in a 1:1 ratio. Both components are shipped to the processor in 20 or 200 l barrels.

7.2.4 Processing components for liquid injection moulding

7.2.4.1 Dosing equipment for liquid silicone rubber

Since the raw material is shipped from the manufacturer in 20 or 200 l barrels, it must be further processed at the production plant using dosing equipment which is specifically designed for the purpose. This type of dosing equipment must fulfil the following requirements:

- Removal and transport of the material from the shipping drums.
- Mixing the two material components in a 1:1 relationship.
- Adds dosages of colour and additives if applicable.
- Conveying the processed mixture to the IJM.

Figure 7.7 depicts a schematic view of the system with a functional overview of a typical commercially available dosage system for liquid silicone.

The removal of A and B components is performed by barrel presses. In these presses a device called the follower plate is pressed onto the material surface so that the material flows in the desired direction of conveyance under the applied pressure. Piston pumps are installed as conveyor and dosing pumps. The two components pass through conduits to a mixing block in which they are brought together for the first time. Both pumps must be set to run synchronously in order to achieve the desired 1:1 mixing ratio. A static mixer, through which the material components must run before delivery to the IJM, is installed downstream for further homogeneous mixing. For pigmentation with colour paste or for dosing of additives, the dosing systems have separate small dosing units available. Portions of 0.3 to 5 % can be added for mixing.

The dosing system for LSR, which was described previously, is designed for only one machine. In order to save cost on space and investment, central material supply units are also available. In this scenario, multiple IJM are supplied from one dosing system. Modern material conveyance systems operate on the concept of central material supply with a decentralised supply by volume at every IJM. Figure 7.8 depicts this type of system in a schematic overview.

A central material delivery system consists of 200 l barrel presses with delivery pumps, which supplies several stand-alone dosing systems that are located directly at the IJM. Decentralised dosing systems may consist, for example, of time-

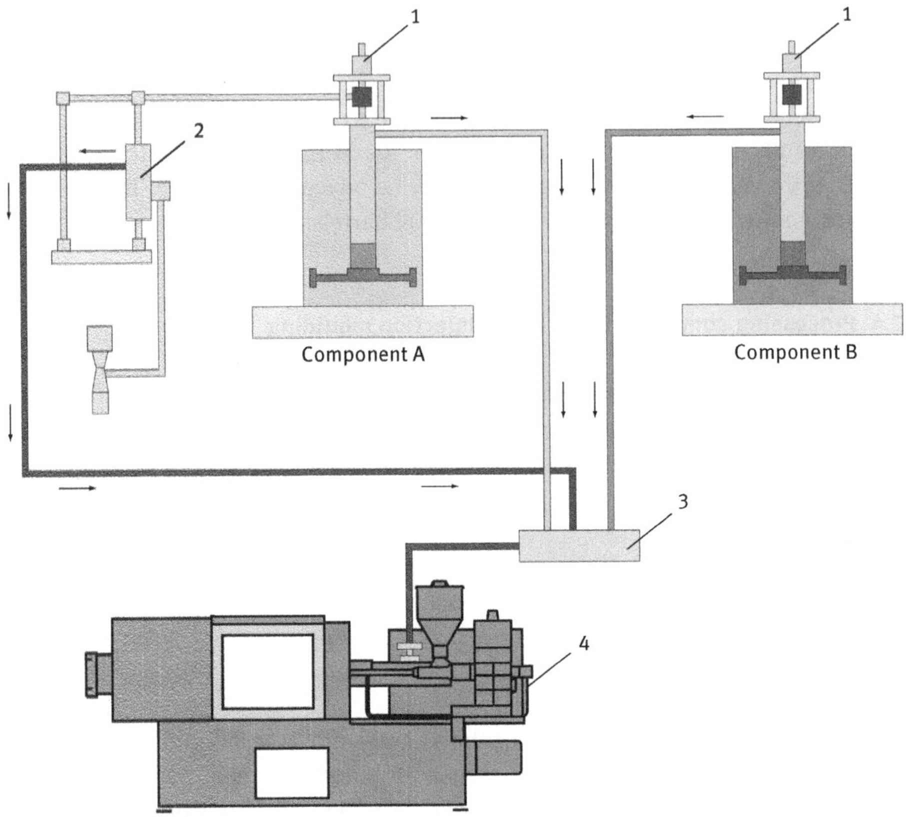

Figure 7.7: Mixing and dosing equipment for LSR. 1) Dosing apparatus for A and B components of the LSR, 2) dosing apparatus for colour dosing, 3) mixing station and 4) IJM with LSR cylinder module.

controlled needle dosing valves with electronic volume gauges, which ensure the exact mixture ratio during dosing. By means of additional electrically controlled dosing units, colours and additives may be freely dosed. Mixers and the inlet to the machine are downstream from the dosing system.

Since every IJM consists of autonomous volume dosing with controls, there is no limitation in regard to the number of machines that can be supplied.

A method for exchanging barrels without interruption may be integrated into the system as an added option in order to avoid unnecessary downtime for the machines.

7.2.4.2 Machine technology

IJM for the processing of LSR are differentiated by a few process-relevant peculiarities from machines that process other materials, and in particular, LSR must pass

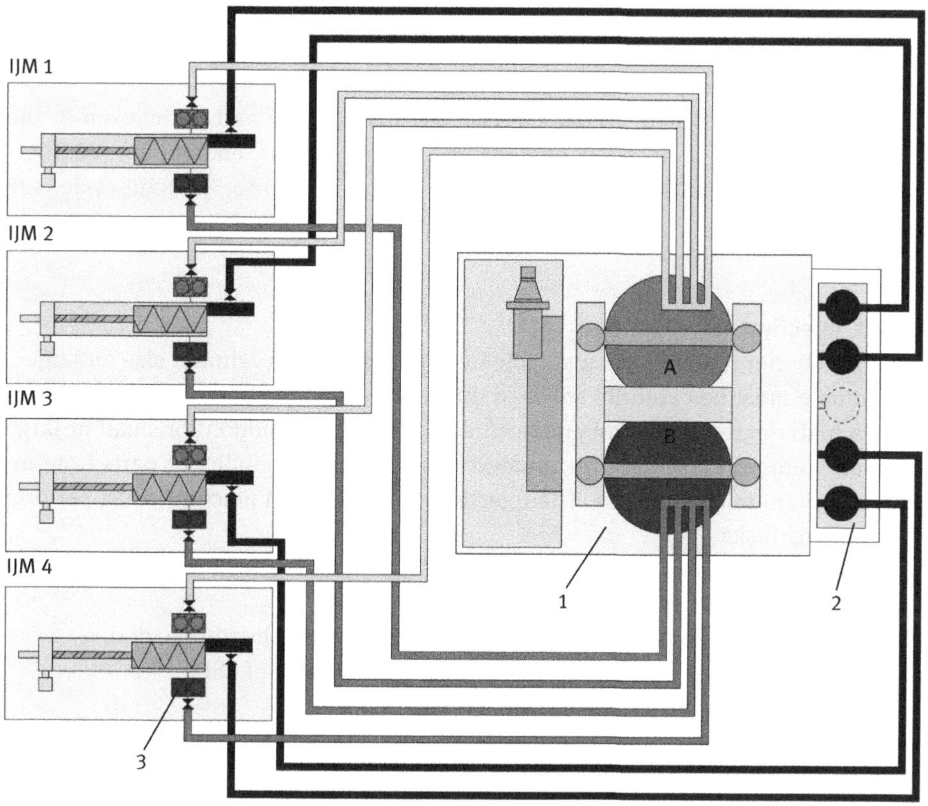

Figure 7.8: Central material delivery with decentralised volume dosing. 1) Central material supply with components A and B, 2) supplemental frame with up to five colour or additive delivery pumps and 3) volume gauge at every IJM.

through a special injection unit. A standard injection unit can be converted relatively quickly and easily for LSR processing by the installation of an LSR cylinder module. Since LSR technology places great demands on the machine in order to ensure problem-free operation, the following points should be kept in mind during the selection of a suitable IJM:

- The machine control unit must be structured simply and flexibly in order to make the connection of different peripheral machines possible. In addition, secondary equipment and mould heating must be operated and monitored directly through the machine control unit.
- The most important motion process phases of the machine should be governed to ensure the highest possible reproducibility of parts.
- A screw with position regulation allows, among other things, the precise maintenance of the defined stroke and pressure profile at injection.

- Ramp-controlled mould sealing enables positioning on the clamping platen which is accurate to the millimetre, providing problem-free operation of the handling devices.
- Mould sealing, which is characterised by high rigidity and parallelism of the mould platen as well as fast opening and closing speeds, enables a rapid injection cycle. Due to the fast vulcanisation time of LSR, 4 to 5 second cycles are not rare.

7.2.4.3 Injection unit

Silicone cylinder modules are available in different sizes as cylinder size and injection volume must be carefully adjusted during processing. These cylinder modules allow a high degree of control and accuracy of the screw motion for small or large injection volumes and hence the greatest possible reproducibility of parts from injection cycle to injection cycle. The injection unit of an LIM machine must perform the following tasks:

- Mixing of components as well as additional dosages of colour pastes or additives.
- Precise volume dosage with sufficient speed within the heating period.
- Reproducible and precise injection of a defined volume of LSR.
- Pressure release from cold-runners with open nozzles.
- Avoidance of early crosslinking by tempering or cooling of the LSR in the cylinder and the machine nozzle.
- Thermal separation by retracting the cooled machine nozzle from the heated mould.

7.2.4.4 Cylinder module

Support elements and device drives in customary standard machine versions may be adapted for the installation of a silicone cylinder module. A drive for a hydraulic needle sealing nozzle must also be provided. The use of cold-runner nozzles with needle sealing or with a cold-runner head, means that the needle sealing nozzle on the machine can be omitted. Figure 7.9 depicts an LSR cylinder equipment kit.

The cylinder module is fitted with liquid thermal collars that are connected in series. Cooling is through a water battery or a governor unit. The temperature level of the cylinder should be maintained at a constant temperature of approximately 20 °C, since crosslinking reactions may otherwise occur. The material inlet for the cylinder is placed at the front near the nozzle, as a high L/D ratio is not necessary for dosing. Mixing and homogenising of the individual components has already taken place *via* the dosing system prior to entry into the cylinder. Only a secondary homogenisation of the components with any subsequent additives or colour pigments takes place in the screw.

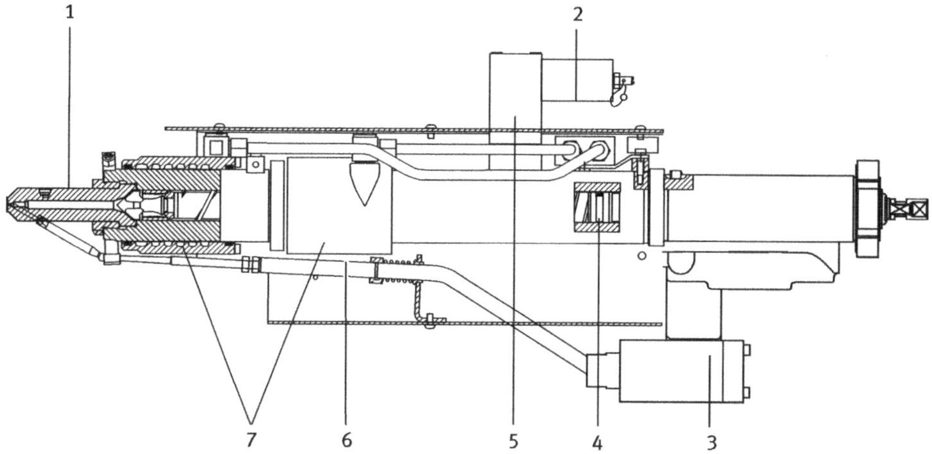

Figure 7.9: LSR cylinder equipment kit. 1) Hydraulic needle sealing nozzle, 2) connection with safety valve, 3) hydraulic cylinder for sealing nozzle, 4) seal from the drive side, 5) insert, 6) rod and 7) heating and cooling mantle.

The screw is designed purely as a mixing and conveyor screw without compression. Due to the low viscosity of silicone, the shaft of the screw is ground cylindrically at the connection of the inlet area and sealed with a radial seal from the drive unit. The coupling of the screw to the drive is *via* an appropriate coupling unit. An important element of this cylinder equipment kit is a high pressure valve at the inlet to the cylinder. This valve serves, on the one hand, to control the material from the mixer of the dosing system and, on the other, to fulfil its function as a high pressure valve.

7.2.4.5 Non-return valve

Because of the low viscosity of LSR, the non-return valve at the point of the screw has a special significance. At start-up and at the beginning of the injection process stage, this valve must close cleanly and without delay in order to maintain good reproducibility of the parts. Typical non-return valves for thermoplastic processing cannot be used since the typical seal rings cannot be positively moved due to the very low resistance of the silicone. The disc non-return valve for 2-K silicone has proven to be beneficial and is shown in Figure 7.10. It has the following features:

- Precise dosage characteristics.
- Positive closing.
- Low backflow.
- Precise injection volumes.
- Is standard equipment for 2K-LSR machines.

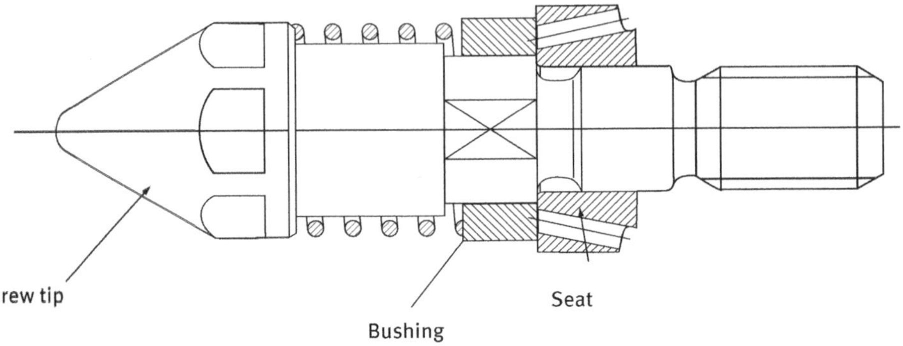

Figure 7.10: Non-return valve for 2K silicone. The bushing slides forward during screw rotation and plastic is pumped in front of the screw tip. During injection the bushing slides back to make a seal on the seat.

With the disc non-return valve, a seal ring is pressed onto the rotating spring in the closed position. During the plasticising process phase, the seal ring opens because of the pressure of the melt from the screw chambers and allows material into the screw prechamber. At the end of the plasticising phase, the spring forces the seal ring back into position. At the start-up of the screw, the non-return valve is thus already locked, and any leakage flow in the screw chambers is very low as a result.

7.2.4.6 Auxiliary heating control circuits

Since LSR moulds are frequently heated electrically, internal heating circuits in the machine for the control and supply of mould heating are beneficial. The essential heating parameters may thus be managed directly with the other machine adjustment data and monitored by the control unit, thereby improving production accuracy. For example, ARBURG machines offer up to 16 such auxiliary heat control circuits.

7.2.4.7 Interface for brush and cleaner units

With LSR moulds, it is common practice to clean the form after every injection cycle or, to facilitate the removal process from the form, to brush the parts from the form. Interfaces for brush and cleaning units are customary for this.

7.2.4.8 Handling interface

Due to the increased automation in injection moulding manufacturing, more and more evacuation devices are being used in production. In order to ensure that these handling devices work reliably, they must be coordinated by the machine controls and therefore, handling interfaces are required.

7.2.4.9 User-programmable inputs/outputs

As moulded parts become more and more complex with many undercuts, the moulds themselves become more complex. As a result, hydraulic or pneumatic core-pullers must frequently be employed for removal of the part from the mould. In order to control them, user-programmable inputs/outputs for the control unit are necessary. The core-pullers in the mould are thus controlled and monitored by the machine.

7.2.4.10 Air blast equipment with pressure reducer

The removal of LSR parts is frequently performed with pneumatic support. For this, air blast equipment is required, including a pressure reducer, which allows the compressed air to be introduced only at the specified pressure.

7.2.4.11 Control unit for vacuum pump

As a result of the precise functioning of LSR moulds, there is very poor airflow. For extremely high injection speeds, and the extremely fast mould filling which is associated with them, or due to extremely tight tolerances, the air does not have adequate time to escape from the cavity. The result is either scorching or moulded parts, which are not filled completely. In order to avoid this, the cavities are evacuated before the injection step. This signal for control of the vacuum pump must come from the machine control unit.

7.2.4.12 Hydraulic/pneumatic cold-runner control

Ready-to-use cold-runner systems are offered as standards in the marketplace. These function especially well with needle sealing nozzles that are controlled hydraulically or pneumatically. In order to allow integration with the process sequence of the IJM, interfaces were developed for control of this type of cold-runner. The machine control unit also monitors the position of the needles.

7.2.4.13 Protective screens made of polycarbonate

In most instances, LSR moulds are heated electrically to temperatures of 170 to 210 °C. Due to the level of heat that is radiated from the mould, heat-resistant material must be installed for the protective screens of the machine (polycarbonate meets this requirement).

7.2.4.14 Simple cold-runner nozzle

The simple cold-runner nozzle with hydraulic needle sealing system, as shown in Figure 7.11, enables either direct or indirect injection of silicone parts. The advantages of using this system are good thermal separation between the cold nozzle and the heated mould, and the flat gate location on the moulded part.

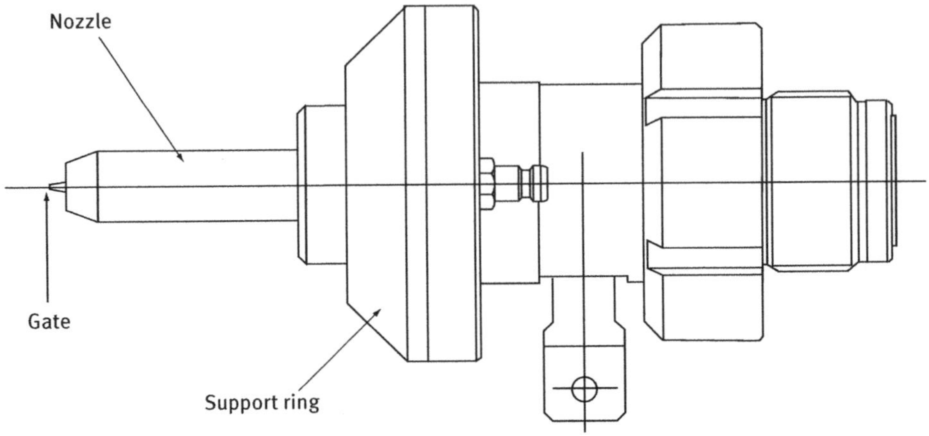

Figure 7.11: Single cold-runner nozzle.

7.2.4.15 Cold-runner head

The cold-runner head shown in Figure 7.12 is a machine-integrated system and hence applicable for different moulds. It is attached to the cylinder with a split coupling and enters into the mould through the fixed platen. The individual nozzles are fitted with a hydraulic needle sealing system, thus ensuring precise sealing of the sprue. As a result, the injection locations are absolutely clean and the surface quality is flawless. With the open design concept and modular construction, manifold systems with up to six sprues may be easily realised.

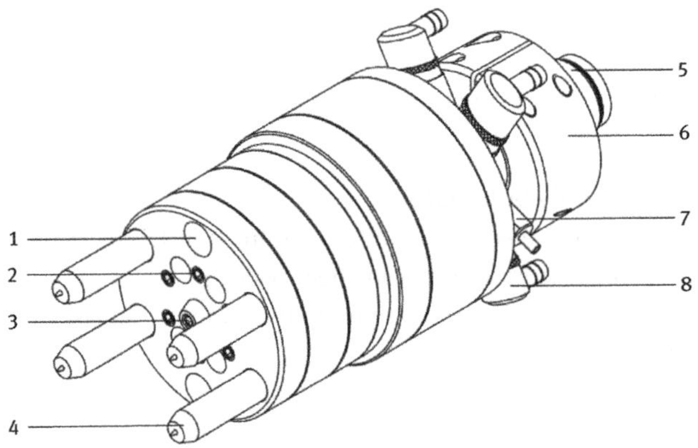

1. Insert cartridge
2. Volume flow limiter
3. Cone centring
4. Nozzle with shut-off needle
5. Cylinder connector
6. Split coupler
7. Hydraulic connection
8. Temperature control connection

Figure 7.12: Cold-runner head.

The mould can be completely separated from the cold-runner head due to a quick-coupling system, thus ensuring an optimal separation between hot and cold

modules during stationary periods. Along with the technical benefits of this system, its application has a cost-reducing effect in the construction of moulds, because different moulds can be matched to the same cold-runner head. The following significant features characterise the cold-runner head:
- A component of the machine and not of the mould.
- Set-up and assembly as with a special nozzle.
- Retraction from the mould possible at any time.
- Opening and closing of the needle sealing nozzles through the machine controls.
- Arrangements with up to six nozzles.
- Each individual nozzle with a needle sealing system.
- Volume flow may be regulated.
- Good cleaning and maintenance features.
- Simple and sturdy construction of the exchangeable needles and nozzles.
- Optimal thermal separation between hot and cold modules.
- Uniform force application to all nozzles through the central hydraulic cylinder.
- Reduced mould costs, since installation with different moulds is possible.
- Easy matching of new moulds to the same cold-runner head.
- Cycle-time savings.
- Cost reductions through material savings.
- Absolutely clean injection locations.

7.2.4.16 Mould technology

The mould is heated electrically and it is recommended to control both mould halves separately. The heating power for small moulds should be between 1,600 and 2,200 Watts per mould half, which varies according to the mould size (e.g., 215 × 300 mm). The mould temperatures vary according to the LSR material and standard values range from 180 to 250 °C. A general guide of 50 Watts per kg of metal can be used.

Making LSR moulds requires expert know-how and it is highly recommended that only mould designers and makers who already have experience in this technique are used. In view of the high mould temperatures, insulating platens with a thickness of 10 mm for the mould are required.

Due to the described material-specific characteristics of LSR, correspondingly high demands are placed on the moulds with regard to rigidity, processing precision and temperature control. Even flaws of less than $1/100^{th}$ of a millimetre can form ridges and thus make cost-intensive reworking procedures necessary for the moulded parts. The primary distinguishing feature of the different conceivable mould technologies is the manner of injection. Figure 7.13 provides an overview of the injection systems in use today.

As is the case with the use of other materials, the injection in LSR moulds requires guiding the material flowing from the machine nozzle into the cavity with as little pressure loss as possible and without interference.

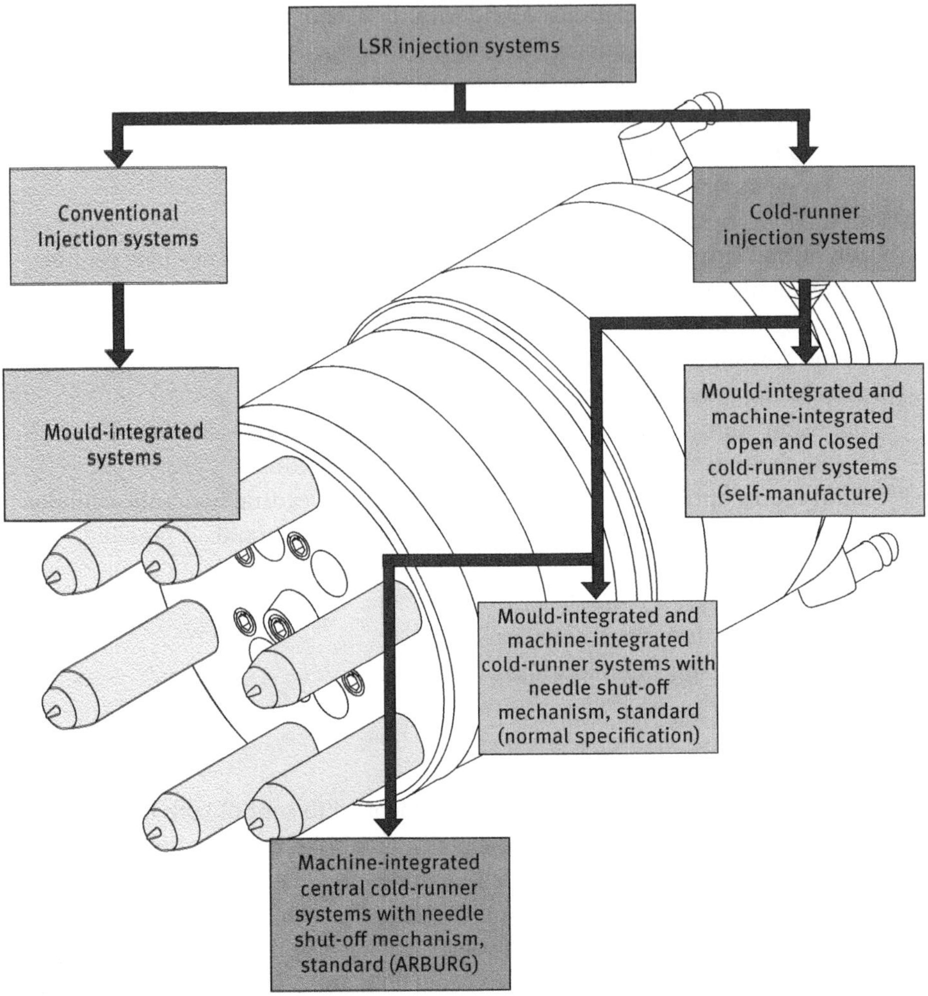

Figure 7.13: LSR mould technologies.

- **Conventional Injection System**

The conventional injection system, that is, the non-temperature controlled, mould-integrated sprue runner, provides the simplest way of achieving the injection phase and can be used for thermoplastic processing.

The cone or pin-gated sprue, whether with or without a submanifold, offers the least resistance to the inflowing material. It is used primarily in simple moulds for thick-walled parts with high-quality requirements. The significant disadvantage of this injection method is the high expense for isolating, separating and recycling the very loosely dimensioned sprue cone and the submanifold.

Point gates for simple or multiple sprues are frequently found in the mould parting line. The runners are guided up to just before the mould cavities with a somewhat larger cross-section, in order to flow into the narrow, short gate runner. The sprue can thus be separated easily from the formed part, although not automatically.

Tape or film gates bind flat parts from one side, which allows material flow that promotes uniform orientation in the part, and thus fewer tendencies for distortion. In order to set this sprue optimally, the part must be injected asymmetrically. Using the injection unit positioned horizontally, where it may be displaced parallel to the fixed platen of the sealing unit, is especially suited for this application. An additional option for injecting this type of part can be realised by locating the injection unit at the parting line.

Tube-shaped parts that must have a high degree of cylindrical accuracy and be free of welding seams are joined with tunnel sprues, which provide the significant advantage of automatic sprue separation from the moulded part when the part is ejected. The sprue runners end shortly before the mould cavities and open into bores which form tunnel shapes leading toward the lateral wall surface of the mould cavity. Figure 7.14 depicts a mould construction with a conventional, mould-integrated runner system (pin-gated sprue with submanifold).
The significant disadvantage of using a conventional system is the fact that material that has been previously vulcanised cannot be returned into the system for reprocessing. Based on the high price of the raw materials for liquid silicone, this is a very considerable economic factor.

As an alternative to the conventional injection system, cold-runner systems are increasingly used for the processing of liquid silicone. These systems may be compared with the hot-runner technology found in thermoplastic processing.

– **Cold-Runner Sprue System**

The cold-runner sprue system may be divided into categories. First of all, there are mould-integrated and machine-integrated, open and closed cold-runner systems, which are primarily self-manufactured. These systems are not generally available as they are utilised for special applications by individual process technicians.

The decisive problem with the application of an open cold-runner system is thermal separation around the gate zone. It has proven to be very difficult to inject a moulded part directly with a cold-runner nozzle. The vulcanisation process is not limited to just the part, but also continues in the gate zone of the cold-runner nozzle, and this vulcanised plug can become fixed in the part at the next injection as an unsightly mark. As an alternative method for this cold-runner system, processing may also be handled with a nozzle sealing system.

– **Mould-Integrated Cold-Runner System**

Manufacturers now offer standardised mould-integrated and machine-integrated cold-runner systems with pneumatic needle sealing nozzles. The advantage of this

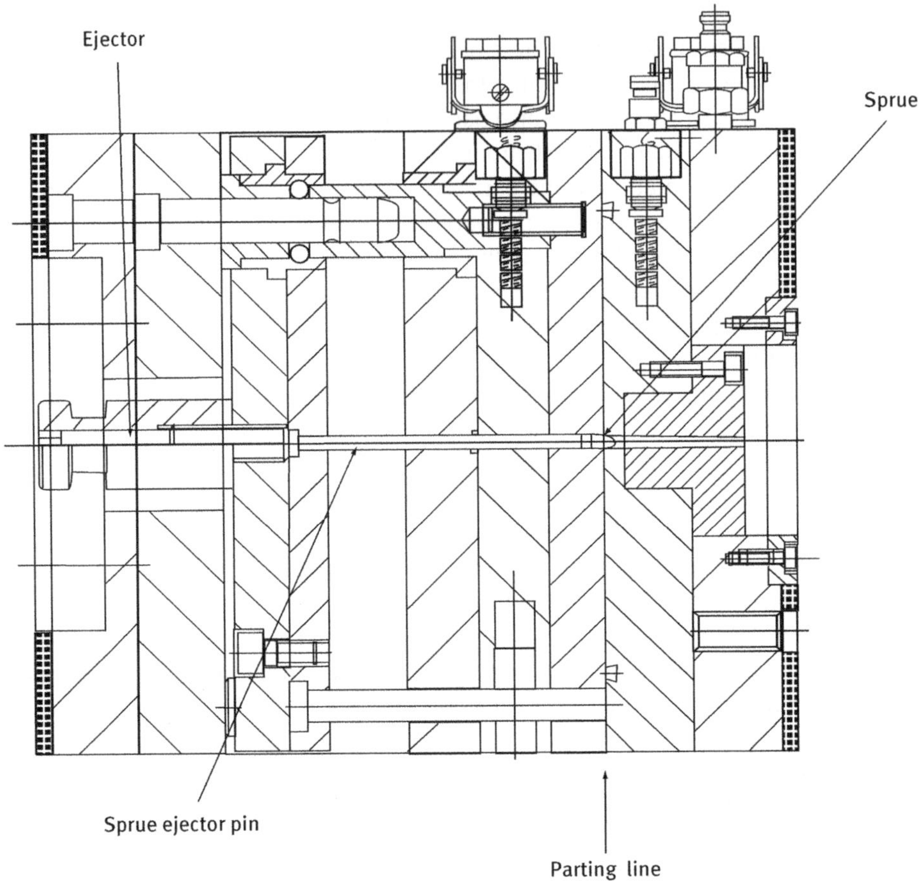

Figure 7.14: Conventional sprue system, mould integrated.

type of standardised cold-runner system is its simple installation into the injection moulds. The cold-runner system may be considered as a standard mould design. It is equally possible to use a cold-runner system for multiple moulds. Figure 7.15 depicts a mould construction with the application of a mould-integrated cold-runner system as a standard specification.

The application of this type of standard specification enables manufacturing moulds with a minimum injection volume by the precise separation of the cold sprue from the hot cavity. The sprue markings are minimal thanks to the pneumatic needle sealing system. Cycle times can be reduced because of the high temperatures of the cavity. As an ARBURG standard, for example, cold-runner systems are offered with up to four gate points.

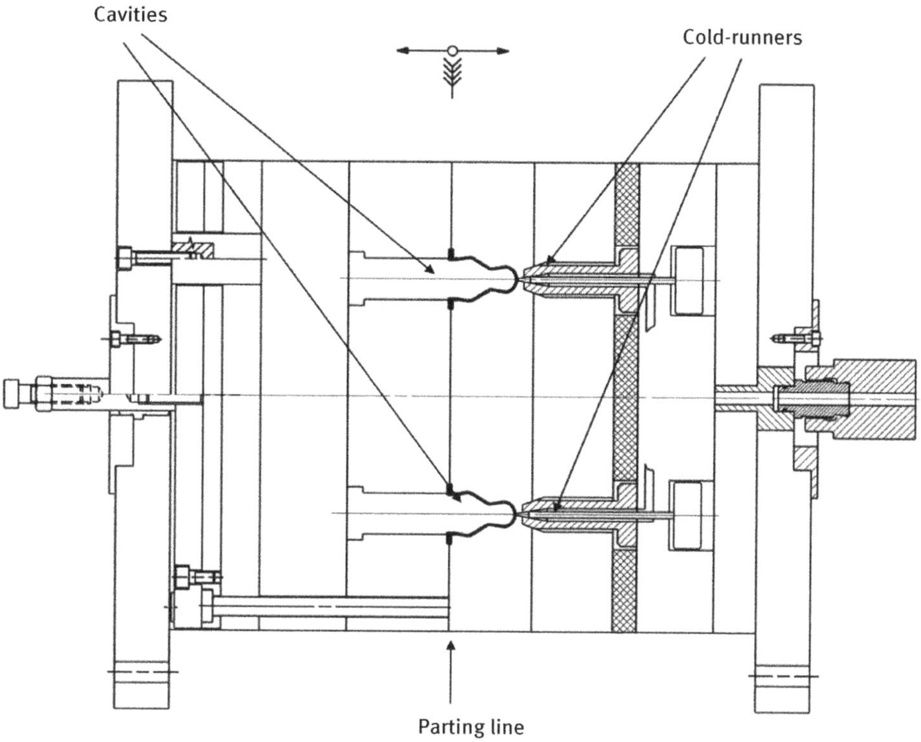

Figure 7.15: Cold-runner system as part of the mould standard specification.

– **Machine-Integrated Cold-Runner System**

The third category of cold-runner system is represented by a machine-integrated system, which can utilise a simple cold-runner nozzle as well as the cold-runner head with up to six gate points. The major advantage of this system is that they are a fixed component of the machine. The needle sealing nozzles may thus be controlled from the machine control system. A machine-integrated cold-runner system may therefore be installed independently of the mould. LSR moulds can hence be produced much more cost-effectively, since it is only necessary to procure the cold-runner once per machine. The simple application of a cold-runner nozzle in an existing design is depicted in Figure 7.15.

The advantages of using this type of cold-runner nozzle in a conventional injection system are clear and include:

- Material savings *via* direct injection into the cavity or *via* injection into a sprue manifold.
- Optimal thermal separation between the cold sprue and hot cavity.
- Absolutely clean injection areas.
- Retraction from the mould is possible at all times.
- Control of the needle sealing nozzle *via* the machine controls.

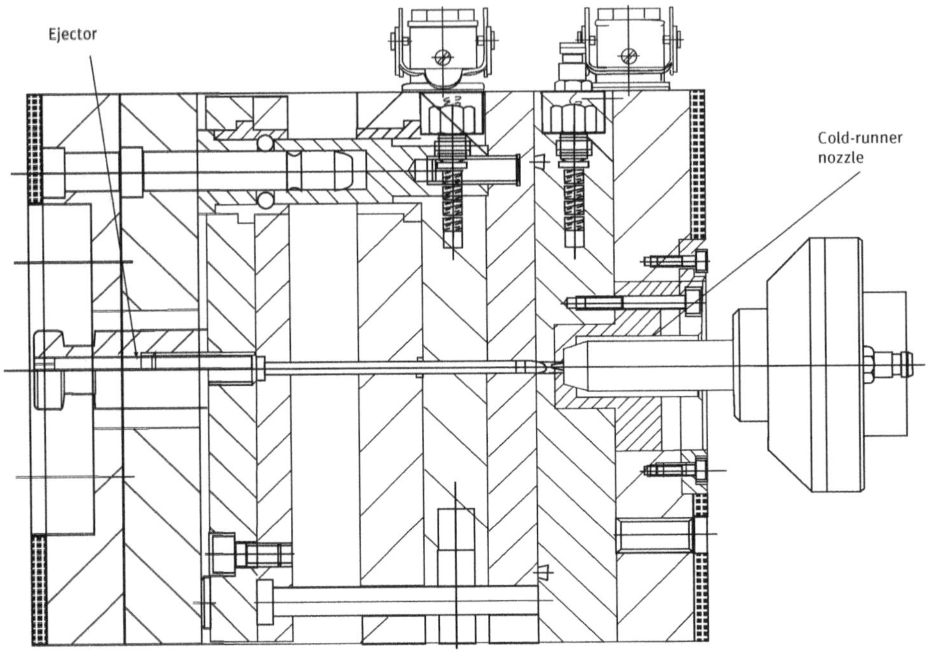

Figure 7.16: Mould design with a simple cold-runner nozzle.

The cold-runner head is installed with LSR moulds with several cavities and is mounted as a forward unit on the cylinder. Through machine contact, the head enters the mould *via* the fixed mould platen so that the nozzle tips reach the cavities. Due to the open concept and modular construction design, application of the head is flexible and it may be fitted with up to six nozzles. Reducing or adding to the number of nozzles is simple and easy to perform with the use of blank adapters. Set-up and assembly on the machine are similar to a special nozzle so that installation work on the mould is eliminated. Figure 7.17 illustrates an installed cold-runner head.

7.2.5 Application of liquid silicone rubber parts in food production

LSR parts can be applied in a great variety of fields; however, for use in food production, for example, the BgVV has only allowed the application of specific LSR materials. The particular raw material supplier should be aware of the suitability of their materials for specific applications if there are concerns.

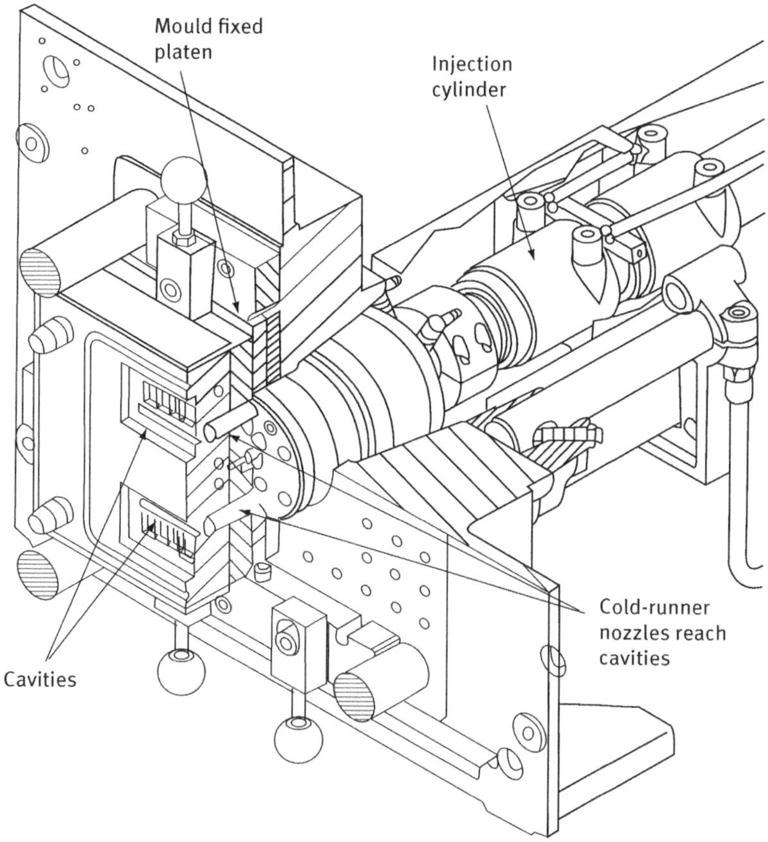

Figure 7.17: Installed cold-runner head.

7.2.6 Summary

Outstanding material properties and good processing factors increasingly make LSR a suitable material for current applications which have high specifications. Among the varied benefits are good electrical insulating properties, excellent heat, cold and light resistance, as well as the harmless physiological properties of many types.

The elimination of preparation techniques is certainly another reason for the rapid acceptance and expanded use of LSR; in particular, for purely thermoplastic process operations, LSR provides excellent potential for elastomer manufacturing, without the necessity of making costly investments in facilities and expertise for material preparation.

For LIM processing, only an appropriately modified IJM and a dosing machine for conveying and mixing in a 1:1 ratio are necessary. In addition, heated moulds, which are especially constructed for the processing of LSR, are used.

Cold-runner systems are the state-of-the-art mould technology; here, mould-integrated or machine-integrated cold-runner systems with needle sealing nozzles are generally used.

Nonetheless, LSR is not a material which can be processed entirely without problems. LSR is a modern high-performance material that must be thoroughly managed during processing. An entire series of process factors must be observed in order to produce LSR moulded parts with quality and high value. Each process factor in the cyclical flow shown in Figure 7.18 must be mastered precisely in order to work successfully with LSR.

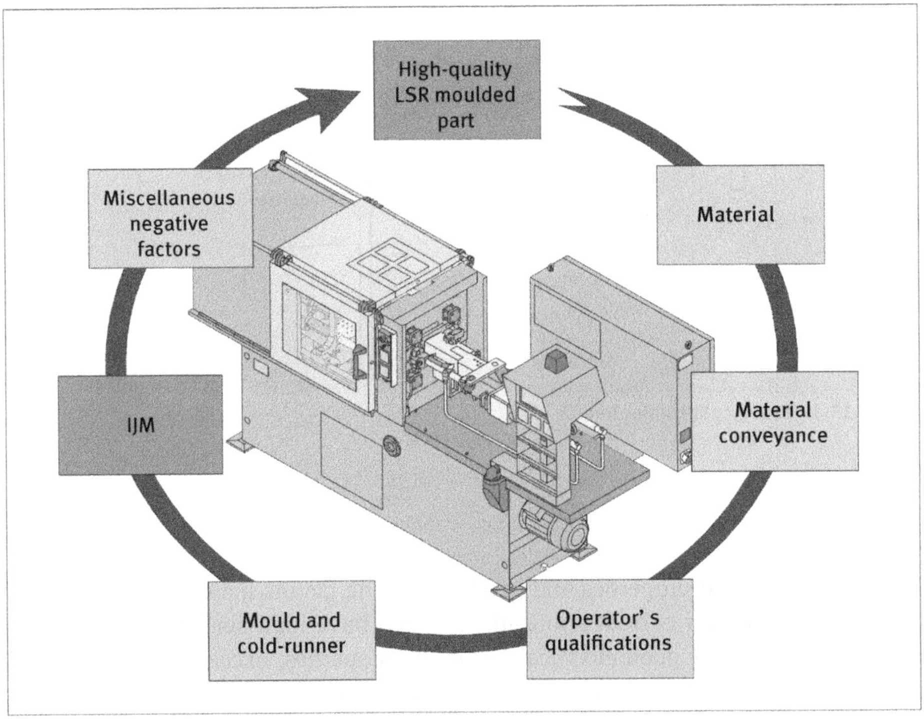

Figure 7.18: Process cycle for a successful LSR part.

Machine technology, mould construction and process management incorporate many specific features which must be clarified in detail and, if required, must be ascertained, before a successful result is achieved.

7.3 Processing of ceramic and metal powder materials

7.3.1 Introduction

In recent years industry users have raised their standards considerably, asking for higher machine output longer production periods to increase operational efficiency and lower servicing costs. The possibilities of conventional materials for machine construction have thus been exceeded.

In this situation, ceramic and metal powder materials have opened new perspectives to solving engineering problems. In the 1960s, industrial products were first made of aluminium oxide and, due to its excellent characteristics, this material has won more and more ground in the technical, chemistry and electronics industries.

Some of the outstanding characteristics of ceramic materials are:
- High mechanical stability, even at high temperatures.
- Good resistance to changes in temperatures (but avoid thermal shocks >100 °C, these will cause cracking).
- Good chemical resistance.
- Very hard and wear-resistant.
- Low specific gravity.

In principle, all materials available in a sinterable powder can be mixed with an appropriate binder and processed on IJM. Therefore, in addition to the traditional oxide ceramics it is also possible, for example, to use metals, carbides and nitrides. Some typical materials are shown in Table 7.11.

Table 7.11: Some colours of ceramic and metal materials.

Material	Colour
Aluminium oxide	White
Zircon oxide	Yellowish
Silicon carbide	Dark grey
Silicon nitride	Dark grey
Magnesium oxide	Yellow–white
Stainless steel	Grey
Copper	Red brown
Hard metal (WC-Co/Ni)	Dark grey

Since mixing and injection units can be subjected to increased wear when using powder materials, it is recommended to keep the grain size of the powder as small as possible. However, the range in which the grain size of the material guarantees optimum production, and therefore the desired properties of specified parts, is relatively narrow.

Table 7.12: Powder properties and performance.

Properties	Coarse powder	Fine powder
Surface of the moulded part	Rough	Smooth
Green strength (strength after binding but before sintering)	Low	High
Injection moulding performance	Poor	Good
Wear on machine and mould	High	Low

Fine powders produce less surface roughness, can be processed with little wear and result in higher green strength. The characteristic spectrum of various powder materials can be seen in Tables 7.13–7.14.

Table 7.13: Material grain sizes.

Material	Grain size
Metals (atomised)	<30 µm, X50~6-7 µm
Carbides	<1.5 µm, X50~0.7 µm
Silicates (e.g., porcelain)	<45 µm, X50~5 µm
Oxides (e.g., aluminium oxide)	<15 µm, X50~0.75 µm
X50: Measure of cumulative particle size	

Table 7.14: Shrinkage.

Grain shape	Shrinkage
Round	Slight, isotropic
Irregular	High, isotopic
Flat	Anisotropic

7.3.2 Binder

The binder must have the following properties:
- Be dimensionally stable during debinding.
- Have a good storage capacity.
- Does not react with the powder materials.
- Has a high green strength (strength after binding).
- Good mould removal properties.
- Thermal stability.
- Decomposes easily and completely upon debinding.

The bonding between the powder and the powder particles should be as great as possible, so that the centrifugal forces arising during the injection process do not give rise to any separation of the two components, thereby resulting in non-homogeneously filled parts.

In order to achieve good injection moulding characteristics and isotopic sintering with a low rate of shrinkage, the use of spherical-shaped particles should be favoured.

During mixing, the binder and powder are combined to form the most homogeneous compound feedstock possible. In order to guarantee that the powder particles are completely surrounded by the binder components, all powder agglomerations must be broken up by shear forces.

The binder component should be kept to as low a concentration as possible, as this reduces shrinkage during sintering, thereby allowing closer tolerances on the finished part to be maintained. The binder component of injection moulding masses is between 35–55% by volume. If the binder component concentration is too low, the quality of the injection moulded part is reduced and the wear on the machine and injection equipment increases greatly.

Various possibilities are available for homogenising the powder and binder, for example, the material can be mixed with a twin screw extruder, a shear roller extruder or a rapid mixer.

The use of powder injection allows the economical manufacture of components that could not be made by employing metal cutting or pressing methods. Injection moulding allows almost limitless design freedom for shaped parts and the production process includes moulding, debinding and sintering of the shaped parts. This means that the final tolerances of the components are determined by the following important factors:
- Binder content.
- Powder characteristics.
- Mixing process.
- Injection parameters.
- Distortion due to gravity.
- Sliding properties of the sintering surface.

7.3.3 Processing

To determine the most economical process one needs to consider the size of the part, its geometry and the quantity to be manufactured.

7.3.3.1 Compression moulding

The ceramic powder is compacted inside a mould under high pressure. Bores, steps, bevels and threads can be pressed into the part in the pressing direction only, since the part has to be ejected after moulding. The compacted parts can be further machined to have additional bores, undercuts and so on (high-tech, high-cost machining is necessary).

7.3.3.2 Injection moulding

Injection moulding is an economical method of processing ceramics. The material is plasticised by a screw. Dosage and plasticising take place simultaneously, while the previously moulded part still cools inside the mould. With automatic removal devices the entire process can run automatically.

Advantages of processing ceramics on IJM are:
- Wall thickness of part can be changed.
- Holes and undercuts possible.
- A higher density is possible.
- Better surface and end quality.
- Shot weights of less than 1 g are possible.

7.3.4 Configuration of injection moulding machines

IJM can be used to process ceramics, providing the following equipment is available:
- Plasticising cylinder, Arbid version with electrical heating.
- Dosage delay unit.
- Thermostatic water valve (feeding yoke).
- Nozzle tip of 4 mm bore.
- 'U'-version of clamping unit (which can move horizontally or vertically).
- Injection unit with adaptor for parting line injection.

N.B.: The 'U'-version of the clamping unit and the adaptor for parting line injection facilitate manual or automatic part removal.

7.3.4.1 Peripheral devices

A cooling device for the mould is necessary. In most cases part removal is (still) achieved manually, however, automatic removal devices are increasingly used.

7.3.4.2 Mould

Moulds for powder injection can have features normally used in manufacturing plastics, such as sliding bars, pull cores and cavity pressure transducers. The major difference is the wear characteristics, powders are far more abrasive and moulds may require hardening or the use of alloy materials. The mould needs sufficient cooling and will be connected to a cooling device; it is recommended that both halves are controlled separately. The temperatures vary according to the ceramic material and standard values range from 6 to 10 °C. The mould should have exchangeable inserts and mould cavities must be high gloss polished. Mould designs should incorporate large ejector pins and sufficient bevelling to facilitate easy part removal. Apply insulating platens to both mould halves to prevent condensation.

7.3.4.3 Wear

There will be wear on the plasticising cylinder and the mould, especially when you are processing inorganic materials.

7.3.5 Process requirements

To facilitate processing on IJM the material needs to be free-flowing and must allow good plasticising; once moulded, the material requires debinding. The process sequence is shown in Figure 7.19.

7.3.5.1 Debinding and sintering

Debinding is carried out in furnaces suitable for the binder system. The binder may be removed by way of catalysis, dissolving in a solution or thermal decomposition. Debinding can be effectively achieved using a suitable furnace temperature and atmosphere. Removing the binder transforms the moulded part into a porous and sensitive moulded body, referred to as a ‘brown compact’. In this condition, the part is only kept stable by minimum binder residues and weak van der Waals forces.

In order to firmly bind the particles of the brown compact to each other, the part is sintered by additional heat of up to 2,000 °C; this process is the same for the pressed sintered parts. The final moulded part is formed by way of diffusion and/or the forming of liquid phases and grain growth.

Injection moulded sintered parts can approach 99.9% of the theoretical material density. With suitable powder they will shrink isotropically during sintering and exhibit homogeneous characteristics. Any injection moulding flaws that arise cannot be remedied by sintering; however, in most cases it may only be necessary to refinish fits or regrind cutting edges. Fine powder and polished surfaces guarantee excellent surface qualities as a result ofthe injection moulding of metal or ceramics.

7.4 A growing market for moulders

Injection moulding technology can be successfully employed to process a wide variety of materials. In all cases, the basic requirements are adequate equipment for the

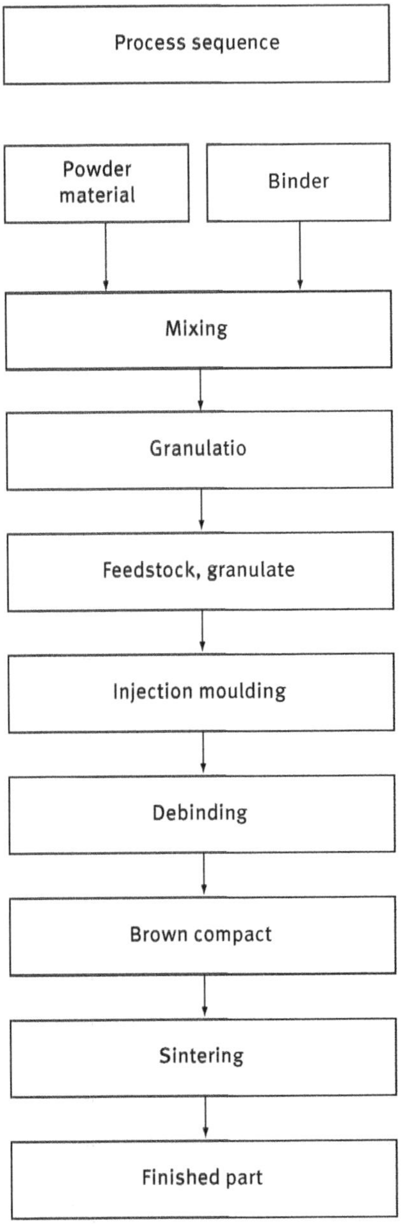

Figure 7.19: Process sequence for injection moulding of powder materials.

IJM and a suitable material. When these requirements are met, injection moulders have a good chance of gaining larger shares of the market.

Our knowledge of the specific properties of materials and adequate processing methods has greatly increased, but almost daily, development seeks and finds new applications.

Further reading

1. *Design and Manufacture of Plastic Components for Multifunctionality*, Eds., V. Goodship,B. Middleton and R. Cherrington, Amsterdam, Elsevier, The Netherlands, 2016.
2. E.J. Barbero in *Introduction to Composite Materials Design*, 2nd Edition, CRC Press, Baco Raton,FL, USA, 2010.
3. D. Gay in *Composite Materials: Design and Applications*, 3rd Edition, CRC Press, Baco Raton, FL, USA, 2014.
4. B. Rodgersin *Rubber Compounding: Chemistry and Applications*, 2nd Edition, CRC Press,Baco Raton, FL, USA, 2015.

Reference

1. *Design and Manufacture of Plastic Components for Multifunctionality*, Eds., V. Goodship, B. Middleton and R. Cherrington, Amsterdam, Elsevier, The Netherlands, 2016.

8 Guide values and processing instructions for the most important thermoplastic injection moulding compounds

This chapter will specify recommended drying and processing settings for the most common plastics, which will include recommended settings for temperature profiles, injection pressure and speeds, holding pressures, back pressures and cooling times. It will also detail the key parameters of the injection moulding process and their effect on the process; however, before some materials can be moulded, they must first be dried. This chapter therefore begins with a guide to drying materials prior to moulding (Table 8.1) as incorrect drying can lead to moulding faults, which will be further discussed in Chapter 9.

8.1 Predrying material

Many plastics are hygroscopic (absorb water) and must be dried before moulding to remove this residual moisture. There are a number of different types of material dryer available on the market, which often come equipped with the ability to deliver the material straight to the hopper to prevent further moisture build-up. Attention should be paid to the maximum permissible water content, as failure to remove moisture to suitable levels can cause moulding problems. It should be noted that Table 8.1 serves as a rough guide only in terms of drying temperature and duration as they can vary depending on the type of dryer used. Attention must also be paid to hopper size, as material left exposed to the atmosphere for extended periods begins to reabsorb water, which may occur even when left for just an hour, dependent on atmospheric conditions and material type. Hoppers should not therefore be overloaded with hygroscopic materials. Once the material has been correctly dried, attention can be turned to plasticising.

8.2 Examples of the moulding parameters of selected materials

A guide to the basic parameters of nozzle temperature, mould temperature, injection pressure, holding pressure and back pressure for selected materials is shown in Table 8.2. It can be seen that injection moulding parameters vary widely from material to material. With this in mind, it is important to understand what these injection parameters are and how they affect the injection moulding cycle. The next section will therefore discuss the injection moulding process.

https://doi.org/10.1515/9783110654813-008

Table 8.1: Drying guide for thermoplastics. Maximum drying temperature and maximum allowable moisture content.

Product	Maximum drying temperature (°C)	Maximum allowable moisture content (granulate) %	Drying time (h)
PA	100	<0.15	4
PBT	120	<0.02	4
PBT + PC/ASA	120	<0.02	4
POM	120	<0.15	4
PES	200	<0.05	4
PSU	160	<0.05	4
PAEK	200	<0.05	4
LCP	80	<0.02	3
SB	80	<0.2	3
ABS	90	<0.2	3
PP + talc	120	<0.03	3
ASA/PC	110	<0.1	3

ABS: Acrylonitrile-butadiene-styrene
ASA: Acrylonitrile-styrene-acrylate
LCP: Liquid crystal polymer
PA: Polyamide
PAEK: Polyaryl ether ketone
PBT: Polybutylene terephthalate
PC: Polycarbonate
PES: Polyether sulfone
POM: Polyoxymethylene PP: Polypropylene
PSU: Polysulfone
SB: Styrene-butadiene

8.3 Parameters of the injection moulding process and the quality of moulded parts

The properties of an injection moulded part depend upon the working material and processing conditions. During the production of a series of parts, a certain deviation in quality features such as weight, dimensional consistency and surface characteristics may occur. The size of this deviation will vary from machine to machine and from material to material. Furthermore, external influences or negative factors have an effect on the quality of an injection moulded part; examples of such negative factors may include changes in the viscosity of the melt, temperature changes in the mould, viscosity changes of the hydraulic fluid and changes in the characteristics of the plastic.

The causes through which these negative factors may arise are, for example, machine start-up after a long period of non-operation, changes in material properties when processing a new lot or a different colour, and environmental influences such as the ambient temperature at the time of processing.

The purpose of regulating the injection process is to eliminate these negative influences, and thus to attain an even higher reproducibility of the parts. The decisive factor for all quality features, which are concerned with dimension and weight, is the internal pressure of the mould. Constant maintenance of the internal pressure curve in every cycle guarantees uniformity of the quality of injection moulded parts, as all of the negative factors mentioned above are eliminated.

Without regulating the injection process during injection moulding, a specified pressure curve can be established for the injection and holding pressure, which can also be maintained with assurance using a suitable machine. However, the mould

Table 8.2: Control values for processing.

Material	Nozzle-side cylinder temperature[1,2] (°C)	Mould temperature (°C)	Injection pressure (bar)	Holding pressure (bar)	Back pressure (bar)	Remarks
PS	160–230	20–80	650–1,550	350–900	40–80	
SB	160–250	50–80	650–1,550	350–900	40–80	
SAN	200–260	40–80	650–1,550	350–900	40–80	
ABS	180–260	50–85	650–1,550	350–900	40–80	
Modified PPO	245–290	75–95	1,000–1,600	600–1,250	60–90	
PVC – hard	160–180	20–60	1,000–1,550	400–900	40–80	3, 5, 6
PVC – soft	150–170	20–60	400–1,550	300–600	40–80	3, 5, 6
CA	165–225	60–80	650–1,350	400–1,000	40–80	3, 4, 6
CAB	160–190	60–80	650–1,350	400–1,000	40–80	3, 4, 6
CP	160–190	60–80	650–1,350	400–1,000	40–80	3, 4, 6
PMMA	220–250	20–90	1,000–1,400	500–1,150	80–120	4
PC	290–320	85–120	1,000–1,600	600–1,300	80–120	4
PSU, PES	320–390	100–160	900–1,400	500–1,100	80–120	4
PE – soft	210–250	20–40	600–1,350	300–800	40–80	
PS– hard	250–300	20–60	600–1,350	300–800	60–90	
PP	220–290	20–60	800–1,400	500–1,000	60–90	
PA6,6	270–295[7]	20–120	450–1,550	350–1,050	40–80	4, 6
PA6	230–260[7]	40–120	450–1,550	350–1,050	40–80	4, 6

Table 8.2 (continued)

Material	Nozzle-side cylinder temperature[1,2] (°C)	Mould temperature (°C)	Injection pressure (bar)	Holding pressure (bar)	Back pressure (bar)	Remarks
PA6,10	220–230[7]	20–100	450–1,550	350–1,050	40–80	4, 6
PA11	200–250[7]	20–100	450–1,550	350–1,050	40–80	6
PA12	200–250[9]	20–100	450–1550	350–1,050	60–90	
PA amorphous	260–300	70–100	900–1,300	300–600	60–90	
POM	185–215	80–120	700–2,000	500–1,200	40–80	3, 6
PET	260–280	20–140	800–1,500	500–1,200	80–120	
PBT	230–270	20–60	800–1,500	500–1,200	80–120	
PPS	300–360	20–200	750–1,500	350–750	40–80	
FEP	340–370	150				5
ETFE	315–365	80–120				5

[1]If no other empirical values are available: nozzle temperature to the set nozzle-side cylinder temperature. Cylinder temperatures falling in the direction of material throat leading to a drop of 5–10 °C for each heating zone; maximum temperature difference between the nozzle-side and throat is 20 °C. For more than two heating zones, set the nozzle-side heating zone and the following to the same temperature.
[2]For heat-sensitive compounds, set higher temperatures only for high cycle times (shorter dwell time in the cylinder).
[3]Heat sensitive.
[4]Process only dry granules.
[5]Do not use shut-off nozzles, only open nozzles.
[6]Work only without non-return valve.
[7]To improve material feed behaviour: set temperature at the same level or slightly rising towards throat.
CA: Cellulose acetate
CAB: Cellulose acetate butyrate
CP: Cellulose propionate
ETFE: Ethylene-tetrafluoroethylene
FEP: Tetrafluoroethylene hexafluoropropylene
PE: Polyethylene
PET: Polyethylene terephthalate
PMMA: Polymethyl methacrylate
PPO: Polyphenylene oxide
PPS: Polyphenylene sulfide PS: Polystyrene
PVC: Polyvinyl chloride
SAN: Styrene-acrylonitrile

internal pressure curve that arises can only be assumed. Pressure losses through the runner manifold as well as the mould-specific filling behaviour cannot be identified.

When the injection process is regulated, the mould internal pressure is first measured and compared with a nominal value. If there is a deviation, a hydraulic valve or the electrical servo system (depending on the control method) that applies pressure to the injection cylinder is actuated. It is thus possible to follow the nominal value precisely and independently of negative factors. The switch-over from injection to holding pressure also occurs as a function of internal pressure; thus, no pressure spikes can occur since the switch-over takes place when a specified threshold value is reached.

The following benefits are achieved through regulation of the injection process:
- Significant reduction in start-up cycles: the required consistency in quality characteristics is achieved after just a few cycles.
- Better reproducibility of the parts: the deviation spread of the various dimensions lies significantly below that of a non-regulated machine.
- Cycle time reduction: the ability to visualise the internal pressure signal enables the sealing point to be determined much more easily and accurately.
- Restarts: if the same internal pressure curve is applied at a restart, the resulting parts are exactly alike.
- Improved quality of the parts through effective speed and pressure profiles: internal pressure profiles without spikes enable the production of parts with low residual stresses. Switch-over as a function of internal pressure prevents over-injection of the part, regardless of the selected dosage stroke.

The enormous significance of a mould internal pressure curve is characterised by the large number of parameters that can influence the appearance of the curve. Only the most important influencing factors are mentioned below:
- Injection phase: the injection speed, the flow resistance as a function of the type of plastic, the material temperature and the mould wall temperature.
- Pressure holding phase: the material temperature, the mould temperature, and the level and duration of the holding pressure.
- Maximum mould internal pressure: the injection speed, the material temperature, the switch-over point and the material flow.

In addition, the nature of the internal pressure curve contains important criteria that influence the following quality data:
- Injection phase: the appearance, the surface characteristics, the orientation and the degree of crystallinity of the moulded part.
- Pressure holding phase: the formation of ridges, the weight, dimensions, shrinkage, shrink holes, sink marks and the orientation are influenced.

The properties and quality of a component are predominantly determined by the moulding process in the mould. The dominating limiting factors here are the pressure and temperature cycles in the mould cavity. It would be ideal if pressure and temperature were uniform at any point in the cavity, and if the temporal pressure and temperature cycle also remained the same from batch to batch. Shrinkage would then be the same in all component batches, there would be no internal stresses and no tendency towards warping in the component, and each component would fall out in the same condition.

This ideal pressure and temperature distribution within the mould, i.e., as uniform as possible, is practically impossible to achieve with injection moulding, as a pressure drop is bound to occur due to the flow resistance, while the mould is being filled. Temperature differences will also arise because filling takes a finite time, even if this is usually very short. To get close to an ideal state, i.e., to aim for the most uniform possible filling process, the flow resistance during the filling of the mould plays a crucial role. The lower the flow resistance, the faster the mould is filled and the smaller the local pressure differences are in the mould.

These factors have corresponding consequences for the design of the component and mould, and the process parameters chosen.

With regards to the influence of mould geometry, the following is generally valid: the flow resistance should be kept as low as possible, e.g., by avoiding sharp edges in the component (pressure losses due to abrupt turning by the flowing compound).

For the process parameter settings, the following points can be generally applied:
- Screw injection speed should be as high as possible.
- Compound temperature should be as high as possible.

High temperatures result in low viscosity of the compound flowing in, low pressure losses, and thus low pressure differences and short filling times.

In practice, there are factors limiting machine settings, which are discussed in more detail when the individual process parameters are dealt with. Here are just a few examples:
- It will not always be possible to take the injection speed right up to the machine's performance capability limit. As the injection speed rises, the tendency of free jet formation and thus the occurrence of surface faults increases. If assistance cannot be provided here by suitable mould design, the machine must be operated at a low injection speed – it may be that two or three speed stages will be required for injection. The compound temperature must obviously not become so high that heat damage occurs. The more heat sensitive the compound, the more advantageous it is to select a larger safety margin from the upper temperature limits. However, it should also be pointed out that the injection speed setting should not fall into the other extreme.

- Too much caution can bring about the exact opposite of the desired effect: low temperatures increase the viscosity, and thus cause higher flow losses due to friction – which reheats the compound as it is injected into the mould. In this way, cylinder temperatures that are too low can actually lead to higher compound temperatures in the mould than in materials where the cylinder temperature was set higher.
- The higher the mould temperature is set at, the longer the cooling off lasts and the longer the cycle time is. Therefore, a temperature should be chosen which is only as high as the desired quality demands in order to be able to produce components as economically as possible. A summary of the most important factors in producing quality components is shown in Figure 8.1.

8.4 Injection and mould cavity pressure

It was mentioned in Section 8.3 that the pressure curve in the mould (as well as the compound temperature and the mould wall temperature) governs the quality of the component. Therefore, the connection between injection pressure and internal mould pressure will now be discussed, which allows the possibility of exerting better control on the pressure cycle in the mould during injection and thus on the quality of the components.

Drive pressure (e.g., in the hydraulic stroke or servo drive of the screw) and injection pressure (e.g., in the screw forward cavity) are practically proportional to one another (independent of the injection conditions) and display the same cycle (if we disregard the friction losses as the screw moves forward) (Figure 8.2). The internal mould pressure, on the other hand, is lower than the injection pressure, as a result of the flow losses during injection, which depend on the viscosity of the melt, the injection speed and the geometry of the flowpath.

Figures 8.3 and 8.4 show the injection pressure and internal cavity pressure curves. The internal cavity pressure can be measured by sensors within the mould, and can be indicated, or visually displayed, using an oscilloscope or pen recorder. The pressure cycle in the vicinity of the gate is the most critical factor here.

As can be seen from Figure 8.4, the internal mould pressure follows the injection pressure, with a time delay. At A, the cavity is volumetrically filled, and between A and B the compound is packed in the mould. The maximum internal mould pressure, at B, is not reached until after the maximum injection pressure is obtained.

Even if the injection pressure stays the same, the internal mould pressure drops slightly, as a result of shrinkage of the compound, down to the curing point, C. From here, the pressure drops rather faster, because now no more compound can be pushed back, until dropping to the residual pressure, Pw remainder, when the mould is opened, at E.

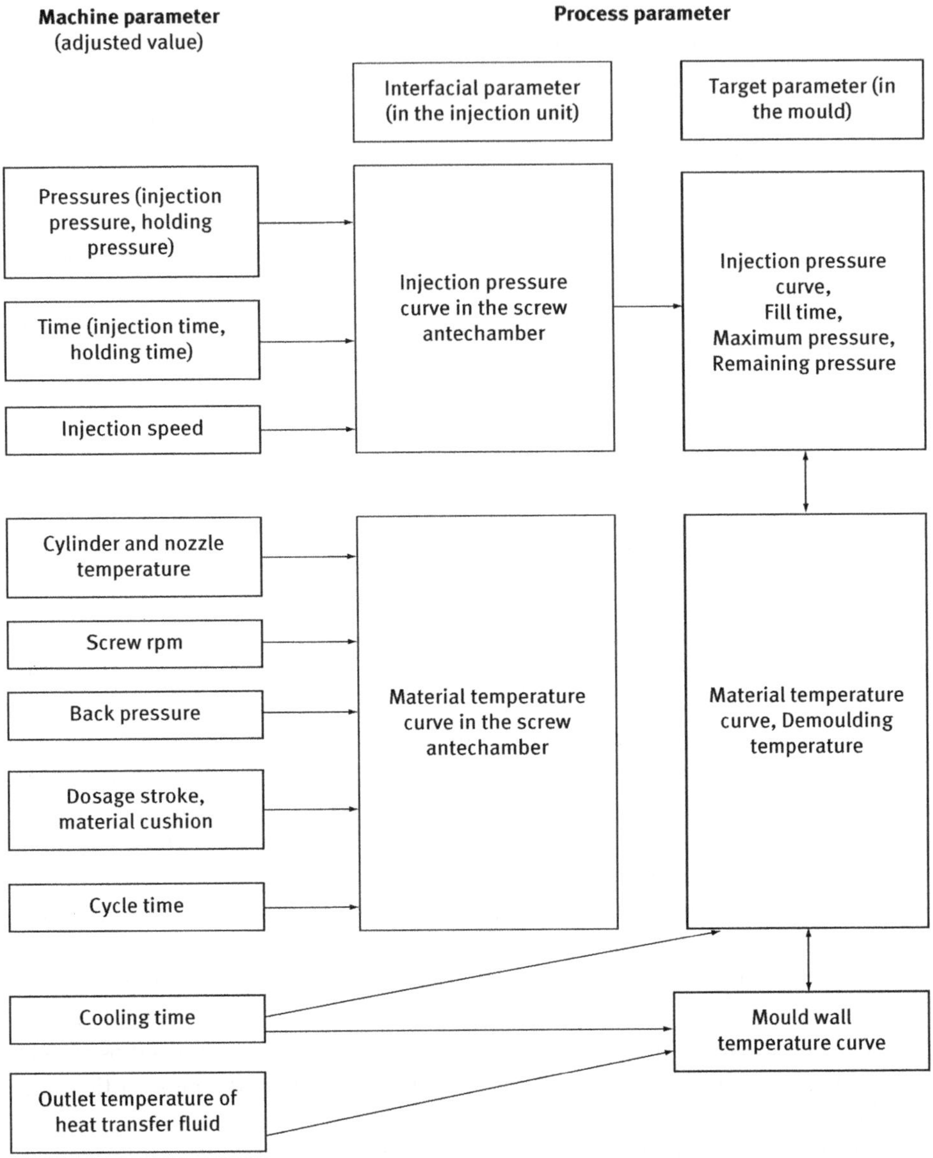

Figure 8.1: Important parameters for part formation (arrows indicate which parameters affect other parameters).

Key:

0–A Filling phase, or more correctly dynamic filling phase, because even though solidification is beginning, the compound is pressed down into the cavity up to the curing point, C.

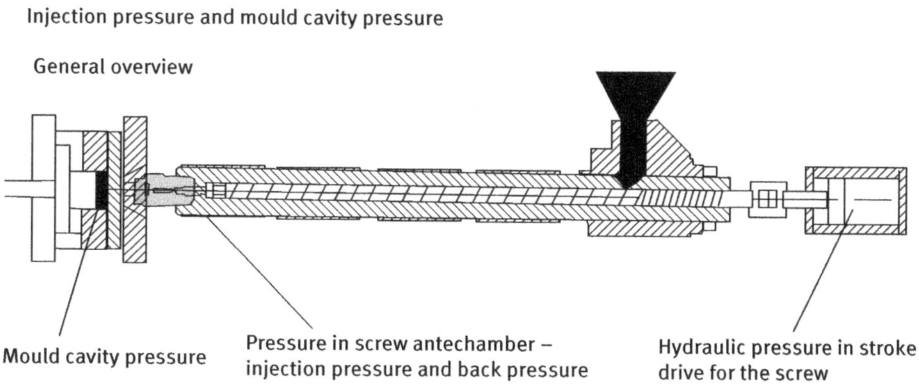

Figure 8.2: Pressure locations during injection moulding.

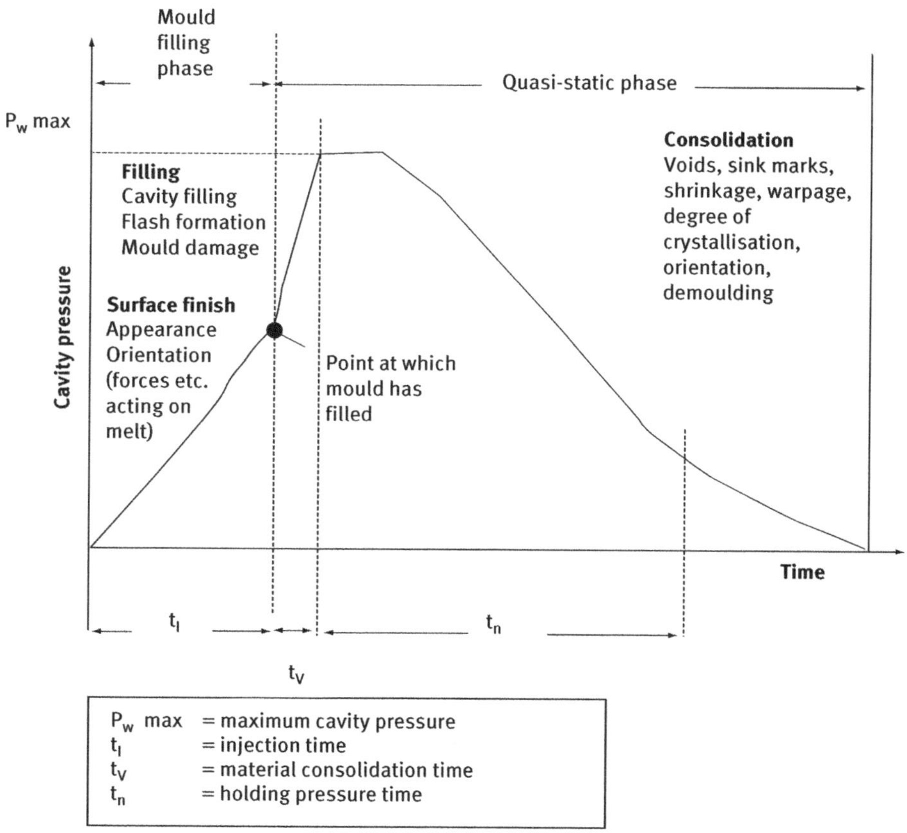

Figure 8.3: Correlation between quality characteristics and pressure profile in the moulding.

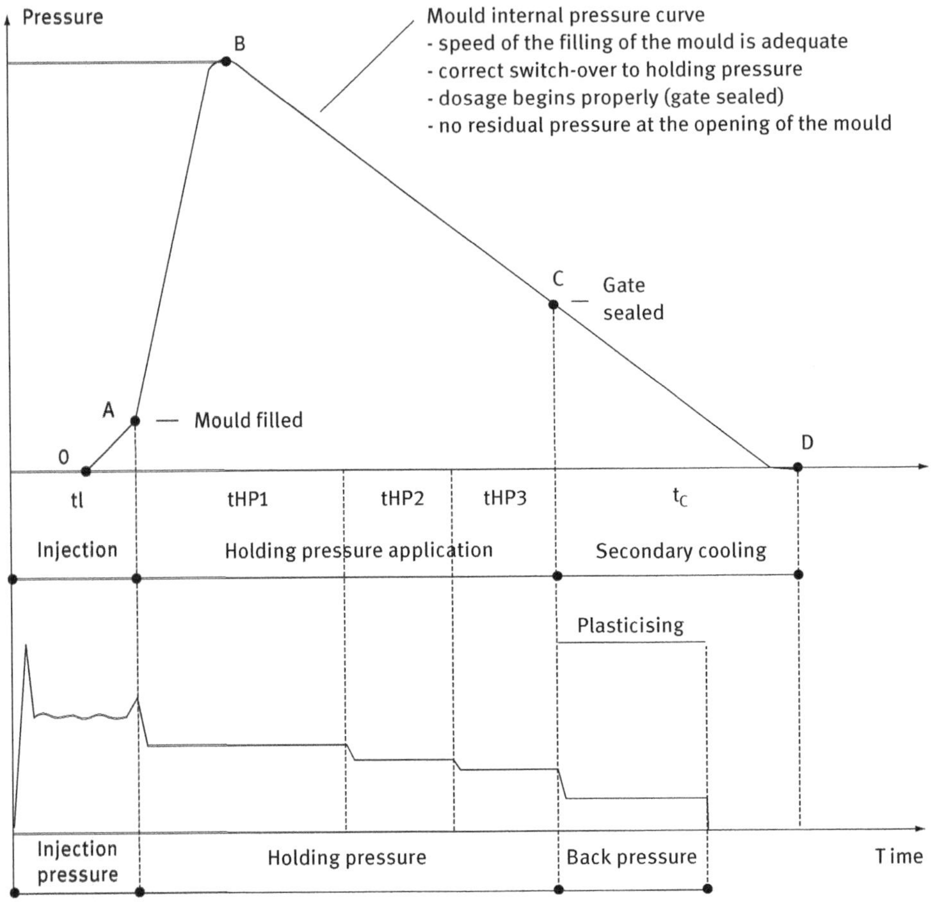

Figure 8.4: Mould cavity pressure curve in relation to injection moulding pressures.

A–D Quasi-static phase: the mould is volumetrically filled. Flow movements of the compound (reserve thrusts from the screw forward cavity) occur only transiently until compacting (up to B) and the balancing out of shrinkage during solidification (up to C).

A–B Compression phase.

B–C Pressure holding phase (up to the sealing point).

C–D Shaping phase (cooling off to a specific temperature at which components are ejected from the mould).

HP Hydraulic pressure. t_k Cooling time.

8.5 Injection pressure and injection time

The selected injection pressure and holding pressure must be as high as necessary to fill the cavity sufficiently fast, completely and efficiently, but on the other hand, as low as necessary to produce low-stressed injection moulded components and avoid difficulties when the components are ejected from the mould.

The injection time (injection time and holding time), i.e., the duration of the effect of the injection pressure, must be selected to be just long enough to solidify (seal) the gate. If the injection time is too short, the compound can flow back out of the cavity, sink marks occur and in general there are larger tolerance variations. Unduly long injection times are uneconomic and increase the internal stresses of the injection moulded component, especially close to the gate. The correct injection time can be determined by weight measurement.

With injection times greater or equal to the solidification time, the weight of the injection moulded component remains practically the same (does not increase). With injection times less than the curing time, the weight of the injection moulded component decreases. The occurrence of sink marks is also a sure indication that the injection time (duration of the effect of pressure) is shorter than the curing time.

A reduction of the holding pressure is necessary with amorphous thermoplastics, in order to avoid difficulties when ejecting the parts and obtain low-stress injection moulded parts.

A constant holding pressure is recommended with semi-crystalline thermoplastics, in order to ensure an undisturbed crystallisation process.

8.6 General information on filling speed

8.6.1 Initial injection speed

The smaller the flowpath cross-section is in relation to the screw/piston surface, the higher the filling speed needs to be. A larger injection cylinder in the same injection unit thus produces a higher filling speed for the same initial injection speed. The initial injection speed, and with it the filling speed, should be selected to be as high as possible, so that the mould is filled as quickly as possible with compound which has as uniform a temperature as possible. As a result, the temperature and pressure variations in the mould are slight, and stress-free components can be obtained. In this way, the component should be filled as uniformly as possible, with the flow head moving away from the gate. Free jet formation can be avoided through suitable mould design.

The optimum filling speed is higher for thin-walled parts than for thick-walled parts, so as to obtain uniform filling of the mould through the flow head. Too low a filling speed causes a greater temperature variation between those parts of the preform near the gate and those far from it, due to increased cooling off of the compound while the cavity is being filled. The higher viscosity of the colder compounds also requires higher injection pressures, which require stronger locking pressures.

Too long a filling speed can also lead to surface faults. If compound which has already solidified onto the mould wall is displaced by a subsequent filling, cross-grooves occur vertically to the direction of flow (gramophone record effect). The effects of too high a filling speed are described in Chapter 9.

8.7 Filling speed and orientation

Orientation effects can arise during the filling of the mould, especially *via* the influence of friction. The molecules initially lying randomly in the compound are now stretched and orientated in the direction of flow. Such orientations lead to non-uniform shrinkage and non-uniform preform properties (anisotropy). The higher the shear rate, the greater the orientation of the polymer chains. In addition, the higher the filling speed and the greater the viscosity of the compound flowing in, the higher the shear rate; thus, higher filling speeds are bound to lead to an increased tendency to orientate.

However, the higher the temperature of the compound is, and therefore the lower the viscosity value, the less negative effect a high filling speed will have. A high compound temperature, in connection with a high mould wall temperature, will cause the orientated molecules to lose their orientation after the filling process has ended (relaxation), which reduces the negative effects of orientation. Therefore, before reducing the filling speed, a check should be made on whether orientation phenomena can be reduced by increasing the compound temperature and the mould temperature. In this situation, preference should be given to compound temperatures and mould temperatures that are as high as possible (as already stated when dealing with the most favourable injection pressure).

In terms of the flowability of the materials themselves, material suppliers sometimes illustrate the relationship between the wall thickness and injection speed as a flowpath/wall thickness ratio (L/s). If a ratio of 100:1 is given, this means for a wall thickness of 1 mm, then the length of flow from the gate will be 100 mm. Because flow is dependent upon wall thickness, a variety of mould wall thicknesses may be quoted. If the material is required to flow further, e.g., 125:1, more pressure will be required to fill the cavity leading to more orientation in the material. The relationship between pressure and wall thickness is shown in Figure 8.5. Therefore ideally, moulds should be designed with the consideration of flowpath lengths and wall thickness ratio in mind

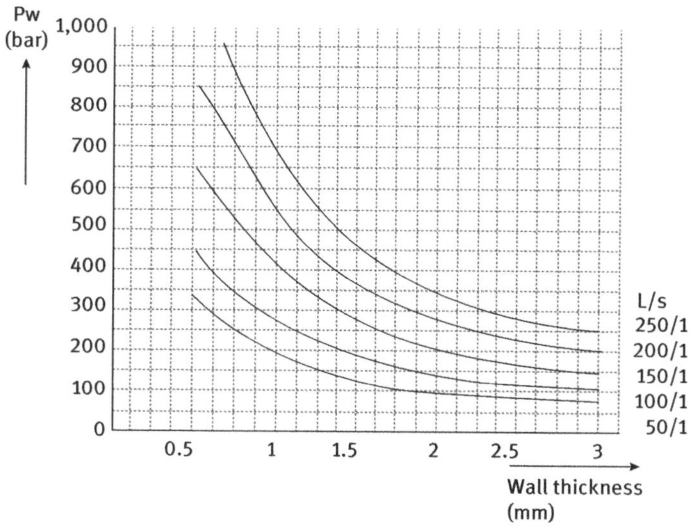

Figure 8.5: Injection pressure/wall thickness relationship.

8.8 Effects of excessive filling speed

A high filling speed also causes a high shear rate between the compound in the core and the compound on the mould wall. The shear stress arising under such conditions can lead to impairment of the plastic (shear fracture). Particularly high shear stresses arise if the compound has to turn sham corners, especially with abrupt changes in the cross-section, which should be taken into account when designing the moulding.

Many plastics display particularly high shear fracture sensitivity, e.g., fluoroplastics such as Teflon. However, PMMA or PC also display more pronounced tendencies to shear fracture than, for example, PS.

If the filling speed is high, the air must be removed from the cavity sufficiently quickly. If this is not the case, the compression, and thus the heating of the air, increases (diesel effect, see Chapter 9), which can lead to heat damage of the material or can even cause burns. So care should be taken to ensure that the air removal system works well in the parts of the mould which are filled last. Under certain circumstances, it is sufficient to reduce the mould clamping force to the necessary level if the air can be extracted through the parting plane.

A high filling speed can cause non-uniform mould filling due to:

- Free jet formation at the gate ('sausage injection moulding'), which results in surface faults; and/or
- Splitting of the compound flow, which leads to unnecessary joint line formation; and/or

- Tearing loose of batches which are already solidified, causing deterioration in the surface finish and usually in the mechanical properties as well.

These faults can usually be avoided by suitable construction design, even at high speed.

Finally, high filling speeds can lead to uncontrolled and excessive heating of the compound, due to strong constrictions in runner cross-sections that are too small. This is particularly the case if the working compound temperature is too low, and thus the melt viscosity is higher. In certain circumstances, heating due to constriction losses can become so great that, when the set cylinder temperatures are reduced, the compound temperature in the mould cavity does not fall, but rises. This matter will be referred to again in connection with cylinder temperatures.

8.9 Setting the initial injection speed

The initial injection speed is set, on the overwhelming majority of the injection moulding machines (IJM) in operation today, using a throttle in front of the drive cylinder of the piston or screw. A higher consistency of initial injection speed from batch to batch can be obtained using a quantity controlling device instead of the throttle. If the initial injection speed is set using a throttle, the connection between the throttle position and the initial injection speed is not linear. For this reason – and also to guarantee a repeatable speed setting in case of a change of machine – the initial injection speed for each machine should be determined independently of the throttle setting (by measuring the initial injection time).

A prerequisite for the chosen initial injection speed actually being achieved is that a sufficiently high pressure head must be available during the initial injection. This is true for both kinds of initial speed setting (throttle or quantity control). Thus, the selected first pressure stage for the initial injection must be sufficiently high. The higher the flow resistance during filling, the higher the pressure selected should be. The initial injection pressure can be raised or lowered to check whether the injection pressure head is sufficiently high. If the initial injection speed no longer changes when the pressure changes (measure the initial injection time to at least 0.01 s), the pressure head is satisfactory. If possible, the initial injection pressure should be set so high that it lies 10–15% above the value at which the pressure influence on the speed becomes noticeable. A guide to injection times relative to the viscosity of materials is given in Table 8.3, and Table 8.4 is a guide to the viscosity type of a number of common materials.

Table 8.3: Recommended injection times for low, medium and high viscosity materials.

Injection volume (cm^3)	Injection time in seconds (s)		
	Low viscosity	Medium viscosity	High viscosity
1–8	0.2–0.4	0.25–0.5	0.3–0.6
8–15	0.4–0.5	0.5–0.6	0.6–0.75
15–30	0.5–0.6	0.6–0.75	0.75–0.9
30–50	0.6–0.8	0.75–1.0	0.9–1.2
50–80	0.8–1.2	1.0–1.5	1.2–1.8
80–120	1.2–1.8	1.5–2.2	1.8–2.7
120–180	1.8–2.6	2.2–3.2	2.7–4.0
180–250	2.6–3.5	3.2–4.4	4.0–5.4
250–350	3.5–4.6	4.4–6.0	5.4–7.2
350–550	4.6–6.5	6.0–8.0	7.2–9.5

Table 8.4: Viscosity of various plastic types.

Viscosity	Plastic type
Low	PE – soft, PA4.6, PA6, PA6.6, PA6.10, PA11, POM, PET, PBT, PPS, TPE
Medium	PS, SB, SAN, ABS, modified PPO, PVC – soft, CA, CAB, CP, PE – rigid, PP, PA12, PA amorphous
High	PVC – rigid, PMMA, PC, PSU, PES, PEI, PAI, PVDF, FEP, ETFE

PAI: Polyamide imide
PEI: Polyetherimide
PVDF: Polyvinylidene fluoride
TPE: Thermoplastic elastomers

8.10 Plasticising

8.10.1 Compound temperature – general information

The smaller the spatial temperature variation of the compound in the mould and the temporal temperature variations from batch to batch are, the higher and more uniform the quality of the component is. The optimum compound temperature is a parameter specific to each material. As mentioned earlier, the compound temperature distribution in the mould becomes more uniform, the tendency toward orientation in the components is reduced. However, the temperature of the compound must in no case be so high that damage occurs *via* decomposition. For this reason, it should not be forgotten that, as well as the

cylinder and nozzle temperature settings, the following parameters also affect the compound temperature:
- Screw speed.
- Back pressure.
- Dwell time of the compound in the cylinder (essentially determined by the cycle time).
- Friction (friction losses) during the initial injection.
- Mould wall temperature.

1, 2 and 3 affect the compound temperature in the screw forward cavity.

4 and 5, like the first factors, influence the compound temperature in the mould.

Too low a compound temperature, in addition to causing the component drawbacks already mentioned, also bring the risk of non-uniform and/or incomplete plasticising (melting) of the compound. It should be taken into account that a low cylinder temperature does not necessarily mean a low compound temperature. Strong friction, as a result of higher viscosity, leads to higher pressure losses, even during the initial injection, and thus a steeper rise in temperature. Apart from the danger of local thermal overloads in the compound, there is also the risk of mechanical damage.

8.10.2 Cylinder temperatures

The temperature ranges provided by the plastic manufacturers can be regarded as the most suitable practical guides. If the manufacturer's data is not available, the guidelines in Table 8.2 should be used. As a first step, these temperatures can be selected as setting values for the nozzle-side cylinder heating band. The influences referred to in the previous section are not taken into account here.

The following data is of assistance in determining the actual compound temperature in the screw forward cavity. The nominal temperature set is the temperature of the cylinder wall, or of the measuring point which is close to the cylinder wall. If the additional heating caused by friction is disregarded, the plastic (as a poor heat conductor) will only reach the wall temperature after a long dwell time. The friction during plasticising causes a rise in temperature, and the higher the screw speed and back pressure, the greater the friction.

Methods of measuring the actual compound temperature will be discussed in Section 8.10.6.

The setting values for the cylinder heating bands should, as far as possible, be graded in descending order going towards the material feed area. The cylinder temperature at the material feed area should not be too low, to ensure that the frictional heat does not become too great. Thus, in practice, a 20 °C temperature difference between the first cylinder heating area and the last has proved to be effective for almost all thermoplastics.

For PA6.6, an even smaller temperature spread gives better material feed behaviour and more uniform plasticising. Under certain circumstances for unplasticised PVC and PMMA, it is preferable that the cylinder temperatures do not decrease going towards the material feed area, but actually increase, in order to keep the frictional heat low. With this in mind, the cooling of the cylinder yoke for the plasticising cylinder must also be mentioned and supporting body temperatures of approximately 30–40 °C (warm to the touch) are recommended. Too low a temperature results in the formation of condensation, which can lead to material feeding difficulties and component faults due to moisture absorption by the granules.

8.10.3 Nozzle temperature

The effect of the nozzle heating should result in the plasticised compound maintaining its temperature while flowing through the nozzle. The nozzle temperature should be set accordingly, usually to the same level as the cylinder heating on the nozzle-side. The nozzle temperature – at least on small- and medium- sized IJM – is essentially determined by the cylinder heating on the nozzle-side, because the nozzle's heat capacity is very small in relation to that of the cylinder. Thus, cases can occur where it is not absolutely necessary to control the nozzle temperature, and it is enough to balance out the heat losses of the nozzle using a power-controlled heater band (cheaper than a controlling device). The power setting for the nozzle heater band should then be set, as stated above, so that the compound does not alter its temperature as it flows through the nozzle. However, controlling the nozzle heating can only lead to satisfactory results with heat balance, and thus with sufficiently long and continuous production. When the machine is started up, and during any production interruptions, the nozzle temperature control should be transferred to a control system. However, it is currently not possible to function without a nozzle temperature control system, especially for the thermoplastics that are more difficult to process. It has already been stated that the nozzle-side cylinder temperature should be taken as a guideline for the nozzle temperature, but the optimum temperature can be slightly lower or higher, in accordance with the material concerned.

If an additional temperature measuring point is established in the nozzle, it is relatively easy to determine the optimum setting temperature; it is the temperature at which the nozzle temperature measured during injection undergoes practically no change. If the temperature is set too high, the nozzle cools off while the compound is flowing through, and if the temperature is set too low the nozzle temperature rises again. A nozzle temperature lower than the last cylinder heating temperature will often turn out to be effective (even if it is not optimum), in order to avoid a leak at the nozzle. Deviating from the optimum value is, in this case, usually justified because this causes the thickness of the material cushion to change during the initial injection and, when two pressure stages are being used, this can have a significant detrimental effect (due to a strong influence on pressure cycle in the mould).

8.10.4 Temperature profile guideline for the plasticising cylinder

There are four different temperature profiles that can be used for injection moulding: horizontal, ascending, descending and finally, ascending with descending at the nozzle. An example of each of these set-ups is shown in Figure 8.6.

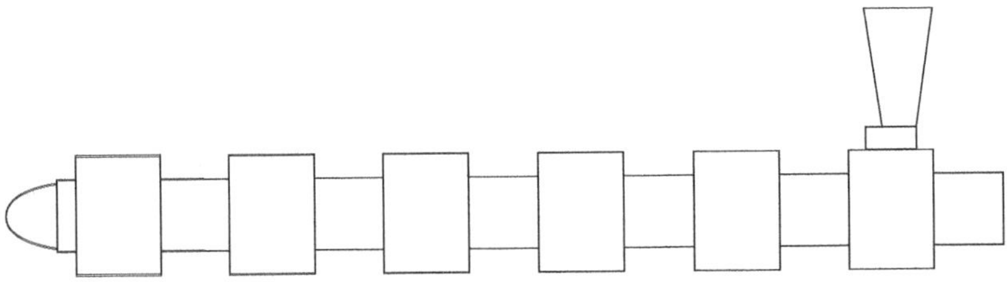

Heater zone	6	5	4	3	2	1	
Temperature guideline (°C)	280	280	280	280	280	280	Horizontal
	280	275	270	260	250	240	Ascending
	280	280	285	290	295	300	Descending
	270	280	270	260	250	240	Descending-ascending

Figure 8.6: Examples of types of temperature profiles.

8.10.4.1 Horizontal

This temperature profile is recommended when fast heating of the material is required, e.g., fast cycle times, long (entire) screw stroke or screws with a deep flight. This profile setting ensures that the material is evenly heated and is completely molten before injection. When using glass-filled material this profile can reduce wear and tear of the cylinder assembly.

8.10.4.2 Ascending

This is the most common temperature profile and is especially recommended for heat-sensitive materials like POM and PES, and flame-retardant materials such as flame-retardant PA and PVC.

8.10.4.3 Descending

This special temperature profile is necessary in exceptional cases, e.g., when working with PA66.

8.10.4.4 Ascending and descending at the nozzle

This temperature profile is normally used with an open nozzle for the prevention of thread formation or drooling. There can be drawbacks with this profile setting as the colder material (which may be hardened) at the nozzle may be injected into the mould to form the product.

The most suitable temperature profile setting can only be achieved through practical experience with the particular material in question. Tables 8.5 –8.7 give recommendations for various setting values.

8.10.5 Temperatures for the cylinder and mould

There are recommended temperatures for the cylinder and mould for thermoplastic and thermoset materials, as shown in Tables 8.5 –8.7. Special instructions specified by the manufacturer must be observed.

Table 8.5: Temperatures for the cylinder and mould: amorphous thermoplastics.

Injection material	Nozzle-side cylinder temperature (°C)[1,2]	Feed yoke temperature (°C)[1,2]	Mould temperature (°C)	Approximate transition temperature (°C)	Remarks
PS	160–230	30–35	20–60	90	
SB	160–250	30–35	20–60	85	4
SAN	200–260	30–35	40–80	100	4
ABS	180–260	30–35	40–85	105	4
PVC – rigid	160–180	30–35	20–60	80	3, 5, 6, 7, 8, 10
PVC – soft	150–170	30–35	20–40	55–75	3, 5, 8, 10
CA	185–225	30–35	30–60	100	3, 4, 8
CAB	160–190	30–35	30–60	125	3, 4, 8
CP	160–190	30–35	30–60	125	3, 4, 8
PMMA	220–250	35–45	60–110	105	4
PPE (PPO) modified	240–290	35–45	70–120	120–130	3, 4, 11
PC	290–320	35–45	60–120	150	4,11
PAR	350–390	45–65	120–150	190	4, 11
PSU	320–390	45–65	100–160	200	4, 11
PES	340–390	45–65	120–200	260	4, 11

Table 8.5 (continued)

Injection material	Nozzle-side cylinder temperature (°C)[1,2]	Feed yoke temperature (°C)[1,2]	Mould temperature (°C)	Approximate transition temperature (°C)	Remarks
PEI	340–425	45–65	100–175	220–230	4, 11
PAI	340–360	45–65	160–210	275	4, 9, 11
PA amorphous	260–300	35–45	70–100	150–160	4

If no other practical values are available: set the nozzle temperature to the nozzle-side cylinder temperature. Cylinder temperature reduction approaching the feed zone by a maximum of 5–10 °C per heating zone; maximum temperature difference between the nozzle and feed-side of 20 °C. Set the nozzle-side heater band and the following one to the same temperature with more than two heater zones.

[1]Only set the upper temperature with a high shot count when using a thermally sensitive material (shorter dwell time in the cylinder).

[2]Thermally sensitive.

[3]The granulate must be dried before processing.

[4]Do not use shut-off nozzles, only open nozzles. [5]Injection without check valve recommended.

[6]Only operate with screw tips without check valves. [7]Only operate with low back pressure.

[8]To improve the feed performance, set the same or a slightly higher cylinder temperature approaching the feed-side.

[9]A corrosion-protected cylinder unit (Arbid) is recommended.

[10]An abrasion-proofed cylinder unit (Arbid) is recommended for the processing of reinforced materials (e.g., fibreglass).

[11]With thermostat cooling water valves: 30...35 °C 2...3 scale markings.

35...45 °C 3...4 scale markings.

45...65 °C 4...5 scale markings.

PR: Polyacrylate

PPE: Polyphenylene ether

Table 8.6: Temperatures for the cylinder and mould: crystalline thermoplastics.

Injection material	Nozzle-side cylinder temperature (°C)[1,2]	Feed yoke temperature (°C)[1,2]	Mould temperature (°C)	Approximate crystalline T_m (°C)	Remarks
PE soft	210–250	30–35	20–40	105–115	9
PE rigid	250–300	30–35	20–60	125–140	
PP	220–290	30–35	20–60	156–168	
PA4.6	210–330	45–65	60–150	295	4, 6, 7, 9
PA6	230–260	45–65	40–100	215–225	4, 6, 7, 9
PA6.6	270–295	45–65	50–120	250–265	4, 6, 7, 9
PA6.10	220–260	45–65	40–100	210–225	4, 6, 7, 9
PA11	200–250	34–45	40–100	180–190	6, 7, 9

Table 8.6 (continued)

Injection material	Nozzle-side cylinder temperature (°C)[1,2]	Feed yoke temperature (°C)[1,2]	Mould temperature (°C)	Approximate crystalline T_m (°C)	Remarks
PA12	200–250	35–45	40–100	175–185	7, 9
POM	185–215	35–45	80–120	165–175	3, 6
PET (PETP)	260–260	45–65	50–140	255–258	3, 4, 9
PBT (PBTB)	230–270	45–65	40–60	220–225	3, 4, 9
PPS	300–360	45–65	20–200	280–288	4, 9
Perfluoroalkoxyalkane	350–420	45–65		300–310	5, 8
FEP	340–370	45–65	150–200	285–295	5, 8
ETFE	315–365	45–65	60–120	270	5, 8
PVDF	220–300	35–45	70–90	171	5, 8
Polyether ether ketone	350–380	45–65	150–180	340	
Polyether ether ketone ketone	390–420	45–65	150–180	363	
Polyether ketone	400–430	45–65	150–180	365	
PAA	250–290	45–65	120–150	235–240	
PPA	320–350	45–65	135–165	310	
LCP	280–450	45–65	30–160	270–380	
		TPE			
TPE-amide type	200–260	30–35	20–50		
TPE-ethylene type	200–250	30–35	20–50		
TPE-styrene type	180–240	30–35	20–50		
TPE-urethane type	190–240	30–35	20–40		
TPE-olefin type	110–180	30–35	15–40		

If no other practical values are available: set nozzle temperature to the nozzle-side cylinder temperature. Cylinder temperature reduction approaching the feed zone by maximum 5–10 °C per heating zone; maximum temperature difference between nozzle and feed-side of 20 °C. Set the nozzle-side heater band and the following one to the same temperature with more than two heater zones.
[1]Only set the upper temperature with a high shot count with a thermally sensitive material (shorter dwell time in the cylinder).
[2]Thermally sensitive.
[3]The granulate must be dried before processing.
[4]Do not use shut-off nozzles, only open nozzles. [5]Only operate with low back pressure.
[6]To improve the feed performance, set the same or a slightly higher cylinder temperature approaching the feed-side.
[7]A corrosion-protected cylinder unit (Arbid) is recommended.
[8]An abrasion-proofed cylinder unit (Arbid) is recommended for the processing of reinforced materials (e.g., fibreglass).
[9]With thermostat cooling water valves: 30...35 °C 2...3 scale markings.
35...45 °C 3...4 scale markings.
45...65 °C 4...5 scale markings. PAA: Polyacrylic acid
PPA: Polyphthalamide T_m: Melting temperature

Table 8.7: Temperatures for the cylinder and mould: thermosets.

Resin type	Temperatures guidelines		
	Mould (°C)	Plasticising cylinder	
		Nozzle-side (° C)	Feed-side (°C)
Phenol formaldehyde	145–175	75–95	45–60
Urea formaldehyde	140–165	80–95	45–55
Melamine formaldehyde	150–180	75–100	50–60
Melamine-phenol formaldehyde	150–180	70–100	45–55
Unsaturated polyester	140–165	80–105	40–65
Diallyl phthalate	150–190	85–110	50–60
Epoxide	155–200	85–115	50–60
Silicone	180–250	5–20	5–20

8.10.6 Measuring the compound temperature

There is still no completely satisfactory solution with regards to the measurement of the compound temperature (which is distinct from measuring the compound pressure). The simplest method is injection into the air, or into a small container made of a heat-insulating material, and measuring the compound temperature with an insertion thermometer. Naturally, this measurement cannot be continuous – the cycle must be interrupted, which is the crucial disadvantage of this process as the heat balance of the machine is disturbed. The accuracy of the measurement here depends on the quantity of the compound that is collected, as the smaller this is, the more inaccurate the measurement is because a small quantity cools off more rapidly.

With small quantities, the heat transfer to the thermometer, which is usually cooler, causes a noticeable cooling off of the compound, and thus gives a false reading. In this case, the temperature should be brought up to somewhere near the compound temperature before measurement, which can be done during the cylinder heating stage.

It is scarcely possible to obtain repeatability of the measurement to within less than around 5 °C using this method. Significantly higher repeatability, and thus significantly higher accuracy, can be obtained by incorporating a thermometer into the cavity in front of the screw.

In this scenario, the thermometer must be mounted in such a way that no (colder) plastic can be deposited on it, or that the plastic can be continually rinsed off. The disadvantage of this process lies in the necessity of adding structures that lie in the compound flow and increase the flow resistance, increasing the risk that dead corners will form. This causes more difficulties on smaller machines than on larger ones, as the quantity of compound which is still in front of the screw tip after the initial injection (which should certainly be kept as small as possible) is generally increased by the addition of these structures. The effects involved are more negative with smaller injection equipment or a smaller sprue fraction than with larger injection equipment.

Temperature measurements using ultrasound are certainly also possible, but the costs involved are still very high; using this method, the compound temperature is measured continually as the compound flows through the nozzle.

8.10.7 Screw speed

An adjustable screw speed is a prerequisite for suitable processing of materials, and the optimum screw speed is a parameter that depends on the material. For example, optimum rotations per minute (rpm) speeds for a range of plastics will be given by the manufacturers, from which the optimum speed for the screw used can be determined.

'Optimum' here should be understood as follows: the melt is satisfactorily homogenised and the friction is not high enough to cause any damage to the plastic due to heat stress or mechanical stress. If the screw speed is too high, not only does the friction heat rise, but the homogeneity of the melt is usually impaired. With fibreglass-reinforced compounds, the wear on screw and cylinder also increases significantly.

It is not only the higher plasticising yield, i.e., a shortening of the plasticising time, which favours raising the screw speed. In borderline cases, the time from the start of plasticising to the opening of the mould is no longer determined by the cooling off time required for the preform, but by the plasticising time. Here, it would be better to use a cylinder fitting with a larger screw diameter, so that processing could continue at a lower speed.

If there is no data on the optimum range of screw speed or rpm, or if these ranges are very wide, the screw speed should be matched to the cooling-off time. Once the cooling-off time is obtained, i.e., the time required for the mould to open (determined by the permissible temperature at which components can be ejected from the mould or the permissible residual pressure in the mould), the screw speed can then be selected in such a way that plasticising is completed shortly before the cooling off time is completed (before the opening of the mould). In this way, the time that the compound is standing still in front of the screw tip is kept short, and any possible interference factors arising should be of little significance.

8.10.8 Back pressure

The back pressure is intended to ensure that a sufficiently high level of homogeneity is attained by the melt during the plasticising stage and relies on uniform temperature distribution, and the absence of air or gas blisters.

The level of back pressure required depends on the melt viscosity of the plastic and/or its heat sensitivity (Table 8.8). Care must be taken to set the back pressure (the pressure present in the compound in front of the screw tip) through the hydraulic pressure or electrical servo, depending on the drive unit used.

Table 8.8: Recommended values for back pressure.

Injection material	Specific weight (g/cm³)	Viscosity	Back pressure (bar)	Screw circumference speed (m/min)	Remarks
		Amorphous thermoplastics			
PS	1.05	M	40–80		
SB	1.04	M	40–80		
SAN	1.08	M	40–80		
ABS	1.03–1.07	M	40–80		
PVC – rigid	1.38–1.40	H	40–80	4–6	1
PVC – soft	1.20–1.35	M	40–80		1
CA	1.26–1.32	M	40–80		
CAB	1.16–1.22	M	40–80		
CP	1.19–1.23	M	40–80	4–6	
PMMA	1.18	H	80–120	4–6	
Modified PPE	1.06–1.10	M	60–90		
PC	1.20–1.24	H	80–120	6–10	
PAR	1.2	H	80–120		
PSU	1.27	H	80–120		
PES	1.37	H	80–120		
PEI	1.87	M	40–80		
PAI	1.38	H	40–80		
		Semi-crystalline thermoplastics			
PE – soft	0.91–0.93	L	40–80		
PE – rigid	0.94–0.96	M	60–90		
PP	0.9	M	60–90		
PA4.6	1.18	L	40–80		
PA6	1.13	L	40–80		
PA6.6	1.14	L	40–80		
PA6.10	1.06	L	40–80		
PA11	1.04	L	40–80		
PA12	1.02	M	60–90		
PA amorphous	1.12	M	60–90		
POM	1.41–1.42	L	40–80	10–15	
PET	1.3–1.37	L	60–90		
PBT	1.26	L	40–80		
PPS	1.34	L	40–80		

Table 8.8 (continued)

Injection material	Specific weight (g/cm³)	Viscosity	Back pressure (bar)	Screw circumference speed (m/min)	Remarks
FEP	2.14–2.17	H	80–120	5–10	1
ETFE	1.70	H	80–120	5–10	1
PAA	1.43–1.64	L	40–80		
PPA	1.26–1.56	L	40–80	10–15	
PAEK	1.27–1.49	M			
LCP		L	40–80		
TPE					
TPE-ethylene type					
TPE-urethane type	1.14–1.26	L			
Thermosets and elastomers					
Thermosets	1.2–2.0	H	40–80	4–6	
Elastomers		M	20–60	2–5	
Liquid silicone rubber	1.86–1.88	L	20–60		
Powder					
Metal, ceramic		M	10...50		
Water	1.00				
Carbon fibre	1.75–1.9				
Fibreglass	2.49–2.52				

1 Only operate with low back pressure.
L: Low
M: Medium
H: High

When starting up the machine, use the lower values from the manufacturer's data. For dry pigmented granules, the back pressure must be increased long enough for the colour distribution in the injection equipment to be satisfactorily uniform.

If the back pressure is too low, this will become noticeable through wide variations of the compound's performance during holding pressure from batch to batch (as a result of air blisters). If the weight of the components is measured, too low a back pressure shows up through underweight components which appear from time to time.

As the back pressure increases, the plasticising time increases (and the plasticising yield sinks), resulting in an increase in back heating because of the increasing and longer-lasting friction effect. Too high a back pressure leads to a greater extension of the plasticising time and drawing-in difficulties also usually occur. The risk of compound damage through overheating also increases.

8.11 The injection stage

8.11.1 Filling to packing

The importance of the pressure cycle on the quality of the moulded part has already been discussed.

The injection pressure ensures that the injection velocity (speed of filling) can be carried out as fast as possible; however, at some stage the filling phase switches to packing. The switch from injection pressure (filling) to holding pressure (packing) is called the switch-over position. This position on the injection ram can be set in a variety of ways, for example, by:

- Screw position.
- Hydraulic (line) pressure.
- Melt pressure.
- Cavity pressure.

To begin with, the switch-over position can be set at about two-thirds of the total shot volume and altered accordingly from there.

A residual amount of material is generally left in the screw after each shot, which is called the melt cushion. A melt cushion is used to ensure the transmission of injection pressure from the hydraulics (or servo) to the mould. Otherwise the screw may bottom out before an adequate holding pressure can be retained on the moulding in progress. If an adequate cushion is not retained, the shot volume should be increased accordingly. The size of the cushion depends on the size of the machine and may vary from 3–10 mm. The length of time that holding pressure is maintained should be kept as low as possible to minimise induced stresses. This minimum time can be found by simply reducing the holding time until sink marks appear; the shortest time in which sink marks do not appear is the optimum setting value. Typical values for injection pressure settings are shown in Table 8.9.

8.11.2 The mould

8.11.2.1 Mould temperature: general information

The mould temperature or mould wall temperature is one of the most important process parameters (in addition to the pressure cycle and compound temperature), which is of vital importance for the quality and dimensions of the components. The mould temperature influences the shrinkage and thus the dimensions of the compound in

Table 8.9: Typical values for injection pressures.

Injection material	Specific weight (g/cm³)	Viscosity	Injection pressure (bar)	Holding pressure (bar)	Mould cavity pressure	
					Relationship with (x) highest holding pressure stage	Expected cavity pressure (bar)
			Amorphous thermoplastics			
PS	1.05	M	650–1,550	300–700	0.75–0.5	150–350
SB	1.04	M	650–1,550	350–800	0.75–0.5	200–400
SAN	1.08	M	650–1,550	350–900	0.75–0.5	250–450
ABS	1.03–1.07	M	650–1,550	400–900	0.75–0.5	300–550
PVC – rigid[1,2]	1.38–1.40	H	1,000–1,550	500–900	0.6–0.4	250–500
PVC – soft[2]	1.20–1.35	M	400–1,550	300–600	0.75–0.5	150–300
CA	1.26–1.32	M	650–1,350	300–650	0.85–0.7	250–450
CAB	1.16–1.22	M	650–1,350	300–900	0.75–0.5	250–450
CP	1.19–1.23	M	650–1,350	400–700	0.75–0.5	200–350
PMMA	1.18	H	1,000–1,400	500–1,150	0.6–0.4	350–550
Modified PPE	1.06–1.10	M	1,000–1,600	600–1,200	0.75–0.5	350–600
PC	1.20–1.24	H	1,000–1,600	600–1,300	0.6–0.4	350–650
PAR	1.2	H	1,000–1,600	600–1,300	0.6–0.4	350–650
PSU	1.27	H	900–1,400	500–1,100	0.6–0.4	400–600
PES	1.37	H	900–1,400	500–1,100	0.6–0.4	400–600
PEI	1.87	M	750–1,550	400–750	0.85–0.7	350–650
PAI	1.38	H	750–1,550	500–1,050	0.85–0.7	450–750
			Semi-crystalline thermoplastics			
PE soft	0.91–0.93	L	600–1,350	300–800	0.85–0.7	200–600
PE rigid	0.94–0.96	M	600–1,350	300–800	0.75–0.5	200–600
PP	0.9	M	800–1,400	500–1,100	0.75–0.5	300–650
PA4.6	1.18	L	650–1,550	550–1,050	0.85–0.7	450–750
PA6	1.13	L	450–1,550	400–750	0.85–0.7	350–550
PA6.6	1.14	L	650–1,550	550–1,050	0.85–0.7	450–750
PA6.10	1.06	L	450–1,550	350–750	0.85–0.7	300–500
PA11	1.04	L	450–1,550	400–800	0.85–0.7	350–550
PA12	1.02	M	550–1,550	400–1,000	0.75–0.5	350–550
PA amorphous	1.12	M	900–1,300	450–800	0.75–0.5	350–450
POM	1.41–1.42	L	800–2,000	700–1,500	0.85–0.7	550–1,050
PET	1.34–1.37	L	800–1,500	550–1,050	0.85–0.7	450–750
PBT	1.29	L	800–1,550	500–1,000	0.8–.0.7	400–700

Table 8.9 (continued)

Injection material	Specific weight (g/cm³)	Viscosity	Injection pressure (bar)	Holding pressure (bar)	Mould cavity pressure	
					Relationship with (x) highest holding pressure stage	Expected cavity pressure (bar)
PPS	1.34	L	750–1,500	400–750	–	350–600
FEP[1]	2.14–2.17	H	1,000–1,500	500–1,000	0.6–0.4	300–600
ETFE[1]	1.70	H	1,000–1,500	500–1,000	0.6–0.4	300–600
PAA	1.4–1.64	L	1,000–1,500	350–800	0.85–0.7	300–700
PPA	1.26–1.56	L	700–1,500	350–800	0.85–0.7	300–700
PAEK	1.27–1.49	M	800–1,500	450–800	0.85–0.7	400–700
LCP		L	400–1,500	350–1,000	0.85–0.7	300–800
TPE-urethane type	1.14–1.26	L	400–1,000	300–600	0.85–0.7	200–450
			Thermosets/Elastomers/Other			
Thermoset	1.2–2.0	H	800–2,500	300–1,000	0.6–0.4	200–600
LSR	1.86–1.88	L	300–800	120–359	0.85–0.7	80–250
Metal, ceramic		M				
Water	1.00					
Carbon fibre	1.75–1.9					
Fibreglass	2.49–2.52					

[1]Do not use shut-off nozzles, use open nozzles only.
[2]Only operate without check valve.
L: Low
M: Medium H: High

the mould, the surface finish and the orientation in the injection equipment and also, not least, the cycle time – through the cooling off time – and thus component cost.

Component quality depends on the level of mould temperature as well as their uniformity and repeatability. Economic quality improvement of injection moulding is not possible without correct and uniform temperatures in the mould. Even with a more expensive injection process control or adjustment system, the negative influence of unsatisfactory mould temperatures cannot usually be balanced out. If it is a question of narrowing the tolerances of the components, the first step is to check the mould temperature data, and, if necessary, these must be improved before, for example, investigating injection pressure control on the basis of the internal mould pressure.

Naturally, a prerequisite for evaluating the mould temperature data is the measurement of the mould temperature. Unfortunately, even today this is often not something that is done automatically.

8.11.2.2 Mould temperature levels

The optimal mould temperature level is a parameter specific to the material. If no data from the relevant manufacturer are available, the guidelines in Tables 8.5–8.7 can be used.

High mould temperatures cause the component to cool slowly, which is necessary for example, with the majority of semi-crystalline thermoplastics in order to obtain components that are to size and have constant dimensions. The crystallisation of these compounds must be completed in the mould, i.e., it must be over before the components are ejected from the mould. Otherwise, after-crystallisation occurs over the course of time, which in every case, causes alterations in dimensions and frequently also leads to warping of the component.

High mould temperatures improve the flow behaviour of the compound in the mould, the injection pressure requirements are lower and lead to an improved surface finish of the components. High mould temperatures break down orientations that arise during the filling of the mould, hence there are fewer orientations in the moulded component.

The upper limit for the mould temperature is governed by the maximum temperature at which the components can be ejected from the mould, which is specific to each material, e.g., for amorphous thermoplastics it lies at least 10 °C below the freezing temperature. High mould temperatures lead to slower cooling, which means longer cooling times and therefore, longer cycle times.

Consequently, during the selection of the mould temperature level, a choice often has to be made between higher quality and a more favourable price for the components. If you allow yourself to be guided by the principle 'only as good as necessary, not as good as possible', a compromise between the two, giving a 'semi-optimum' temperature level, will very often lead to an economically acceptable solution.

8.11.2.3 Uniformity of mould temperatures

Uniformity has two aspects here – the spatial temperature distribution in the mould and the temporal temperature behaviour during the production cycle.

The mould temperature influences the shrinkage, and thus the later dimensions of the component, and local temperature differences can cause parts to warp. Mould temperatures that do not remain constant from batch to batch lead to dimensional variations through varying shrinkage.

Uniform temperature distribution in the mould is essentially dependent upon the mould temperature system. Adequate and uniform temperatures are of importance, not only in relation to the warping tendency, but also for economic reasons from the point of view of the time to produce a part. It is necessary to wait until the hottest part of the component has cooled enough before ejecting the component from the mould, thus a uniform intensity of temperature becomes a prerequisite for economic manufacturing. When adequate core temperature patterns cost more to achieve, this is often a sign of omission of this parameter during the mould design phase; the mould certainly becomes cheaper but it means that a substantial increase in the cycle time – often up to 100% and more – must usually be taken into account.

The temperature systems used are essentially responsible for maintaining the required mould temperatures throughout the manufacturing cycle. The temperature systems must be matched to the mould with regards to their production capacity, i.e., they must be in a position to supply or extract the necessary amounts of heat sufficiently quickly.

In order to guarantee a repeatable, uniform temperature distribution in the mould, even after a change of mould or a refit, the inlet and outlet paths of the temperature control fluid at the mould must be unambiguously marked. If the connections are mixed up, this will certainly alter the temperature conditions, which admittedly, need not always have a negative effect, but which can often lead to substantial deterioration, especially in cooling cores.

8.11.2.4 Mould temperature patterns

During the cooling of the equipment, the heat is drawn from it by the mould. The amount of heat involved depends on the component weight, the type of plastic (specific heat content), the ejection temperature and the cycle frequency. The mould transfers this heat to the environment (*via* heat conduction to the machine and radiation or convection to the ambient air), if no special temperature control system is connected up. The higher the mould temperature, the greater the amount of heat released to the environment. So, when the machine starts up, the mould initially heats up to the temperature at which the amount of heat supplied by the solidifying and cooling plastic can be released to the environment.

This temperature is, in practice, scarcely likely to match the optimum mould temperature for the injection moulding process. The temperature reached will only remain constant for as long as the machine carries on running continuously, as long as no cycle interruptions occur and as long as the heat removal conditions, e.g., the ambient temperature, do not alter. Satisfactory consistency and repeatability of the temperature level cannot therefore be obtained without special temperature control equipment.

If the desired mould temperature is lower than the automatically set temperature referred to above, additional heat must be drawn from the mould by cooling: if the required mould temperature is higher, more heat must be supplied to it *via* heating.

These processes are also referred to as hard temperature control, if heat is drawn from the mould, and soft temperature control, if heat is to be supplied; a separate temperature control for the two halves of the mould is important from this point of view. The parts of the mould on which the component is shrunk, such as cores, for example, will accept more heat, and thus will be more heated to a greater degree, if they are not temperature controlled in a suitable fashion. Parts of the mould from which the component shrinks away accept less heat and will therefore remain colder without temperature control. In extreme cases, a situation may arise where one half of the mould, with the core parts, has to be cooled, while the other half has to be heated, so as to obtain the same mould temperature in both halves of the mould.

Again, this clearly shows that a good temperature control system for the mould is crucial (it must be able to quickly exchange the requisite amounts of heat). The temperature control device must also be suitable for the specific system (so as to be able to supply and extract the necessary amount of heat).

If hard temperature control is possible and necessary, its capability has an essential part to play in whether the process is economically viable, because inadequate cooling means the cycle time is extended and can lead to ejection losses of up to 100%.

At this point, a warning should be given that trial injections should not be carried out without monitored mould temperature control. Not only would this have the consequence that later, under production conditions, the dimensions obtained could in certain circumstances be completely different, but also that the cycle times could be detrimentally altered.

8.11.2.5 Temperature control devices

Fluids are usually used for mould temperature control, because both heat extraction and heat supply can be controlled equally well. Electrical heating for soft temperature control is certainly very rapid, but still not fast enough on its own as it would cause excessively high temperature variations. In all cases, electrical heating is used in conjunction with fluid temperature control, the latter being there to provide – if required – sufficiently rapid heat extraction (so as to minimize temperature variations). This combined temperature control system does indeed obtain the best results, but of course costs are significantly higher, and only rarely economically acceptable. So, in what follows, only pure fluid temperature control will be considered.

The cheapest fluid temperature control can be carried out using a water circuit, which of course, is only suitable for hard temperature control. For this purpose, a water distribution system is present in all parts of the IJM, to which the temperature control circuits of the mould can be connected. The cooling intensity is set by the water throughput using valves and, to obtain values with a certain amount of repeatability, throughput gauges and thermometers (in the counterflow) are required in each circuit.

For more demanding hard control requirements and for all cases involving soft temperature control, fluid temperature control units (forming units, heating units and cooling units) are used, which pump a temperature control medium (water or oil) through the temperature control system of the mould. The cycle temperature of the temperature control medium is adjustable and is set to a predetermined value.

Units using water as a temperature control medium generally operate over a temperature range of 20 and 95 °C. At temperatures of over 100 °C, the units must be under constant pressure in order to keep the water liquid, as the heat transfer ratios for steam are considerably poorer. The lower temperature limit is defined by the temperature of the available cooling water. If special cooling units are installed, cycle temperatures down to -10 °C can be reached.

Oil is used as a temperature control medium for cycle temperatures over 95 °C. As a rule, a temperature range of 30 to 250 °C can be covered using these units. They are less suitable for hard temperature control, of course, as the heat absorption capacity of oil is smaller than that of water; thus, for the same amount of circulated fluid, more heat can be transferred using water than oil.

The mould temperature can be controlled using the devices described so far. The temperature control units can be used to regulate the flow temperature, but not the mould temperature. To set the desired mould temperature, it is necessary to measure the actual mould temperature and to alter the flow temperature (or the fluid quantity) until the desired values are obtained. It should be kept in mind here that temperature changes, as a whole, are relatively sluggishly, and therefore it is necessary to wait a sufficient period any change or a state of equilibrium to be reestablished. It is especially important to bear in mind that the flow temperature is not identical to the mould temperature. For hard temperature control (cooling), the flow temperature is always below the mould temperature, while for soft temperature control (heat supply), it is higher than the mould temperature.

As in any kind of control, there are disturbing influences, for example, cycle interruptions, which can affect the process. There is no heat supply from the injection equipment and the open mould has a large heat-radiating surface; hence, when production resumes, the old temperature balance, and with it the original dimensions of the component, will not be reinstated until a few cycles have been completed.

Mould temperature regulation can be used to considerably reduce the effects of these interference factors. A temperature sensor in the mould detects the temperature and relays the actual mould temperature to the temperature control unit. The flow temperature is now no longer kept at a constant value, but is adjusted by means of a regulator until the actual temperature in the mould corresponds to the desired temperature for the mould set on the temperature control unit.

Given the impact of the heating and cooling cycle, more rapid heating and cooling solutions, known as variotherm control can provide significant advantages through the use of dynamic temperature control. The use of faster and more precise control of the heating cycle brings benefits through:

- Reduced cooling times due to faster heat transfer mechanisms, which gives
- Reduced cycle times and therefore can lead to greater productivity.

There are various systems on the market under names such as Rapid Heat Cycle Moulding (RHCM), Electricity Heating Mould (E-mould) and Induction Heating Moulding (IHM). Other solutions bring the cooling solution closer to the specific contours of the surface of the mould using selective heating solutions, an example is the Ball Filled mould system (BFMOLD®).

Numerous quality improvements can be achieved if greater control is gained as a result of understanding the relationships between material and processing mentioned in Chapter 2. This can lead to improvements in both mechanical and aesthetic properties of the subsequent mouldings.

8.11.2.6 Matching temperature control units and mould maintenance

Satisfactory results can only be obtained from mould temperature control if the performance of the temperature control unit is suitable for the quantities of heat required to be exchanged in the mould, i.e., if the temperature control unit not only operates efficiently, but also sufficiently rapidly, only slight temperature variations will occur in the mould.

If the dimensions of the temperature control unit are too large, this would not cause any damage to the moulded part, but it would not be economical. If the dimensions are too small, this will always increase the temperature variations in the mould. The temperature control units must therefore possess a sufficiently high pump delivery (l/min) in association with a sufficiently high transmission pressure level as it must be possible to overcome the flow resistance of the temperature control circuit with a suitable delivery.

Unfortunately, the transmission pressure of a number of temperature control units on the market today is insufficient. A safety valve is frequently built into such units, which opens a parallel circuit within the unit once a limiting pressure has been exceeded; unfortunately this is not always noticeable to the user. Only a fraction of the flow delivery then flows through the mould and the temperature control suffers accordingly. Naturally, the flow resistances of the temperature control circuits should be kept as low as possible, i.e., there should be sufficiently thick hoses, as short as possible, between the temperature control unit and the mould. The temperature control channels in the mould must be of suitable dimensions.

It is also important that the heat transfer ratios of the walls of the temperature control flues do not deteriorate over the course of time. The temperature control channels must be suitably maintained and must be regularly checked for cleanliness, i.e., after each mould change. If water is used for temperature control, then special attention must be paid to rust deposits and also, at higher temperatures, to scale formation.

8.11.2.7 Measuring the mould temperature – checking the uniformity of the mould temperature control

Most information concerning the injection process cycle can be supplied by a temperature measurement point mounted directly in the moulding nest wall (mould wall temperature). During the cycle, this temperature value can be observed during the compression phase and a lower value during ejection. The lower value can give a good guide for the ejection temperature.

A measuring position like this is not suitable for regulating the mould temperature, because of the temperature variations already discussed. The temperature gauge should therefore be sufficiently far away from the mould wall, so that at the measuring point the temperature variations have already been sufficiently dampened. However, it should also be an adequate distance away from the temperature control channels, so as to exclude any interfering effects. A middle position between the mould wall and the temperature control channels is recommended. As the temperature falls as the clamping surface is approached, the measuring point should obviously not be located too close to this. It is in any case advisable to insulate the mould halves from the clamping surface, so as to reduce the interference factors; for high mould temperatures, economic considerations also require this.

If no fixed temperature measuring point is incorporated in the mould, it is possible to use probe thermometers as an aid. To carry out any measurements in the cavity, you will be forced to interrupt the cycle. The negative effects, which thus arise on acquiring the measurement and the production cycle must be considered; however, in no case should the measurement of the mould temperature – at least at one specific comparison point or reference point – be neglected.

A sufficiently uniform spatial temperature distribution can be obtained in the mould if the inlet and outlet temperatures of the temperature control medium do not differ by more than 5 °C. In order to achieve this, a sufficiently high throughput volume of the temperature control medium is required.

For temperature control (regulated flow temperatures), the variations in the cycle temperatures should be less than 5 °C, in order to reach a suitable temporal consistency of mould temperatures.

8.12 Recommended values for holding pressure time and remaining cooling time

The cooling time consists of both the holding pressure time and the residual cooling time. The holding pressure time and cooling time can be calculated as follows:

- For holding pressure with either a single holding pressure time (t_n) or holding pressure times ($t_{n1} + t_{n2}$).
- For injected parts with a medium wall thicknesses of d = 1 mm to d = 4 mm and with mould temperatures less than 60 °C (over 60 °C: 30% extra).

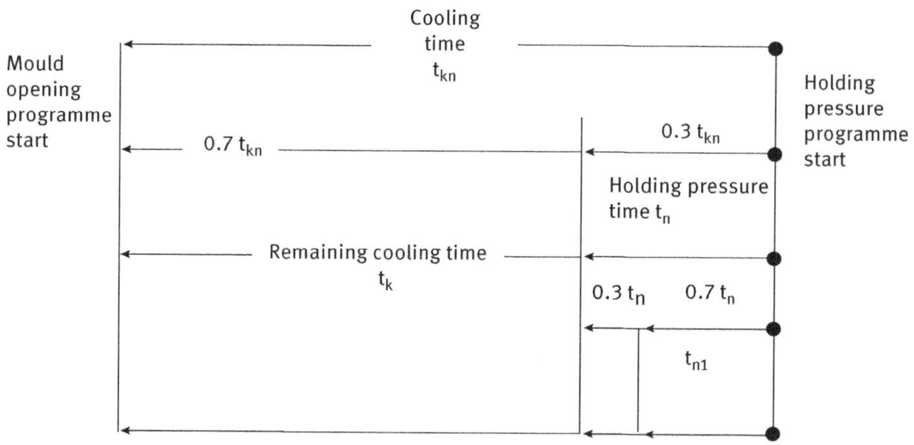

Figure 8.7: Holding time and cooling time.

The following empirical formula for the necessary cooling time (t_{kn}, in seconds) with start-up applies:

$$t_{kn} = d(1 + 2d) \tag{8.1}$$

From this calculation, the next step is to divide the determined cooling time according to Figure 8.7 into the remaining cooling time and holding pressure times. Respective values should be taken from Table 8.10.

The remaining cooling time should be minimised so as the surface temperature of the component is just under the heat distortion temperature of the moulding material, which can be provided by the material supplier.

8.13 Cycle time

The total cycle time consists of:
- Machine and mould movement time.
- Time for mould opening and ejection, mould closing and injection unit movement.
- Injection time.
- Holding pressure time.
- Residual cooling time.
- Plastication time.

The total cycle time obviously has a considerable economic impact on effective production.

Table 8.10: Holding and cooling values.

Mould temperature below 60 °C						Mould temperature above 60 °C					
d (mm)	t_{kn}	t_n	t_{n1}	t_{n2}	t_k	d (mm)	t_{kn}	t_n	t_{n1}	t_{n2}	t_k
1.0	3.0	0.9	0.6	0.3	2.1	1.0	3.9	1.2	0.8	0.4	2.7
1.1	3.6	1.1	0.7	0.4	2.5	1.1	4.6	1.4	1.0	0.4	3.2
1.2	4.1	1.3	0.9	0.4	2.8	1.2	5.3	1.6	1.1	0.5	3.7
1.3	4.7	1.4	1.0	0.4	3.3	1.3	6.1	1.9	1.3	0.6	4.2
1.4	5.4	1.7	1.1	0.6	3.7	1.4	7.0	2.1	1.4	0.7	4.9
1.5	6.0	1.8	1.2	0.6	4.2	1.5	7.8	2.4	1.6	0.8	5.4
1.6	6.8	2.1	1.4	0.7	4.7	1.6.	8.8	2.7	1.8	0.9	6.1
1.7	7.5	2.3	1.6	0.7	5.2	1.7	9.8	3.0	2.1	0.9	6.8
1.8	8.3	2.5	1.7	0.8	5.8	1.8	10.8	3.3	2.3	1.0	7.5
1.9	9.2	2.8	2.0	0.8	6.4	1.9	11.9	3.6	2.5	1.1	8.3
2.0	10.0	3.0	2.1	0.9	7.0	2.0	13.0	3.9	2.7	1.2	9.1
2.1	11.0	3.3	2.3	1.0	7.7	2.1	14.2	4.2	3.0	1.3	10.0
2.2	11.9	3.6	2.5	1.1	8.3	2.2	15.5	4.7	3.3	1.4	10.8
2.3	12.9	3.9	2.7	1.2	9.0	2.3	16.8	5.1	3.5	1.6	11.7
2.4	14.0	4.2	3.0	1.2	9.8	2.4	18.1	5.5	3.8	1.7	12.6
2.5	15.0	4.5	3.1	1.4	10.5	2.5	19.5	5.9	4.1	1.8	13.6
2.6	16.2	4.9	3.4	1.5	11.3	2.6	21.0	6.3	4.4	1.9	14.7
2.7	17.3	5.2	3.6	1.6	12.1	2.7	22.5	6.8	4.7	2.1	15.7
2.8	18.5	5.5	4.0	1.6	13.0	2.8	24.1	7.3	5.1	2.2	16.8
2.9	19.8	6.0	4.2	1.8	13.8	2.9	25.7	7.7	5.4	2.4	18.0
3.0	21.0	6.3	4.4	1.9	14.7	3.0	27.3	8.2	5.7	2.5	19.1
3.1	22.4	6.8	4.7	2.1	15.6	3.1	29.1	8.8	6.1	2.7	20.3
3.2	23.7	7.2	5.0	2.2	16.5	3.2	30.8	9.3	6.5	2.8	21.5
3.3	25.1	7.6	5.3	2.3	17.5	3.3	32.6	9.8	6.8	3.0	22.8
3.4	26.6	8.0	5.6	2.4	18.6	3.4	34.5	10.4	7.2	3.2	24.1
3.5	28.0	8.4	5.8	2.6	19.6	3.5	36.4	11.0	7.7	3.3	25.4
3.6	29.6	8.9	6.2	2.7	20.7	3.6	38.4	11.6	8.1	3.5	26.8
3.7	31.1	9.4	6.5	2.9	21.7	3.7	40.4	12.2	8.5	3.7	28.2
3.8	32.7	9.9	6.9	3.0	22.8	3.8	42.5	12.8	9.0	3.8	29.7
3.9	34.4	10.4	7.2	3.2	24.0	3.9	44.7	13.5	9.4	4.1	31.2
4.0	36.0	10.8	7.5	3.3	25.2	4.0	46.8	14.1	9.8	4.3	32.7

8.14 Setting the injection moulding machine

Now you have read this chapter you should be in a position to set-up an IJM. The following 12-step guide should help you on your way:

1. Set the melt and mould temperature as per specific material instructions.
2. Set-up switch-over position (roughly two-thirds of the potential stroke) to prevent damage to the press or mould and set an initial holding pressure at 0 bar

(the idea at this stage is not to completely fill the tool but to produce short shots).
3. Set the screw rotation and back pressure as per specific material instructions.
4. Set the injection pressure and injection speed to maximum.
5. Set the holding time and cooling time.
6. Set mould opening time (usually 2–5 s).
7. Gradually increase the shot volume to almost a full shot. You may use small increments of perhaps 5% at a time.
8. Switch the machine to automatic.
9. Set the ejectors and minimise the mould opening time.
10. Increase the injection volume to around 99% filled.
11. Increase the holding pressure in steps to optimise.
12. Minimise the holding pressure time.

9 Practical applications using injection moulding technology

9.1 Introduction

Like most manufacturers, the first ARBURG injection moulding machine was born of the necessity to perfect their own products. As a consequence, practical requirements were systematically incorporated into the design of the ARBURG ALLROUNDER from the very outset. Injection moulding machines continue to be modelled on the specific needs of the customers and the objective is to offer machines that function optimally under the demanding use and conditions of day-today operation.

In other words, all machines must work accurately and reliably, providing simple and consistent access to complex technical procedures using technology that is easy to operate. Many modern injection moulding machines meet these requirements day in, day out, all over the world.

The operability of a machine, and thus its efficiency, depends not only on a perfected concept but also on the availability of comprehensive services needed to complement an all-encompassing customer support system or supplement turnkey solutions for integrated systems.

These services may include rapid telephone diagnosis, effective support by a customer service organisation and a computer-controlled spare parts service. These facilities may also be backed by leasing and credit arrangements. Training courses are available as well as advice in the areas of mould design and machine planning.

With the benefit of many years of experience, manufacturers such as ARBURG have process engineers able to assist in the design and production of moulds. Their sound knowledge can save time and money when planning to equip and install new machines. This kind of expert system is increasingly necessary to compete in the competitive world of injection moulding.

9.2 Troubleshooting and guidance

Occasionally, problems occur in producing parts of the desired quality. In most cases, the surface quality of thermoplastic injection moulded parts is the main criterion for their quality. Due to the complex interrelationship between the moulded part and the mould, the moulding compound and processing, it is often very hard to recognise the origin of problems and thus take immediate action.

https://doi.org/10.1515/9783110654813-009

The guide that follows aims to help with practical aspects of injection moulding.

Practical hints will be given on how to optimise the moulded part should surface defects occur. This chapter is based on the experience and knowledge of many experts and was written during a three-year team project, which involved comprehensive work by 30 companies.

This troubleshooting guide is designed to help you analyse surface defects in the injection moulded part and to provide hints on avoiding and/or reducing defects. It consists of descriptions, pictures and notes about the different defects, which will help classify the problem, and provides a short explanation of possible physical causes for the defect. Furthermore, important notes on general faults are supplied, as well as notes on boundary conditions which should be considered.

You will also find flow diagrams containing hints on avoiding or reducing defects. Solutions and hints are given, concerning the process, the moulded part, the mould design and the moulding compound. A blank 'data acquisition record' can be found in this chapter to help evaluate the optimisation process and finally, some case studies of real-life problems are included.

9.2.1 Detection and classification of defects

The defects are classified into 17 sections:

1. Sink marks (Section 9.2.3).
2. Streaks (Section 9.2.4).
   - Burnt streaks (Section 9.2.4.1).
   - Moisture streaks (Section 9.2.4.2).
   - Colour streaks (Section 9.2.4.3).
   - Air streaks/air hooks (Section 9.2.5).
   - Glass fibre streaks (Section 9.2.6).
3. Gloss/gloss differences (Section 9.2.7).
4. Weld line (Section 9.2.8).
5. Jetting (Section 9.2.8).
6. Diesel effect (burns) (Section 9.2.10).
7. Record grooves effect (Section 9.2.11).
8. Stress whitening/stress cracks (Section 9.2.12).
9. Incompletely filled parts (Section 9.2.13).
10. Oversprayed parts (flashes) (Section 9.2.14).
11. Visible ejector marks (Section 9.2.15).
12. Deformation during demoulding (Section 9.2.16).

13. Flaking of the surface layer (Section 9.2.17).
14. Cold slugs/cold flow lines (Section 9.2.18).
15. Entrapped air (blister formation) (Section 9.2.19).
16. Dark spots (Section 9.2.20).
17. Dull spots near the sprue (Section 9.2.21).

Physical Cause

In order to eliminate surface defects as quickly and as economically as possible, knowledge regarding the potential causes of the defects is essential and will allow the processor to decide which solution is most appropriate or what should be changed. Furthermore, knowledge regarding the physical causes of defects helps understand the flow diagrams. Additionally, this chapter details some important notes on economical and quick optimisation of the process and avoiding defects. It should be noted that this guide only contains short explanations of the causes without any detailed description of the physical background.

9.2.2 Flow diagrams for troubleshooting

Eliminating surface defects can be a hard task, due to the different physical causes. In order to help the processor, this chapter contains flow diagrams which systematically show how to eliminate the defect. The aim is to reach the desired quality by varying the process parameters.

When applying this troubleshooting guidance, please note that for one defect, the diagram branches into different solutions, according to the questions. Only one parameter should be changed at a time, in order to avoid collective factors and then several cycles should be completed to ensure stable working conditions. In some cases various solutions are possible, but tendencies (+) or (–) are given. Should one parameter variation fail, go through the questions again and apply, if possible, one solution after another.

Of course, these diagrams, although taking practical conditions into account, only offer suggestions and cannot consider all eventualities that may occur during the process. In addition, they can easily be adapted to unique conditions, or completed using your own knowledge. The diagram will also help determine whether the defect can be eliminated by changing the machine settings, or whether the mould or the part has to be changed.

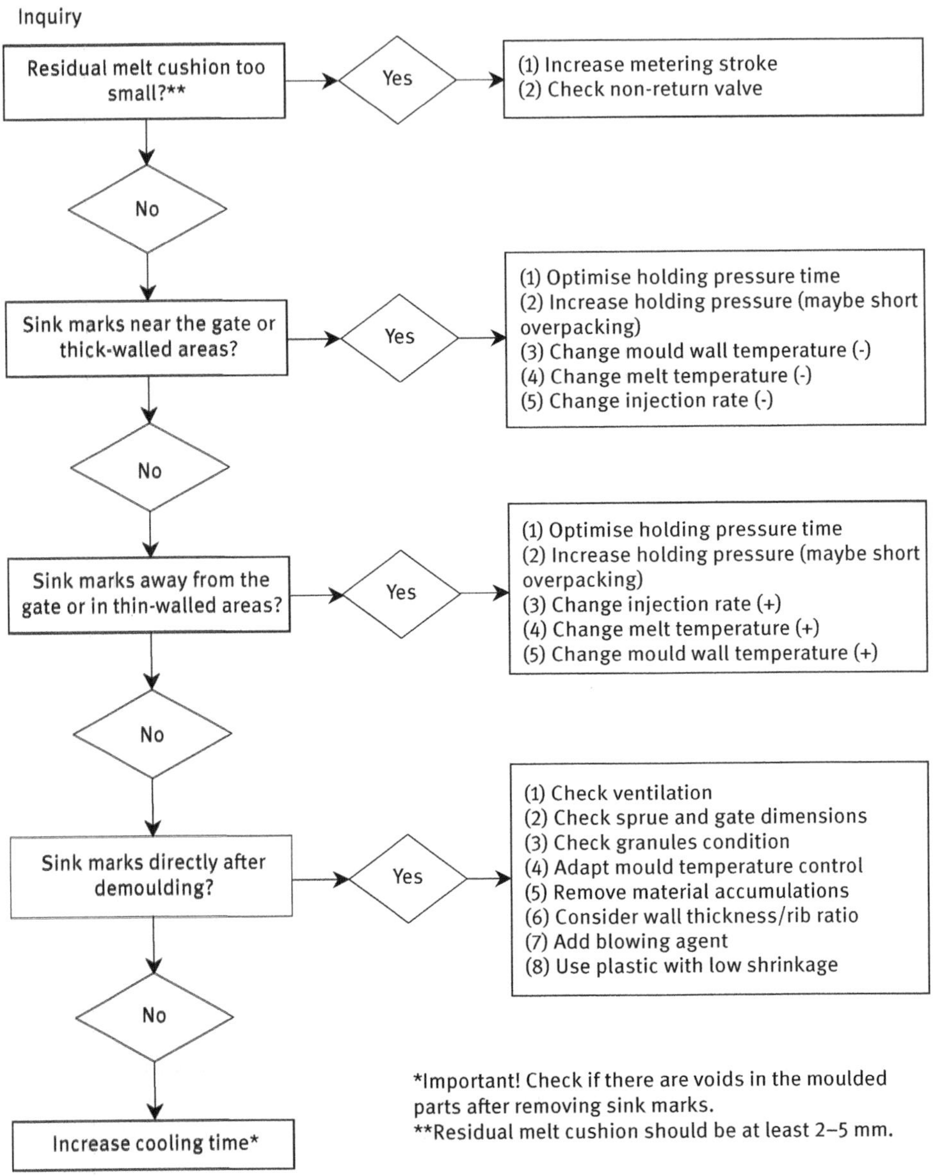

Flow Chart 9.1: Sink marks.

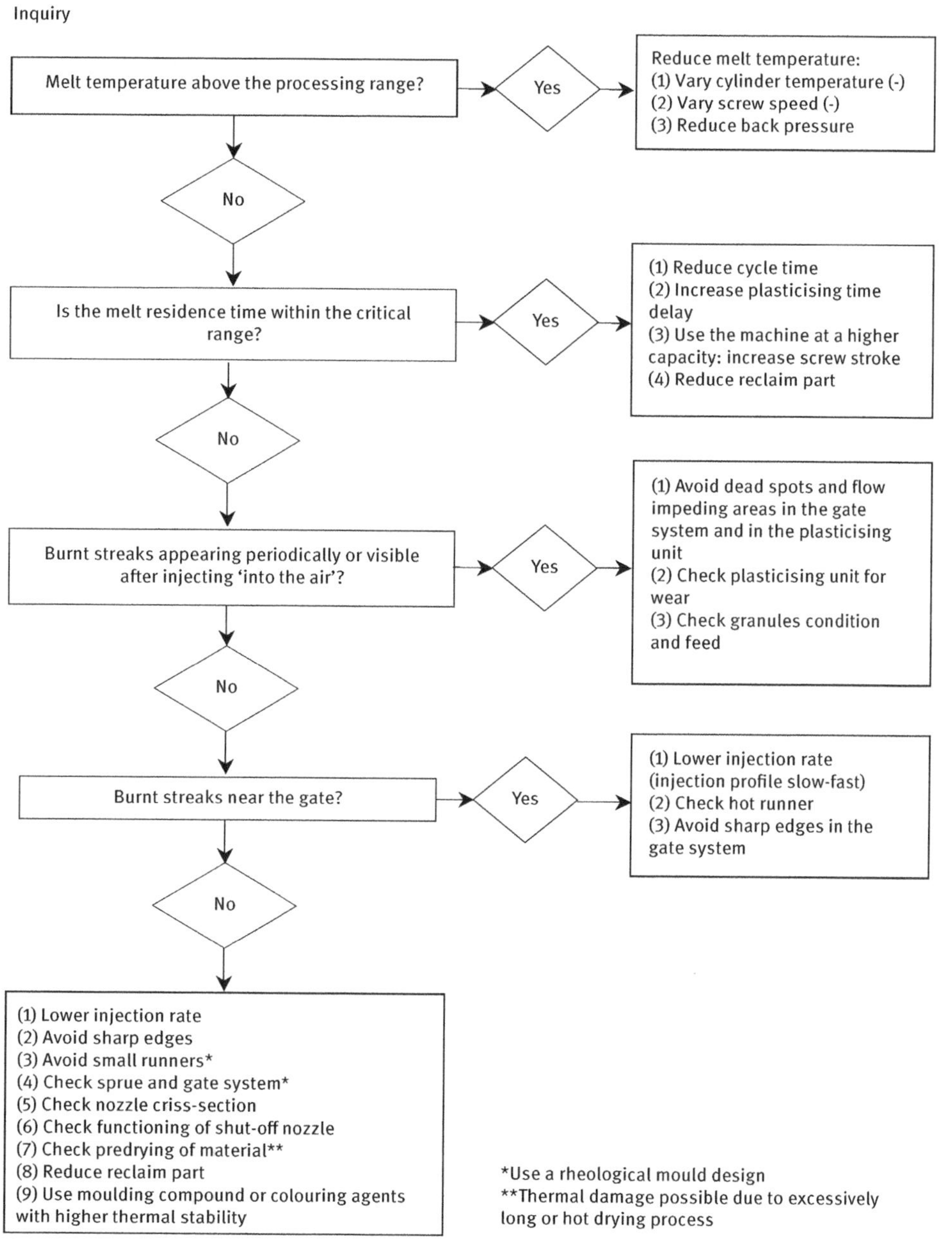

Flow Chart 9.2: Burnt streaks.

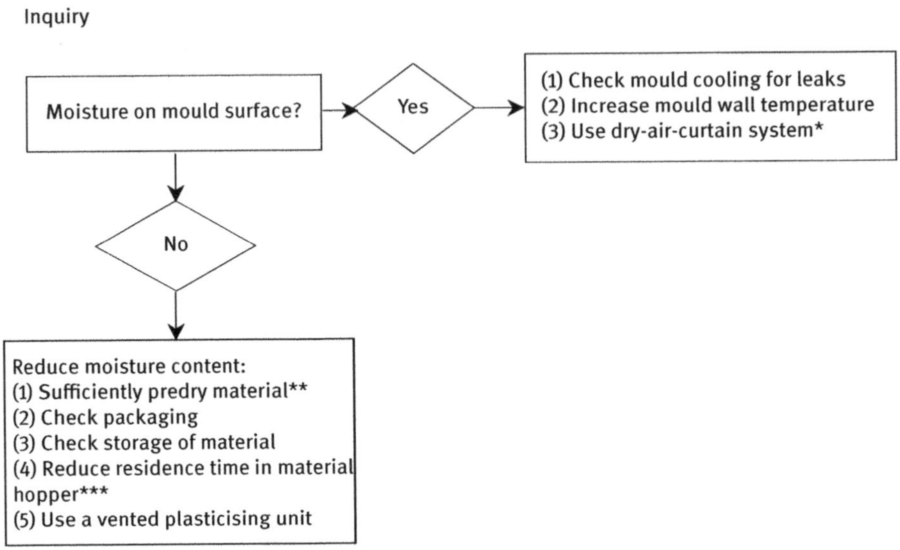

*Application appropriate if:
- Shorter cooling and cycle times can be achieved by lower temperature of the temperature control medium
- Optimum temperature of the temperature control medium cannot be maintained due to condensed moisture

**Pay attention to drying instructions by the raw material producer

***If possible, feed directly from the dryer in a closed system

Flow Chart 9.3: Moisture streaks.

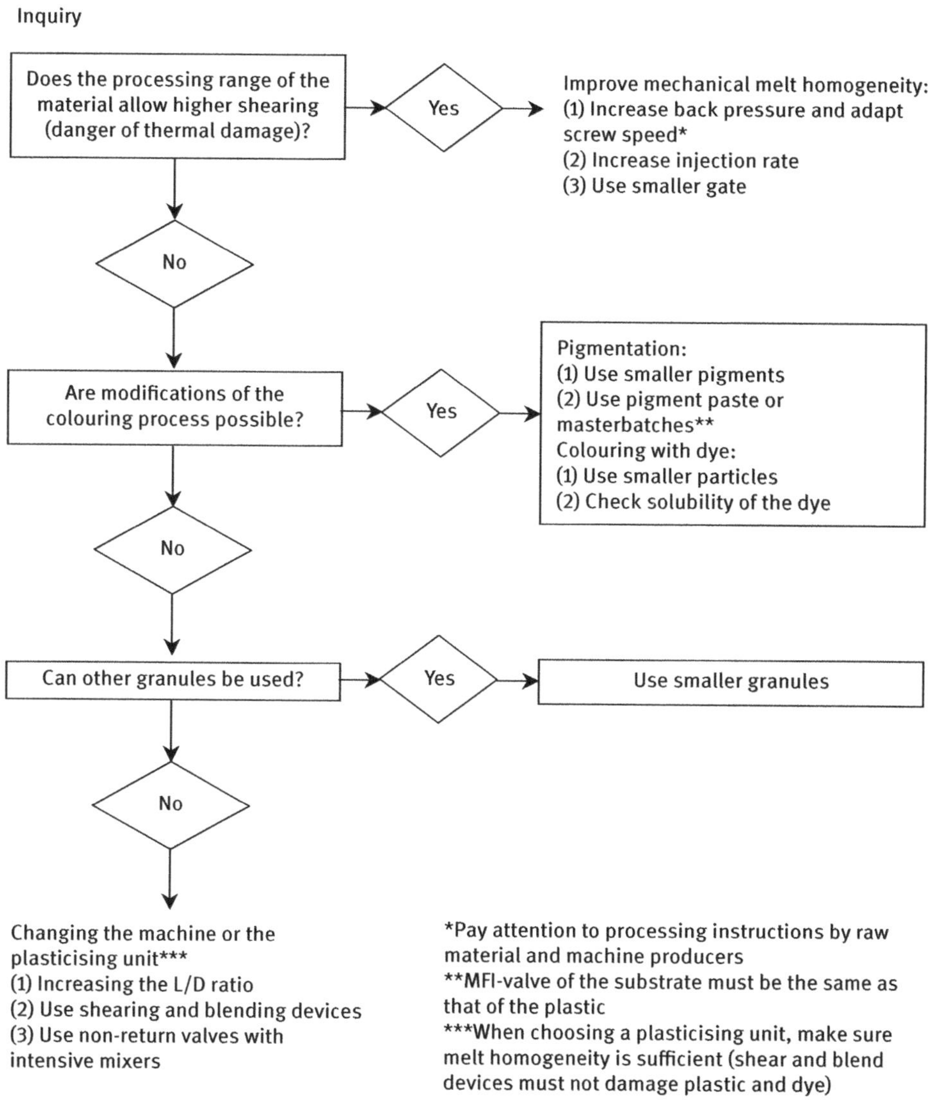

Flow Chart 9.4: Colour streaks.

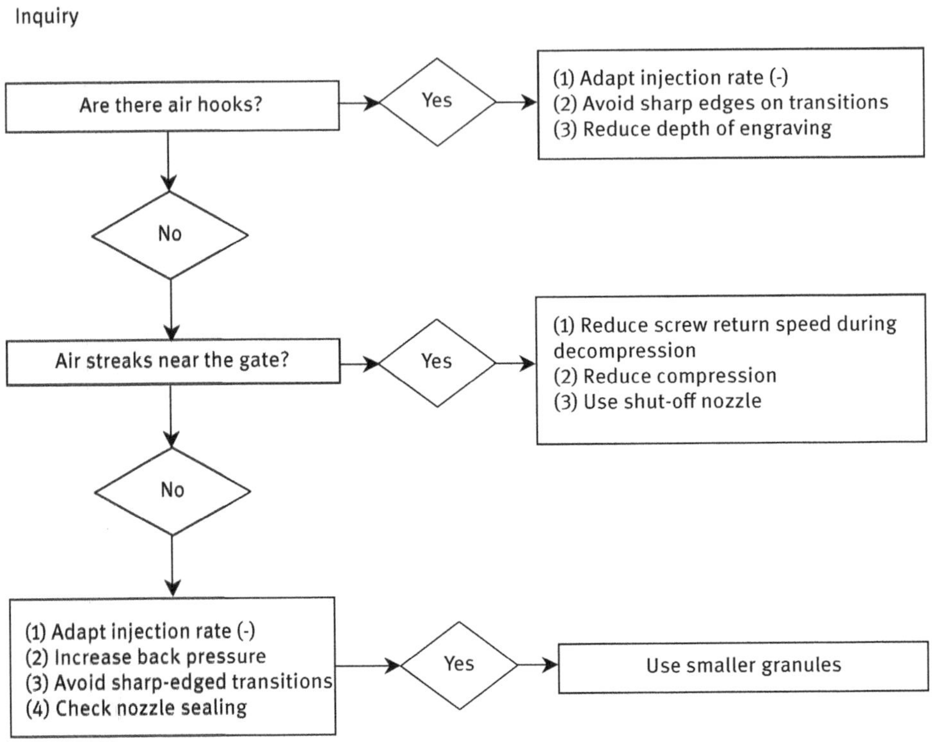

Flow Chart 9.5: Air streaks/air hooks.

Inquiry

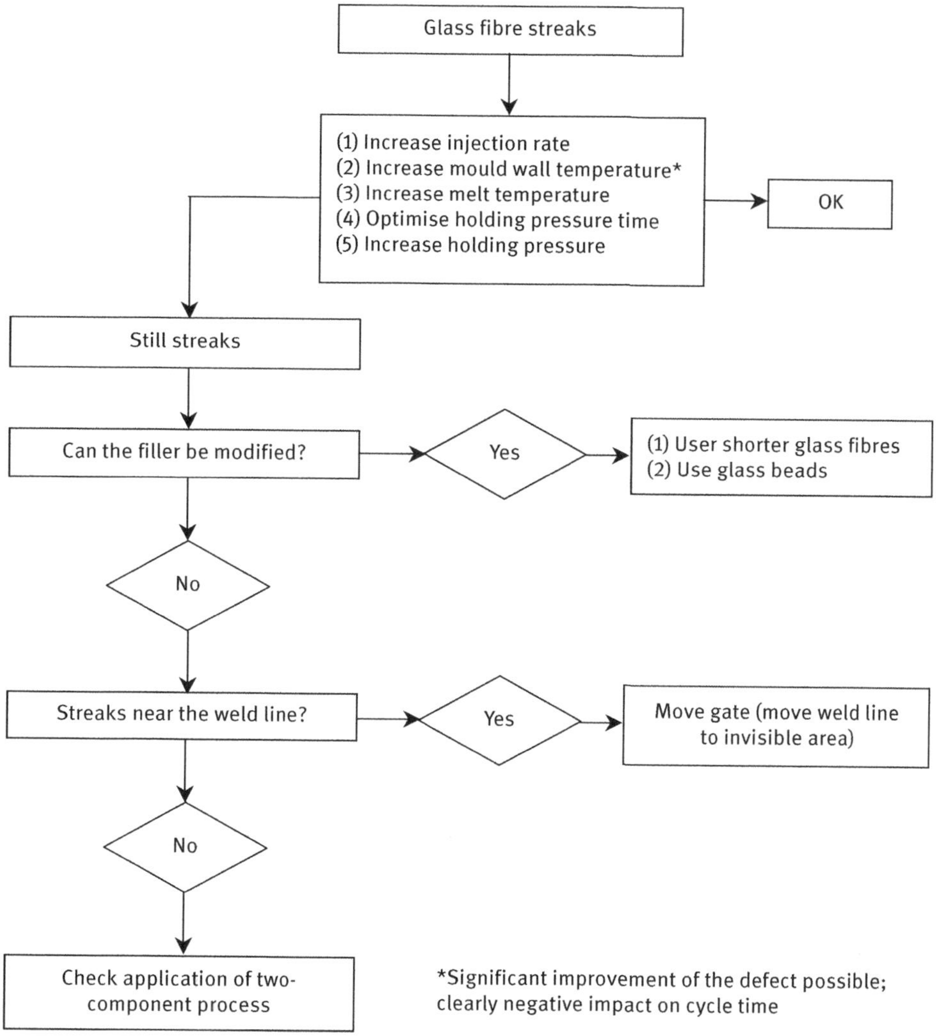

Flow Chart 9.6: Glass fibre streaks.

Inquiry

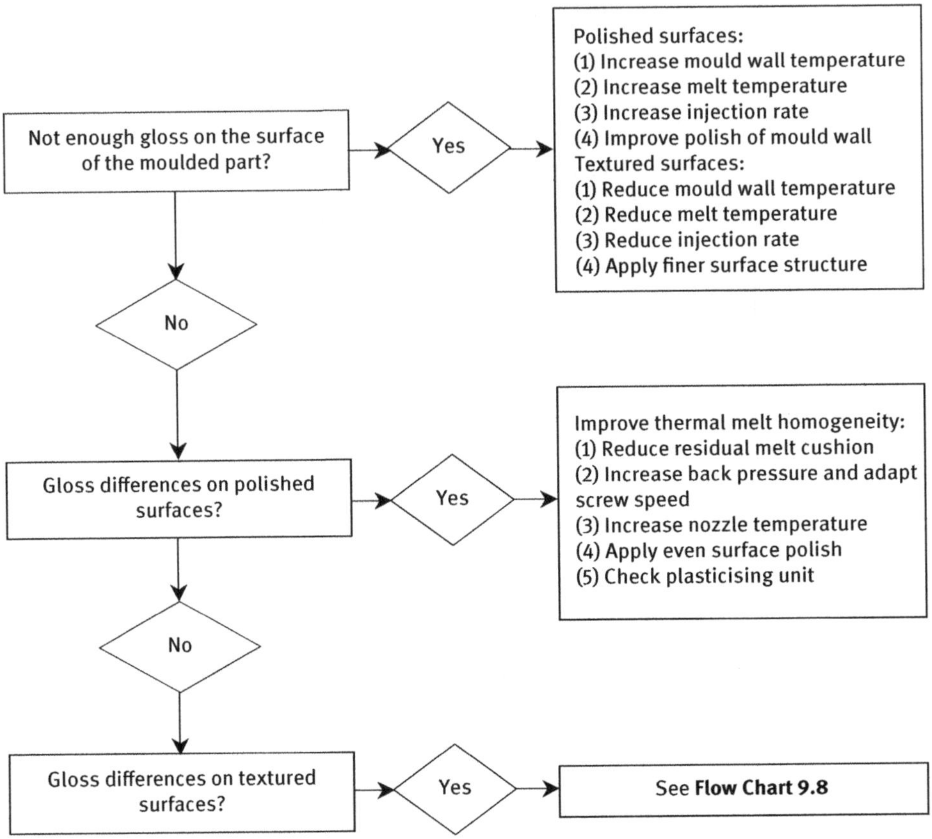

Flow Chart 9.7: Gloss/gloss differences (1).

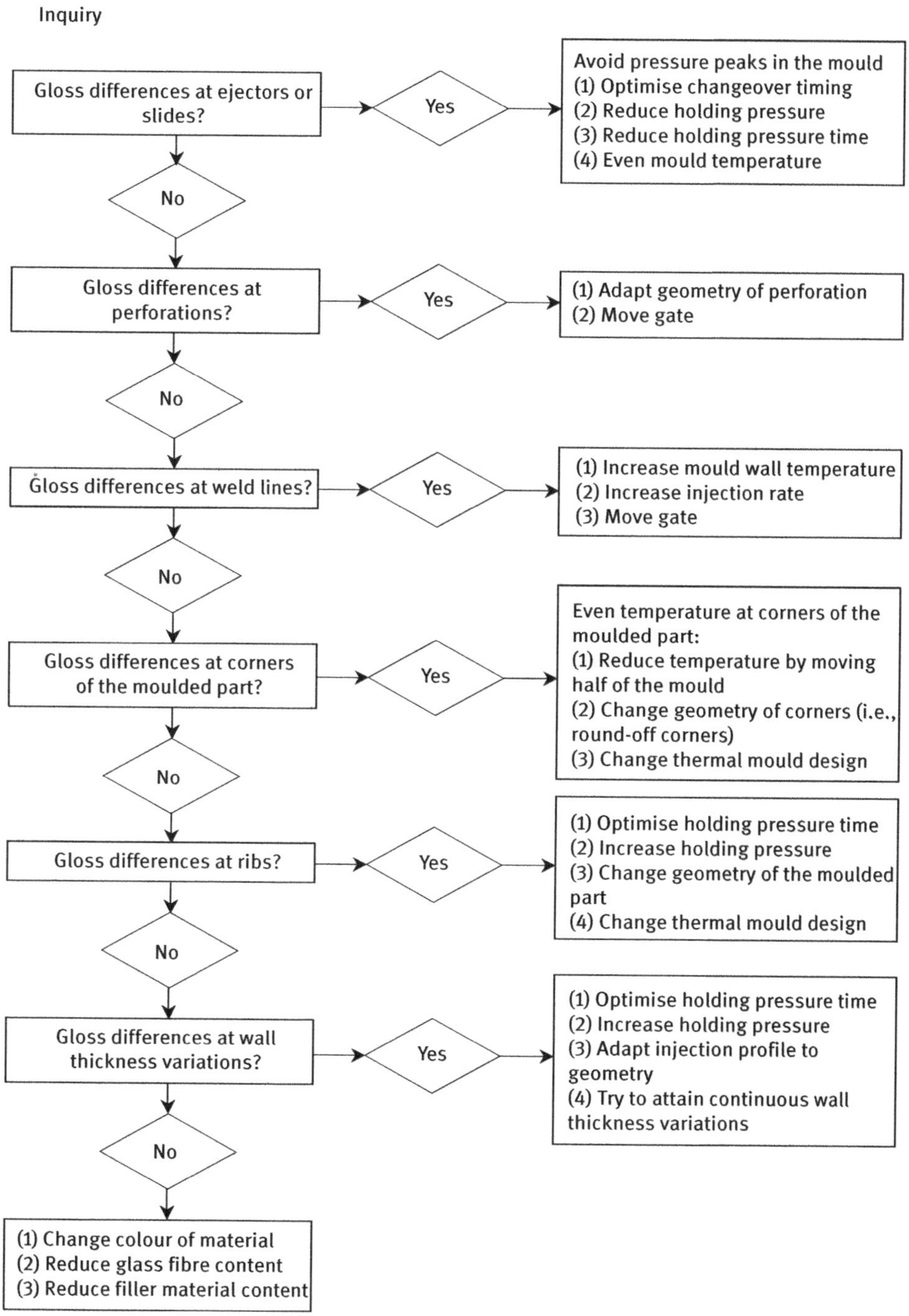

Flow Chart 9.8: Gloss/gloss differences (2).

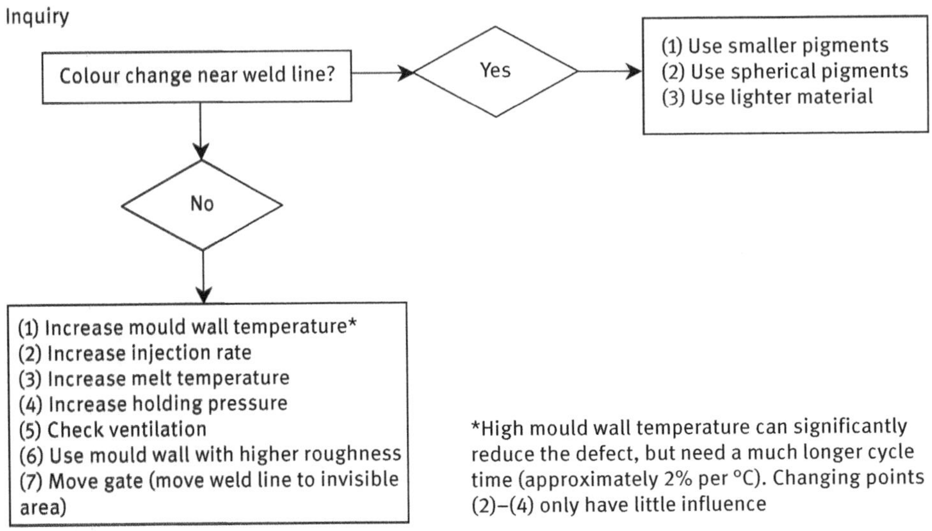

Flow Chart 9.9: Weld line.

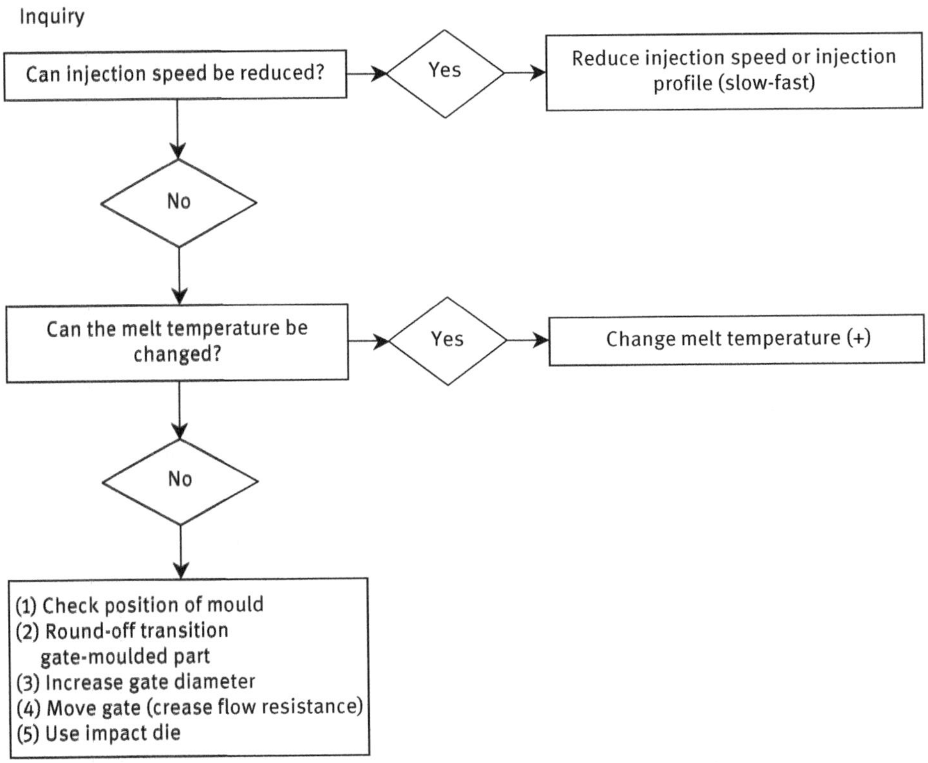

Flow Chart 9.10: Jetting.

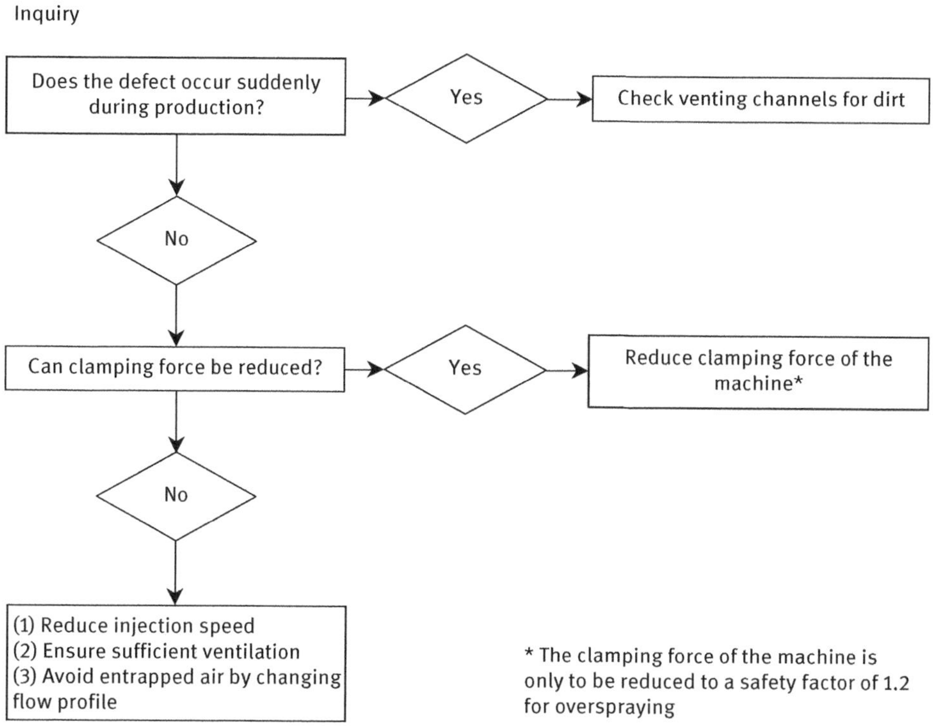

Flow Chart 9.11: Diesel effect.

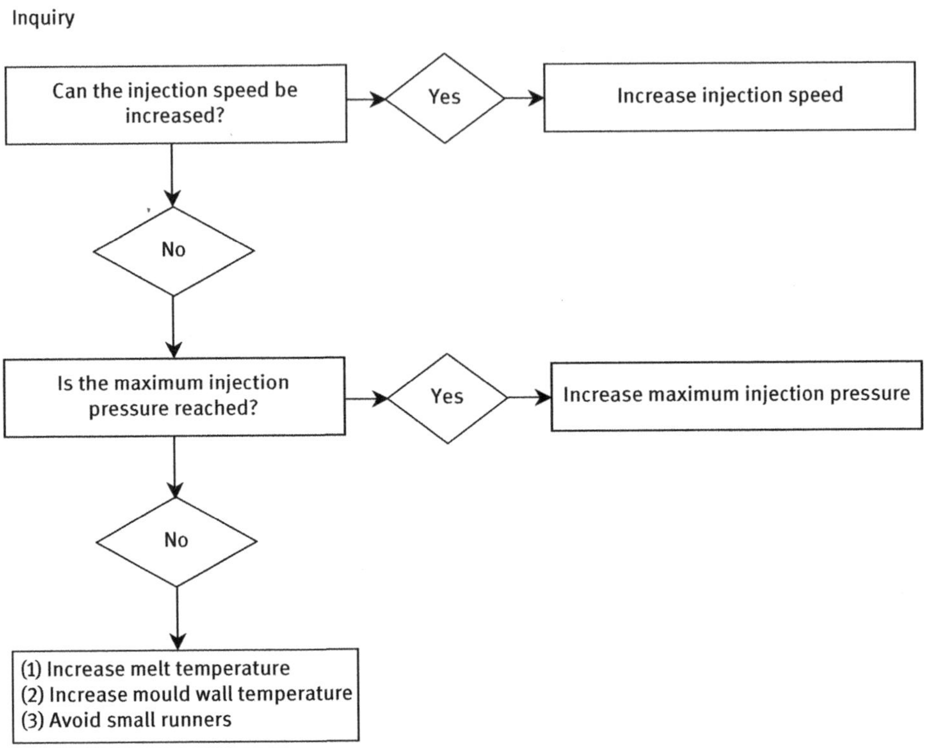

Flow Chart 9.12: Record grooves effect.

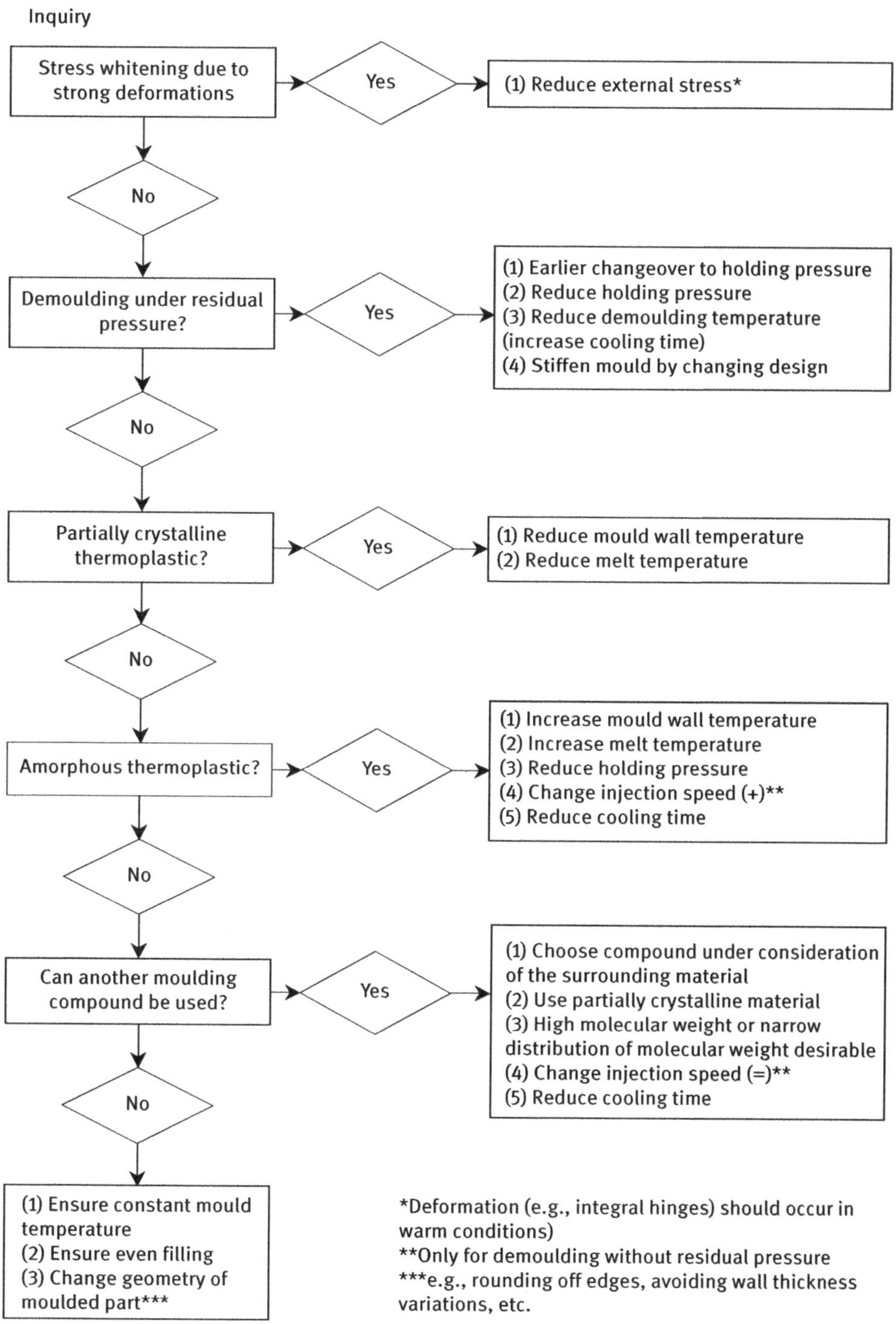

Flow Chart 9.13: Stress whitening/stress cracks.

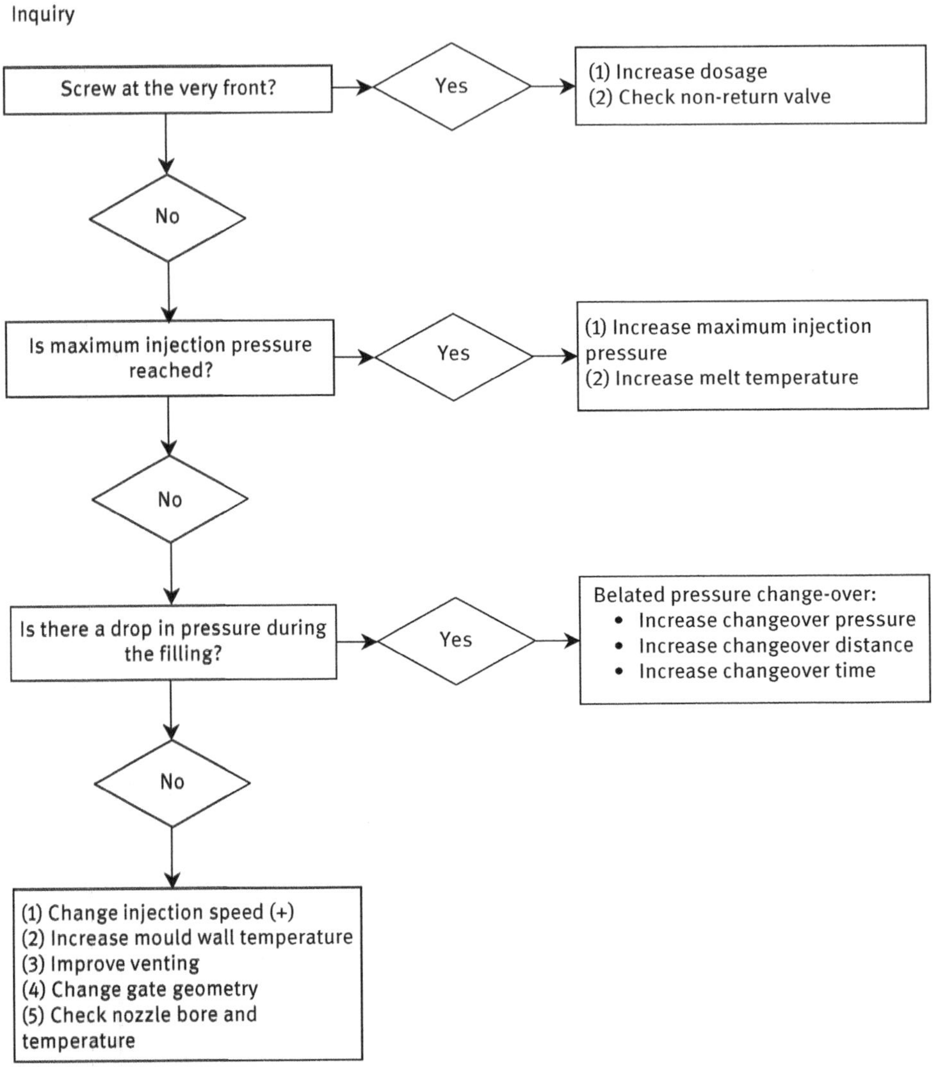

Flow Chart 9.14: Incompletely filled parts.

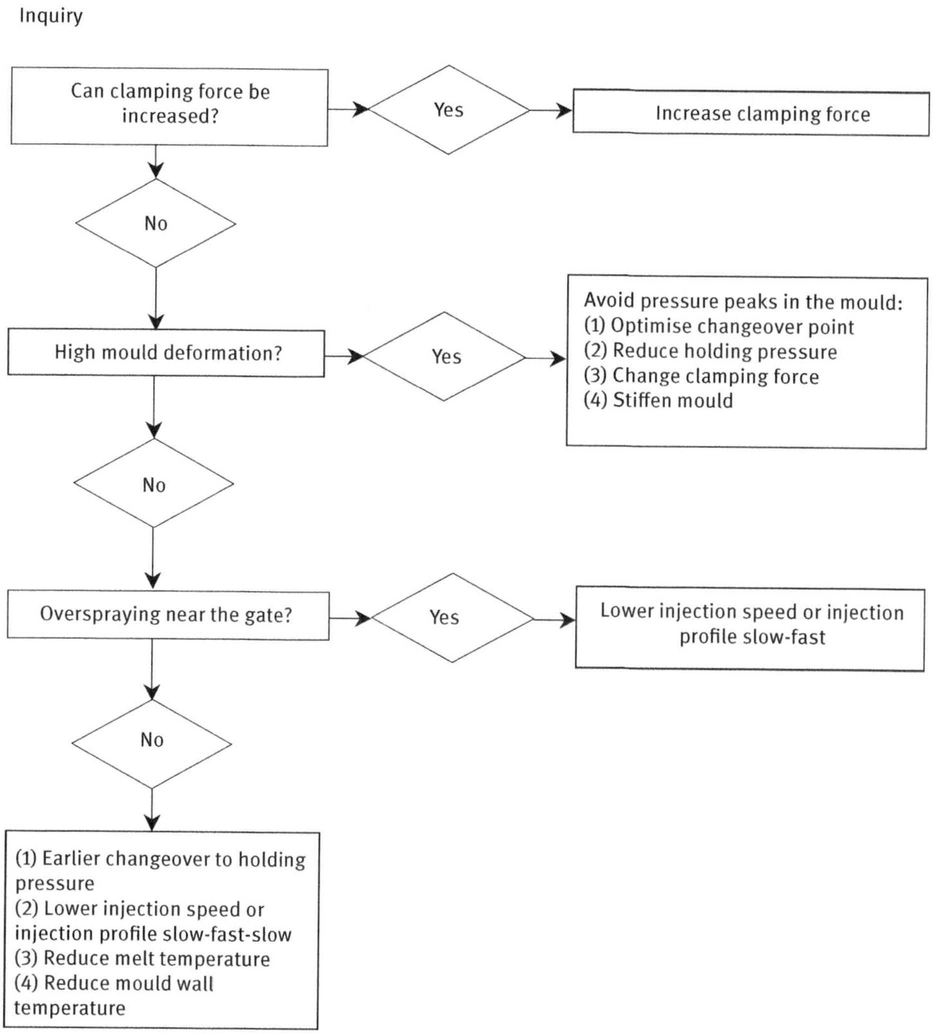

Flow Chart 9.15: Overspraying (flashes).

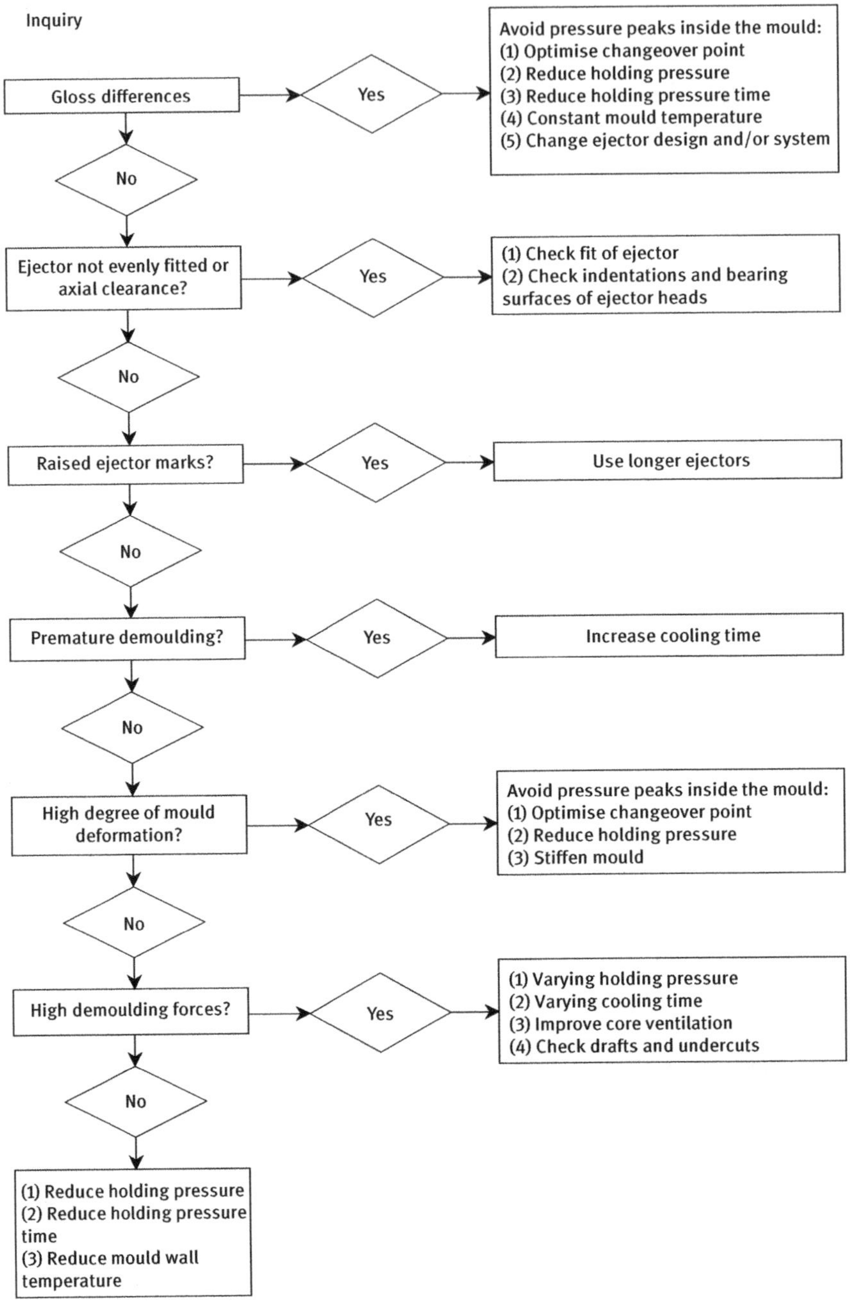

Flow Chart 9.16: Visible ejector marks.

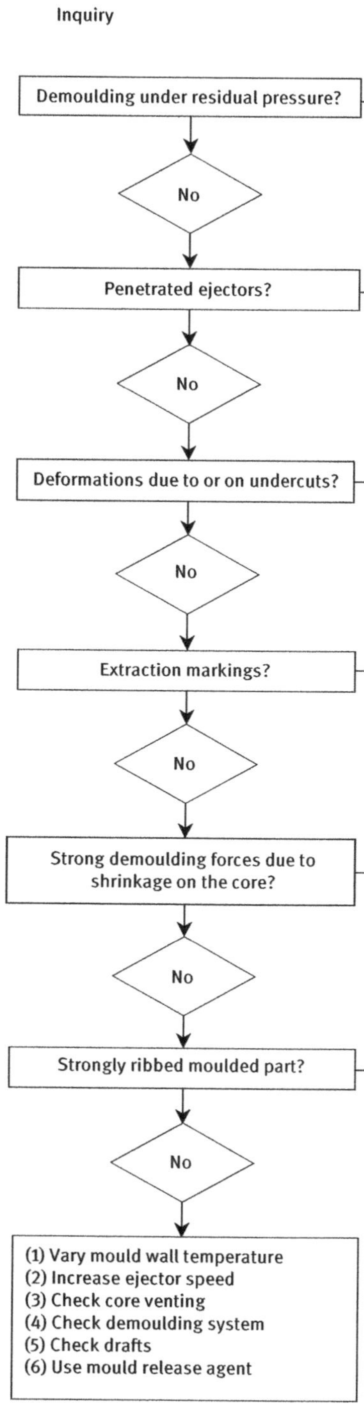

Flow Chart 9.17: Deformation during demoulding.

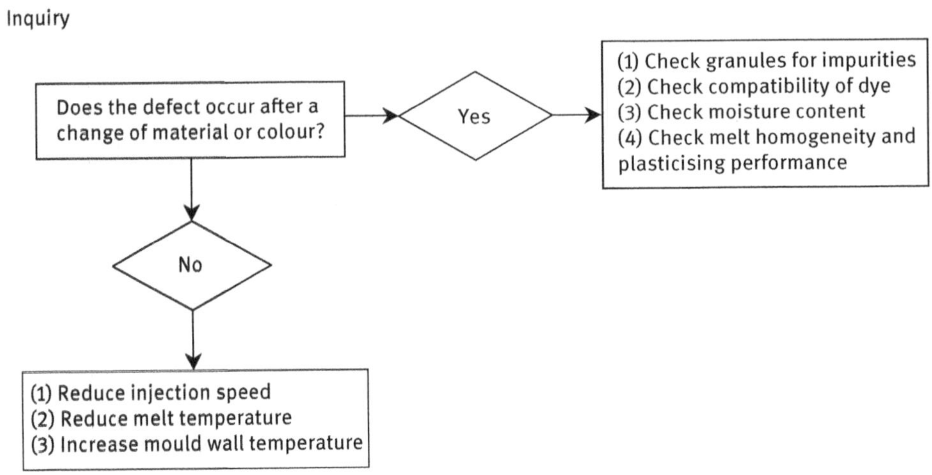

Flow Chart 9.18: Flaking of the surface layer.

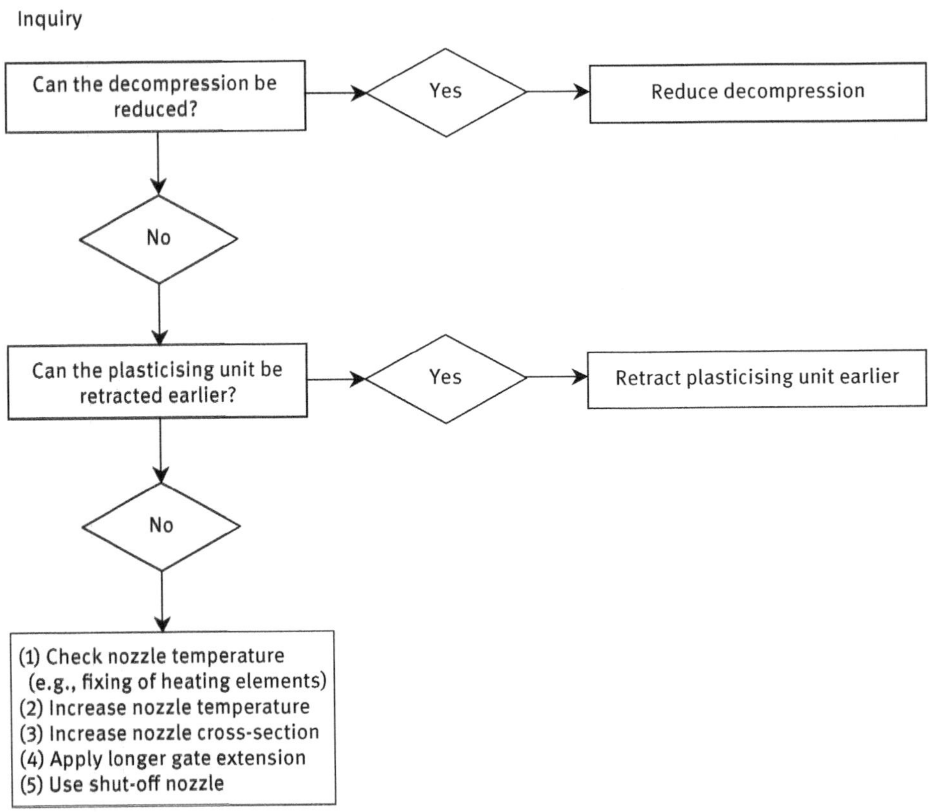

Flow Chart 9.19: Cold slugs/cold flow lines.

Inquiry

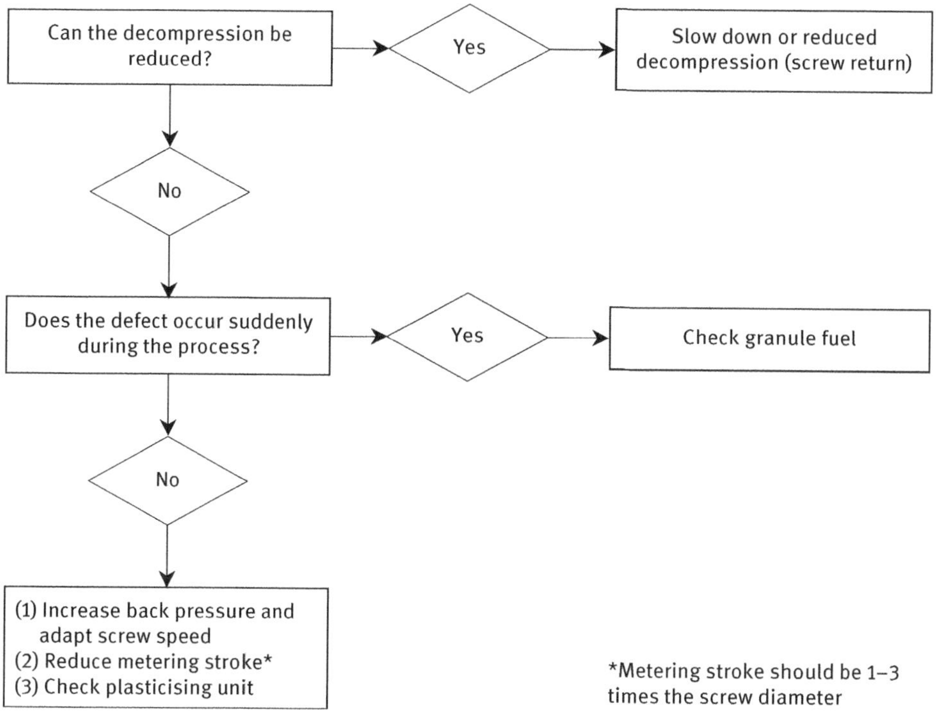

Flow Chart 9.20: Entrapped air.

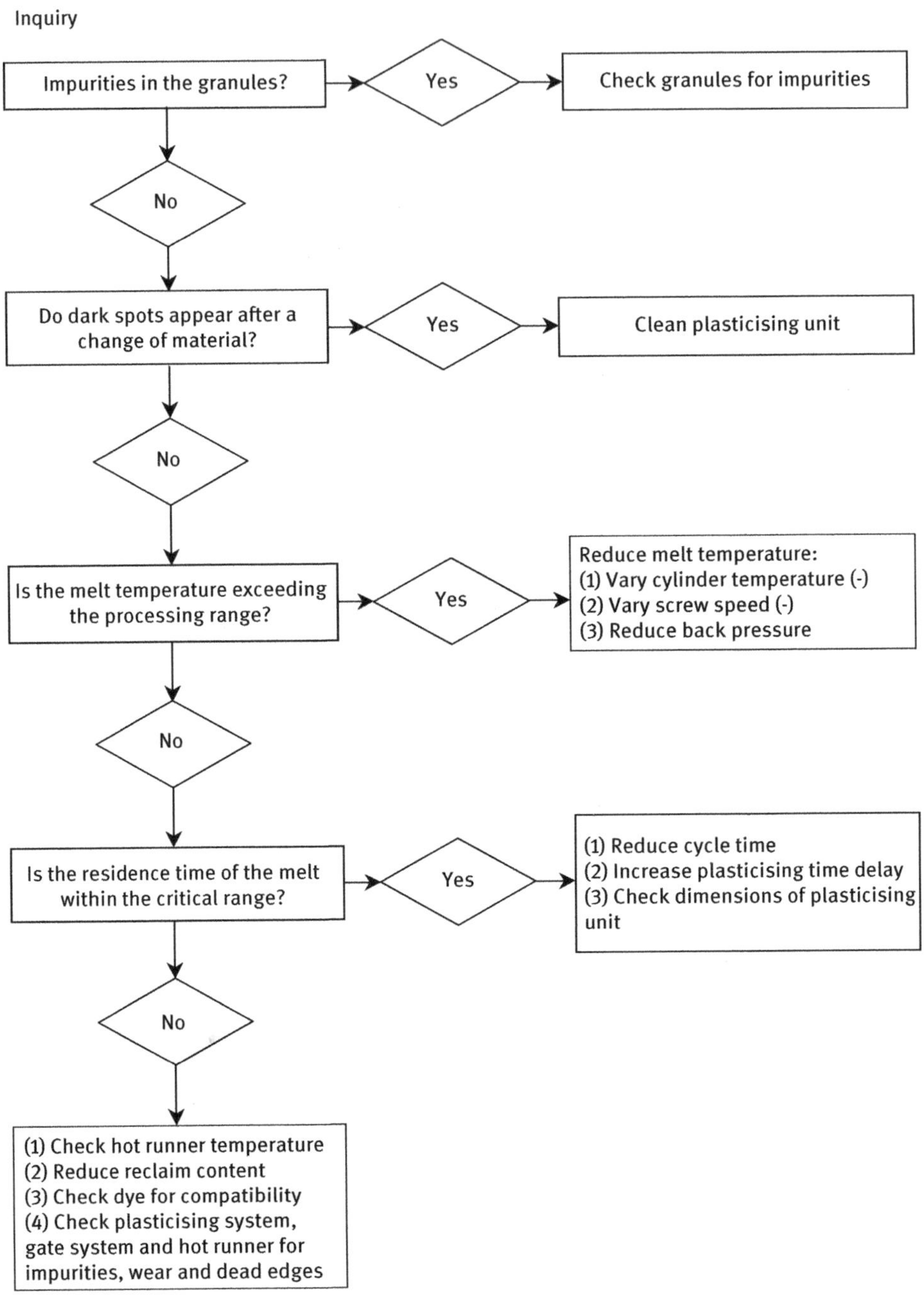

Flow Chart 9.21: Dark spots.

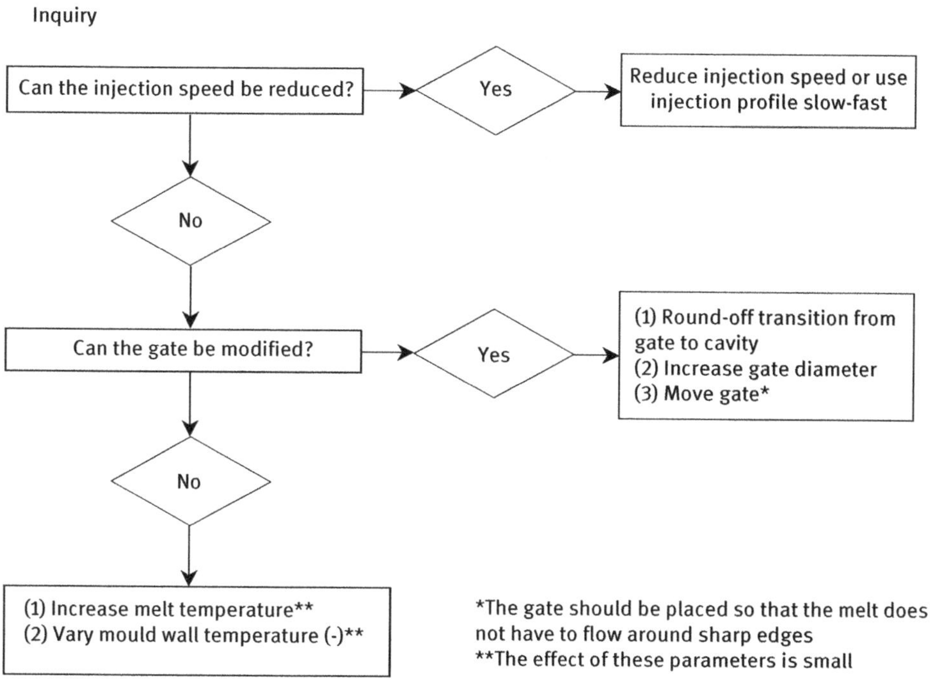

Flow Chart 9.22: Dull spots near the sprue.

9.2.3 Sink marks

Physical cause
Sink marks occur during the cooling process, if the thermal contraction (shrinkage) of the plastic cannot be compensated in certain areas. If the outside walls of the moulded part are not stable enough, due to insufficient cooling, the outer layer is drawn inside by cooling stresses as shown in Figure 9.1.

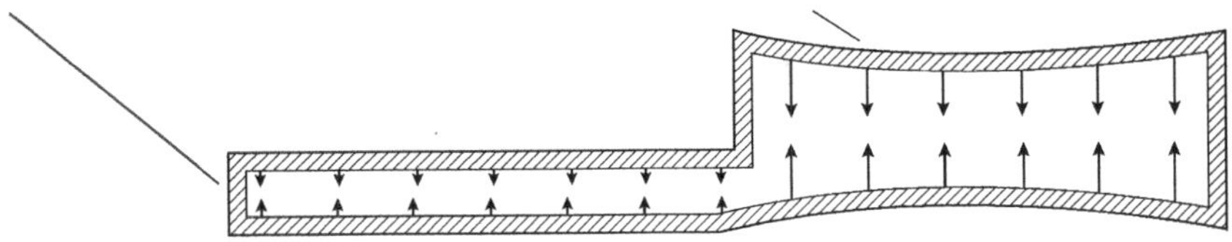

Figure 9.1: Moulded part with sink marks (gating at the thin wall).

There are three fundamental factors:
- Solidification is too slow.
- The effective holding pressure time is too short.
- Not enough holding pressure transfer; because flow resistances in the mould are too high.

N.B.: For optimum holding pressure transfer, the moulded part should be gated to the largest cross-section. In order to avoid premature solidification of the sprue and gate system, sufficient dimensioning is necessary.

Sink marks appear, for example, near material accumulations, as depressions on the surface of the moulded part, if thermal contraction (shrinkage) cannot be compensated as illustrated by Figures 9.2 and 9.3.

Correcting sink marks
Check and/or change the machine settings and change the mould or moulding compound, start a new cycle and go through Flow Chart 9.1.

9.2.4 Streaks

Streaks caused by burning, moisture or air can look very similar making classification difficult, if not impossible. The listed signs do not have to appear, they only give reasons to suspect a certain type of streak.

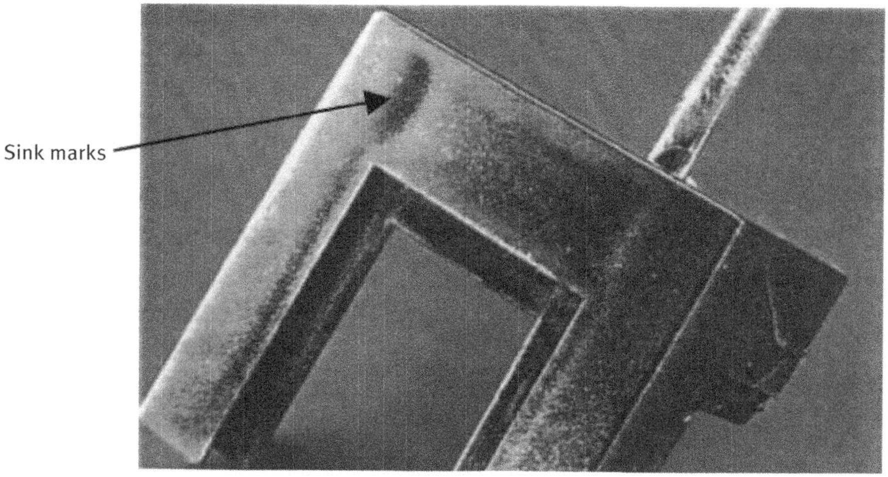

Figure 9.2: Sink marks due to wall thickness variations.

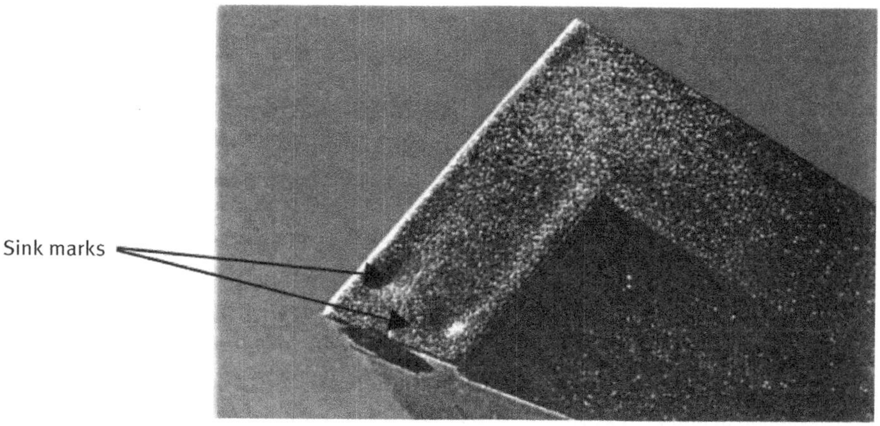

Figure 9.3: Sink marks on the cylindrical core whose temperature was not controlled correctly.

Signs of burnt streaks

The streak appears periodically.

- The streak appears behind narrow cross-sections (shear points) or sharp edges in the mould.
- The melt temperature is near the upper processing limit.
- Lowering the screw advance speed has a positive impact on the defect.
- Lowering the melt temperature has a positive impact on the defect.
- Long residence time in the plasticising unit or the space in front of the screw (due to e.g., cycle breaks or low shot volumes).

- High reclaim content or a part of the material has previously been melted several times before.
- The mould is equipped with a hot-runner.
- The mould is equipped with a shut-off nozzle.

Examples of mouldings with burnt streaks are shown in Figure 9.4.

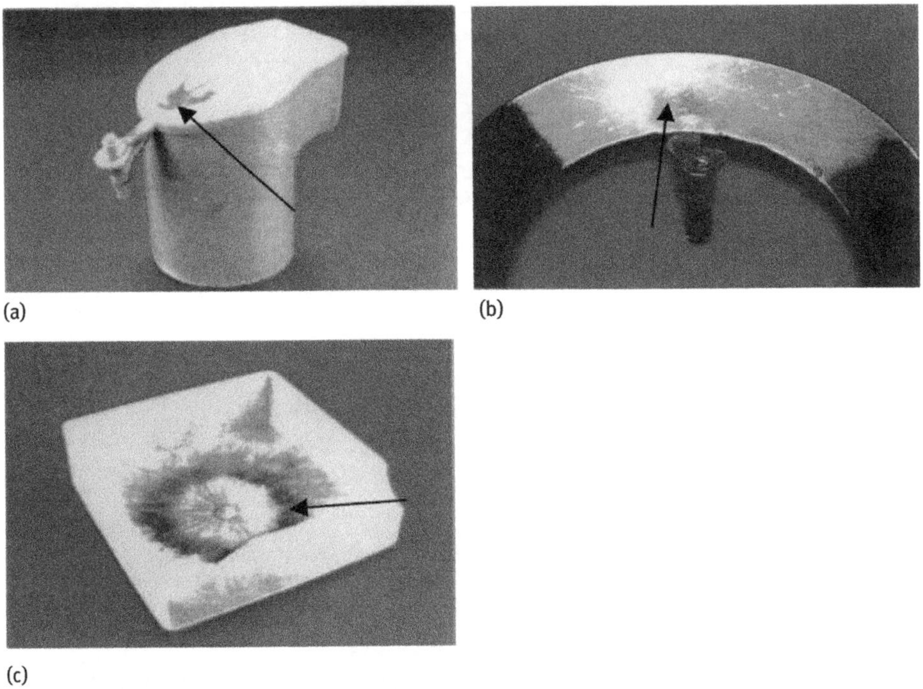

Figure 9.4: Examples of burnt streaks. a) Burnt streaks due to excessive residence time in the plasticising cylinder, b) burnt streaks due to high shearing heat in the gate and c) burnt streaks due to excessive residence time in the plasticising cylinder.

Signs of moisture streaks

- The material tends to absorb moisture (e.g., polyamide, acrylonitrile-butadiene-styrene, cellulose acetate, polybutylene terephthalate, polycarbonate, polymethyl methacrylate, styrene-acrylonitrile).
- When slowly injecting the thermoplastic melt into the air, the melt shows blisters and/or is steaming.
- The solidified flow front of a partial filling shows crater-like structures.
- The moisture content of the material before processing is very high.

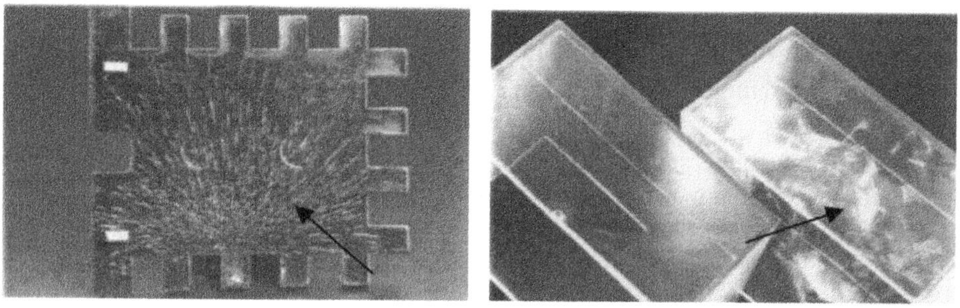

(a) (b)

Figure 9.5: Examples of moisture streaks. a) Streaks due to moist granules and b) streaks due to moisture on the mould surface.

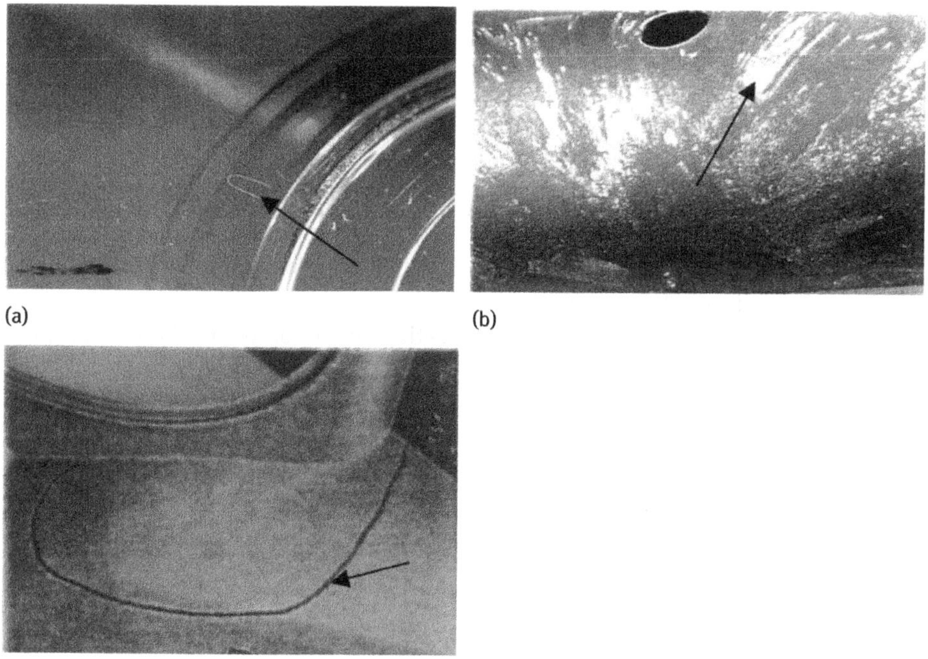

(a) (b) (c)

Figure 9.6: Air streaks/air hooks. a) Air streak behind a wall thickness variation, b) air streak (near the sprue) due to sucked in air during decompression and c) air streak due to entrained and stretched air near rib.

Signs of air streaks

- The moisture content in the environment is very high (especially in combination with cold moulds and cold granules).
- The defect becomes smaller with lower decompression.
- The defect becomes smaller with lower screw advance speed.
- Blisters are visible in the injected material.
- The solidified front flow of a partial filling shows crater-like structures.

9.2.4.1 Burnt streaks (brown or silver)

9.2.4.1.1 Physical cause

Burnt streaks are caused by thermal damage to the melt and can result in a decrease of the length of the molecule chain (silvery discolouration) or a change of the macromolecules (brownish discolouration).

Possible causes for thermal damage:

- Temperature too high or residence time too long during predrying.
- Melt temperature is too high.
- Shearing in the plasticising unit is too high (e.g., screw speed is too high).
- Residence time in the plasticising unit is too long.
- Shearing in the mould is too high (e.g., injection rate is too high).

N.B.: Inject 'into the air' in order to check the melt temperature – measure the temperature using a needle thermometer.

Thermal degradation of plastic has a negative impact on its mechanical properties, even if no damage is visible on the surface.

9.2.4.1.2 Correcting burnt streaks (brown or silver)

Check and/or change the machine settings, change the mould or moulding compound, start a new cycle and go through Flow Chart 9.2 reducing the melt temperature.

9.2.4.2 Moisture streaks

9.2.4.2.1 Physical cause

During storage or processing, moisture is absorbed by the granules, forming water vapour in the melt. Due to the velocity profile at the flow front, gas blisters are pushed to the surface of the melt, as shown in Figure 9.7. As the blisters try to compensate for the pressure, the blisters burst and are deformed by the moving flow front and freeze at the mould wall.

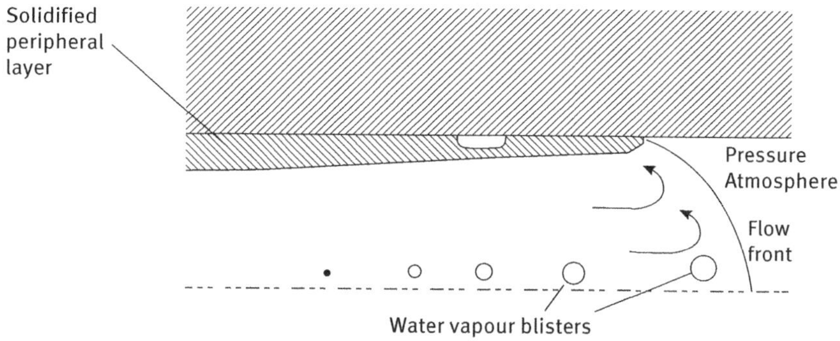

Figure 9.7: Flow of water vapour blisters near the flow front.

9.2.4.2.2 Possible causes for moisture streaks

1. Moisture on the mould surface:
   - Leaky mould temperature control system.
   - Condensation water on the mould walls.
2. Moisture in/on the granules:
   - Insufficient predrying of the material.
   - Incorrect storage of the material.

9.2.4.2.3 Correcting moisture streaks

Check and/or change the machine settings, change the mould or moulding compound, start a new cycle and go through Flow Chart 9.3.

9.2.4.3 Colour streaks

9.2.4.3.1 Physical cause

During the pigmentation stage, pigment agglomerations can lead to differences in the concentration. To some extent this can be mitigated by an increase in shearing, as shown in Figure 9.8, and increases in back pressure can be applied during the plastication stage to increase mixing. This kind of poor distribution can be caused

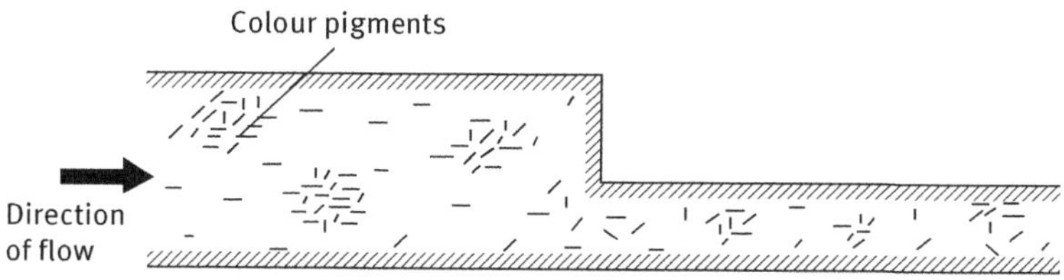

Figure 9.8: Smaller differences in pigment concentration due to higher shearing.

by the plastic, the processing parameters, adhesives and other additives. With the use of inplant colouring dyes, the defect can occur due to incomplete solution of the dye particles in the melt.

Similar to thermoplastics, pigments and dyes are sensitive to excessive processing temperature and residence times. If thermal damage is the reason for colour streaks, they should be considered as burnt streaks.

Extensive stress or warpage can also cause colour differences. The deformed areas break the light in a different manner to other areas.

NB: If using masterbatches for colouring, make sure the substrate is compatible with the plastic to be coloured. The effect of using an incompatible masterbatch is shown in Figure 9.9.

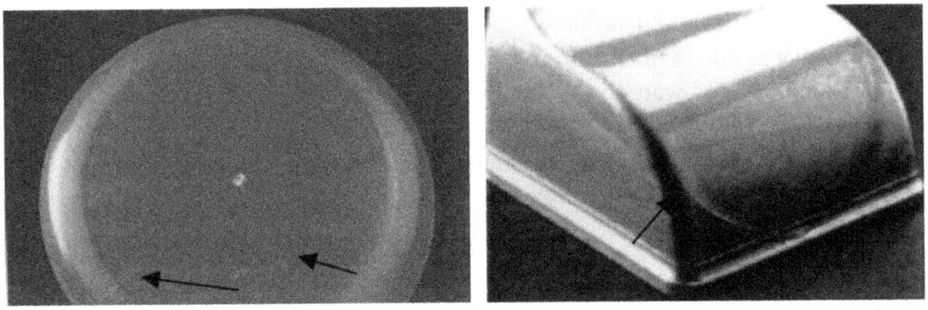

(a) (b)

Figure 9.9: Colour streaks. a) Colour streaks due to an incompatible masterbatch and b) orientation of metal-effect pigments caused by the flow.

9.2.4.3.2 Correcting colour streaks

Check and/or change the machine settings, change the mould or moulding compound, start a new cycle and go through Flow Chart 9.4.

9.2.5 Air streaks/air hooks

Physical cause

Air which cannot escape in time during mould filling is drawn to the surface and stretched in the direction of the flow. Particularly near engravings, ribs, domes and depressions, the air can be rolled over and thus entrapped by the melt, as shown in Figure 9.10. The result is the formation of air streaks or air hooks.

If air is sucked into the area in front of the screw during decompression, air streaks will appear near the gate. Here, the air is transported into the cavity during the injection and then pushed towards the mould wall where it freezes.

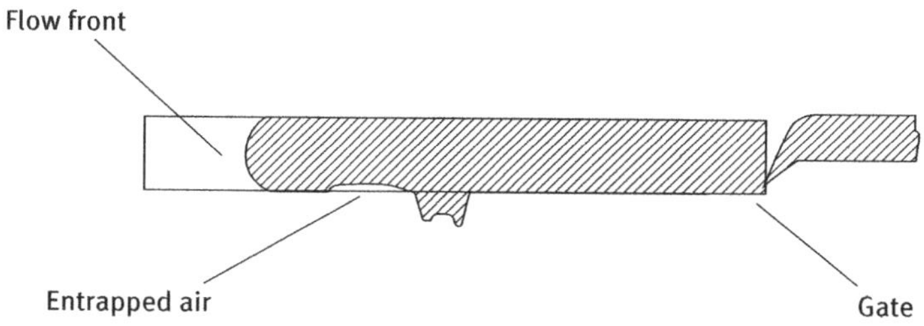

Figure 9.10: Formation of an air streak behind an engraving.

Correcting air streaks/air hooks

Check and/or change the machine settings, change the mould or moulding compound, start a new cycle and go through Flow Chart 9.6.

9.2.6 Glass fibre streaks

Physical cause

Due to their length, glass fibres orientate themselves in the direction of flow during injection. If the melt suddenly freezes when touching the mould wall, the glass fibres may not yet be sufficiently surrounded with melt.

In addition to that, the surface can turn rough because of the large differences in shrinkage (glass fibre:plastic = 1:200). The glass fibres impede the shrinkage of the cooling plastic, especially in the longitudinal direction of the fibre, thus producing an uneven surface, as shown in Figure 9.11.

Figure 9.12 illustrates the effects that fibre orientation can have on moulded parts.

Correcting glass fibre streaks

Check and/or change the machine settings, change the mould or moulding compound, start a new cycle and go through Flow Chart 9.6.

9.2.7 Gloss/gloss differences

Physical cause

The gloss of a moulded part is the appearance of its surface when exposed to light.

If a ray of light hits the surface, its direction will change (refraction of light). While one part of the light will be reflected on the surface, another part will reflect inside the part or penetrate it to different intensities. The impression of gloss is at an optimum, the lower the surface roughness. To achieve this, a mould wall should be used which is as highly polished as possible, a textured mould wall would not

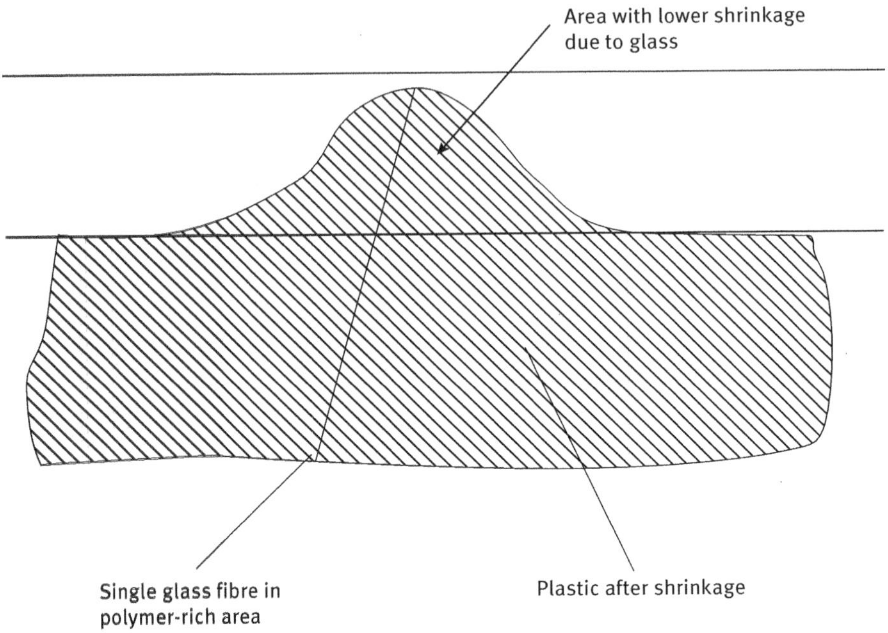

Figure 9.11: Formation of a rough surface due to different shrinkage.

provide high gloss, which is illustrated in Figure 9.13. Gloss differences are caused by different projection behaviours of the plastic at the mould wall, due to different cooling conditions and shrinkage.

Stretching of already cooled areas (e.g., due to warpage) can be another reason for gloss differences. Various examples of gloss-related defects are shown in Figure 9.14.

Correcting gloss/gloss differences

Check and/or change the machine settings, change the mould or moulding compound, start a new cycle and go through Flow Chart 9.7 and/or 9.8.

9.2.8 Weld line (visible notch or colour change)

Physical cause

Weld lines are created when two or more melt flows meet. The rounded flow fronts of the melt streams are flattened and bond when touching each other, which is shown in Figures 9.15 and 9.16. This process requires stretching of the already highly viscous flow fronts. If the temperature and pressure are not high enough,

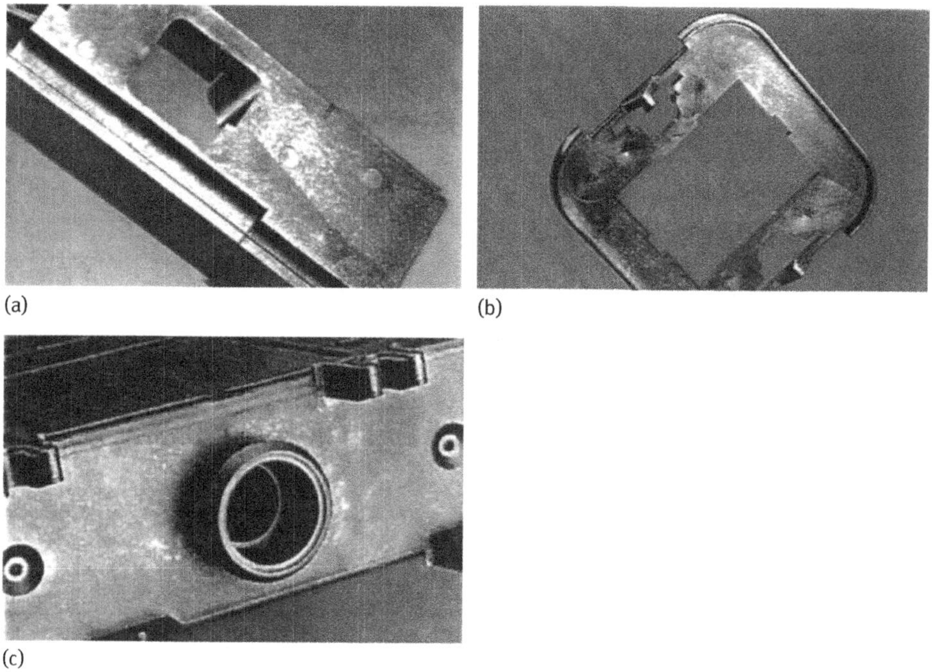

Figure 9.12: Glass fibre streaks. a) Glass fibre streaks clearly visible in the weld line, b) moulded part with rough silvery surface and c) glass fibre streak due to orientation near the sprue.

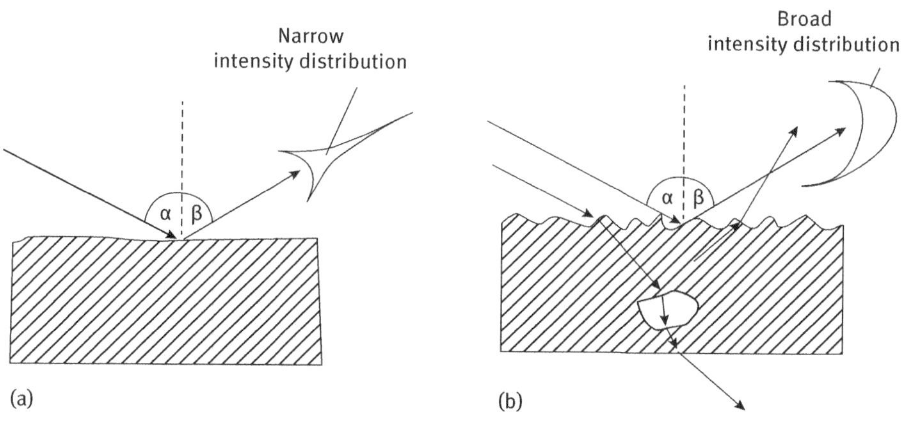

Figure 9.13: Different glosses. (a) Very glossy impression due to reflection on the polished surface and (b) only slightly glossy impression due to reflection on a rough surface and on filler materials.

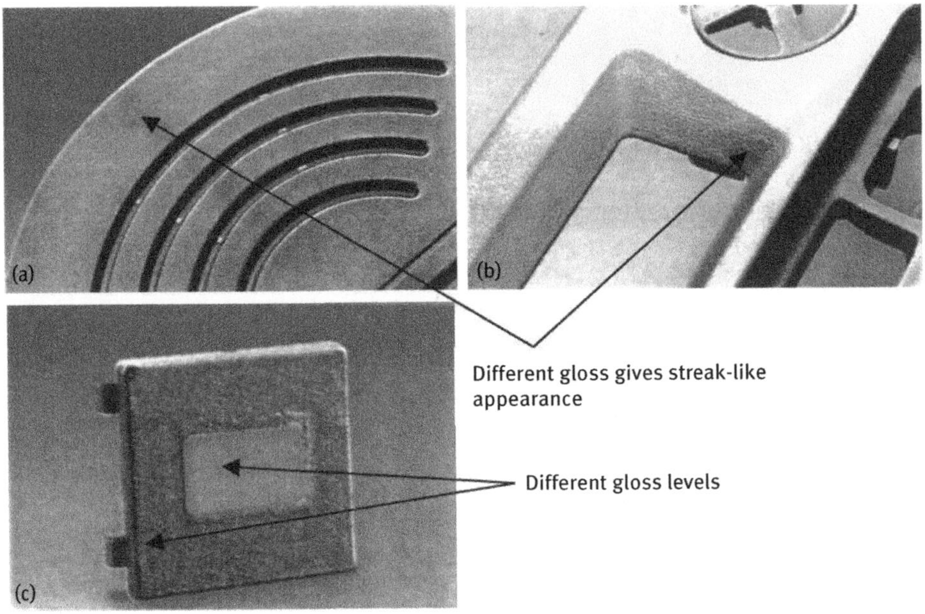

Figure 9.14: Examples of gloss-related defects. a) Gloss differences near weld lines, b) gloss differences due to wall thickness variations and c) gloss differences near ribs.

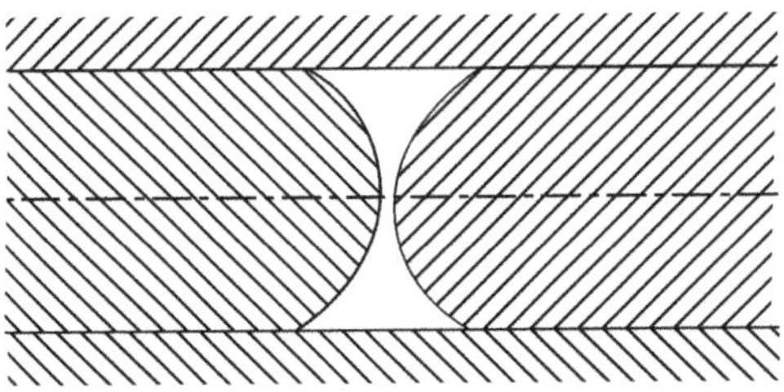

Figure 9.15: Flow fronts before touching each other.

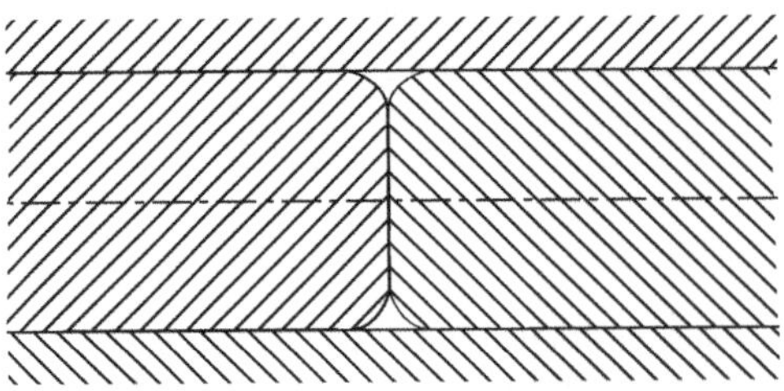

Figure 9.16: Stretching of the rounded flow fronts.

the corners of the flow fronts will not completely develop, creating a notch. Furthermore, the flow fronts no longer melt together homogeneously, possibly producing an optical and mechanical weak spot, as shown in Figure 9.17. If moulding compounds containing additives (e.g., colour pigments) are used, strong orientations of these additives near the weld line are possible, which can also lead to colour changes near the weld line. Notches are particularly visible on dark or transparent parts with smooth, highly polished surfaces. Colour changes are particularly visible on parts with metal-effect pigments.

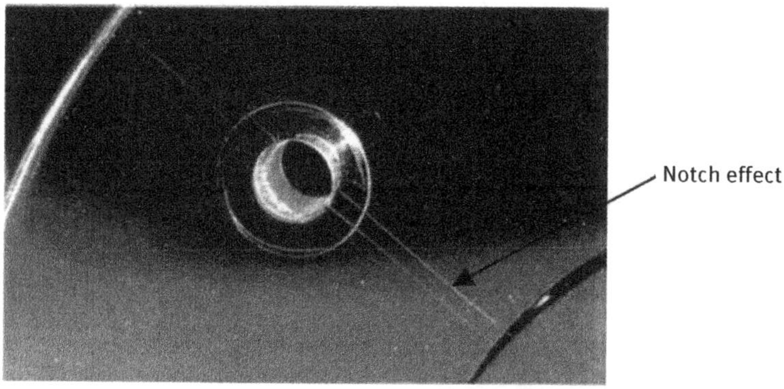

Figure 9.17: Visible notch on the top and bottom side of a transparent part.

N.B.: Significant improvements can only be achieved using high mould wall temperatures. Increasing the mould wall temperature increases the cycle time by approximately 2% per °C.

Improving a weld line (visible notch or colour change)
Check and/or change the machine settings, change the mould or moulding compound, start a new cycle and go through Flow Chart 9.9

9.2.9 Jetting

Physical cause
Jetting is caused by an undeveloped frontal flow of melt in the cavity. A melt strand is developed which, starting at the gate, enters the cavity with uncontrolled movement. During this phase, the melt strand has cooled down to such a degree that it cannot be fused homogeneously with the rest of the moulding compound. This often happens with intermittently increasing cross-sections of the moulding part in conjunction with high injection speeds. Jetting is illustrated in Figures 9.18 and 9.19.

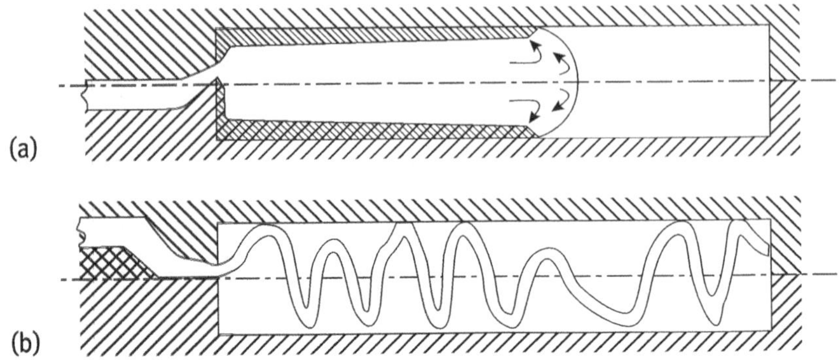

Figure 9.18: Mould filling: a) frontal flow and b) jetting.

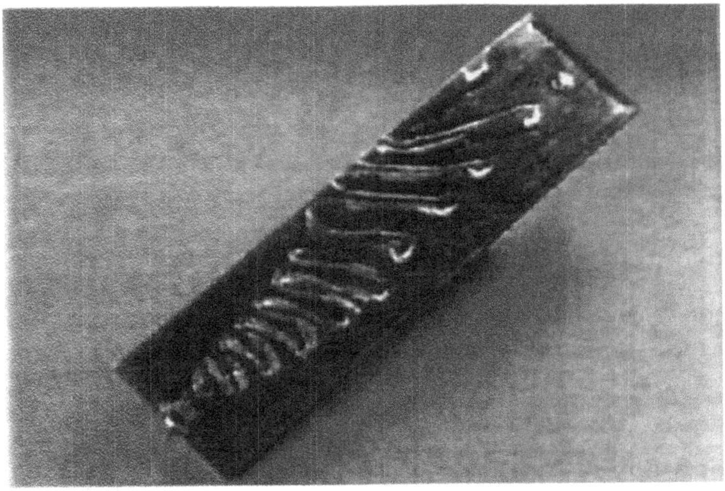

Figure 9.19: Jetting starting at the gate, spreading over the entire part.

Jetting often causes differences in colour and gloss, and in some cases there are similarities to the record grooves effect.

N.B.: Jetting can also be influenced by the position of the mould. In order to avoid defects, the cavity should not be filled from top to bottom.

Correcting jetting

Check and/or change the machine settings, change the mould or moulding compound, start a new cycle and go through Flow Chart 9.10.

9.2.10 Diesel effect (burns)

Physical cause

The diesel effect is purely a venting problem and can occur near blind holes, fillets, the end of flowpaths and near points where several flow fronts fuse. It occurs whenever the air cannot escape or not quickly enough *via* commissures, venting channels or ejector fits. Towards the end of the injection process, the air is compressed and thus heated to a high degree, resulting in very high temperatures which can cause burn marks on the plastic as shown in Figures 9.20 and 9.21.

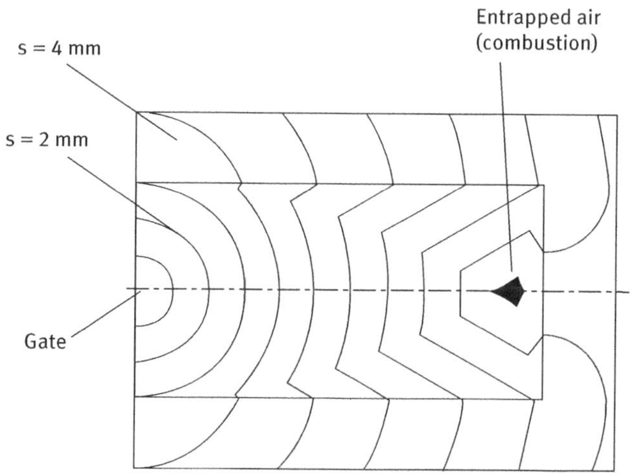

Figure 9.20: Sheet with diesel effect (filling pattern) (s indicates the two different flow fronts and their distance from the sprue).

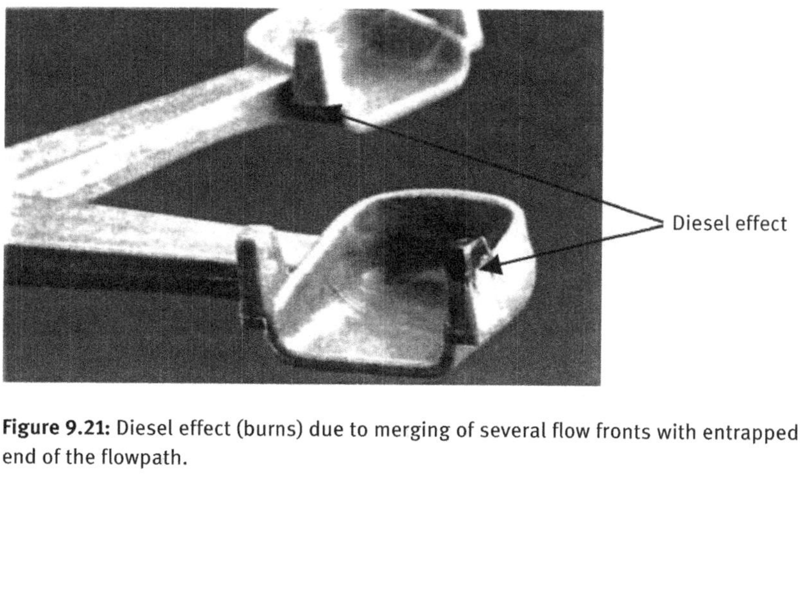

Figure 9.21: Diesel effect (burns) due to merging of several flow fronts with entrapped air at the end of the flowpath.

N.B.: Due to the burning of the plastic, aggressive decomposition products may be created, which often attack the mould surface, leading to destruction.

Correcting the diesel effect (burns)
Check and/or change the machine settings, change the mould or moulding compound, start a new cycle and go through Flow Chart 9.11.

9.2.11 Record grooves effect

This effect results in very fine grooves on the moulded part, which are very similar to those found on records. Concentric rings appear near pin-point gates, while markings are parallel towards the end of the flowpath and/or behind the gate (Figure 9.22).

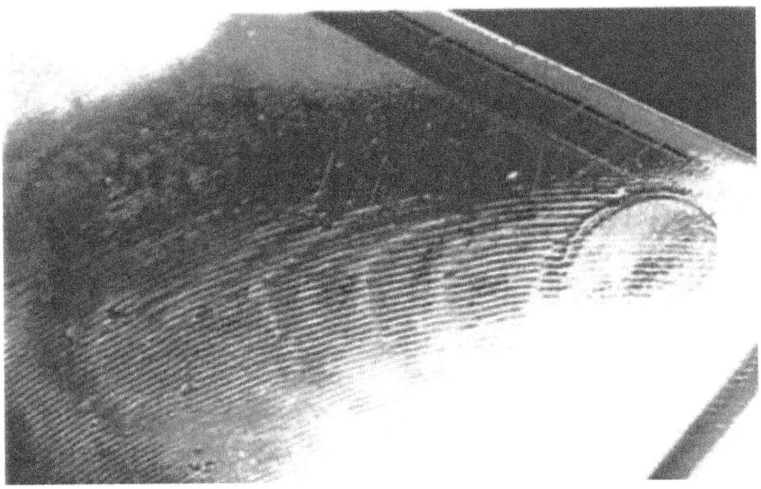

Figure 9.22: Concentric record grooves.

Physical cause
- High cooling velocity.
- Melt temperature is too low.
- Injection speed is too low.
- Mould wall temperature is too low.

When injecting the moulding compound into a cold mould, a solidified peripheral layer is formed behind the flow front due to the high cooling rate. The cooling of the peripheral layer also causes cooling of flow front areas near the mould wall. If this cooling is very high (especially with low injection speeds) these very high viscosity or frozen flow front areas can impede the direct frontal flow of the melt to the wall. Thus, the following hot melt will not be pushed towards thewall as expected, but will cause an elongation of the flow front in the middle. At a certain pressure the flow front will again touch the wall. The cooled down peripheral areas of the flow front have no contact with the wall (Figures 9.23–9.25).

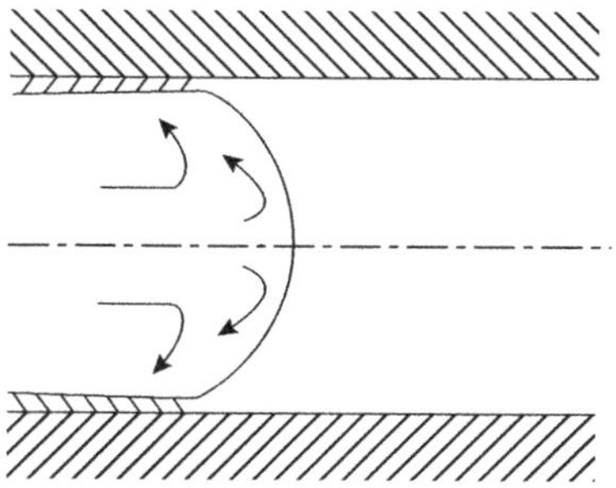

Figure 9.23: Flow front has cooled down near the wall.

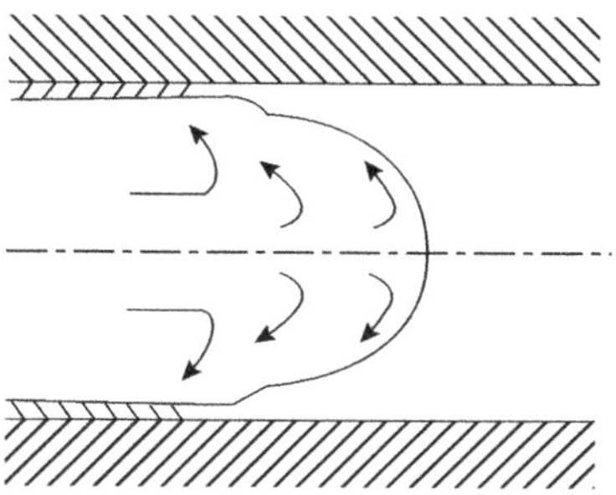

Figure 9.24: Cooled down peripheral layer impedes direct frontal flow to the wall.

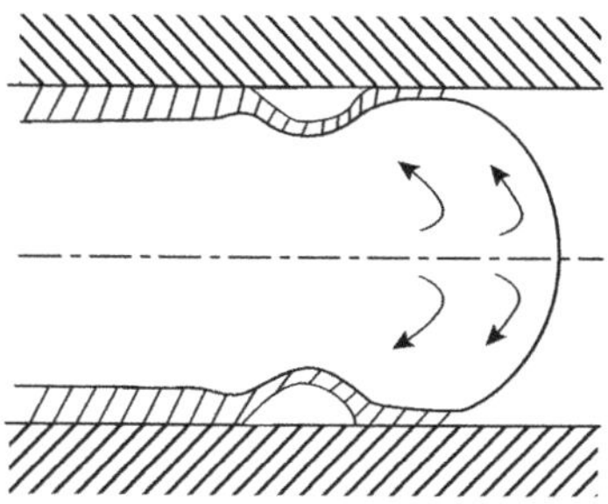

Figure 9.25: Flow front touches the mould wall again.

Correcting the record grooves effect

Check and/or change the machine settings, change the mould or moulding compound, start a new cycle and go through Flow Chart 9.12.

9.2.12 Stress whitening/stress cracks

Physical cause

Stress whitening or stress cracks occur when the maximum deformation is exceeded (e.g., due to external stress or warpage). The maximum deformation depends on the type of material used, the molecular structure, the processing and the surrounding environment of the moulded part. The magnitude of external and internal stresses can be drastically reduced through physical processes, which are dependent upon time and temperature. In this case, the bonding forces of the molecules are reduced through wetting, diffusion and swelling processes, which particularly favour stress cracks. Besides internal cooling stresses and stresses due to flow, internal stresses due to expansion are another main reason for internal stresses. External expansion stress is created by demoulding under residual pressure, when the moulded part suddenly shifts from residual pressure to atmospheric pressure, the inner layers of the moulded part put stress on the outer layers. The main reasons for demoulding under residual pressure are insufficiently dimensioned moulds and/or high cavity pressures. The formation of stress is shown in Figure 9.26 and the physical manifestations on the moulding in Figure 9.27.

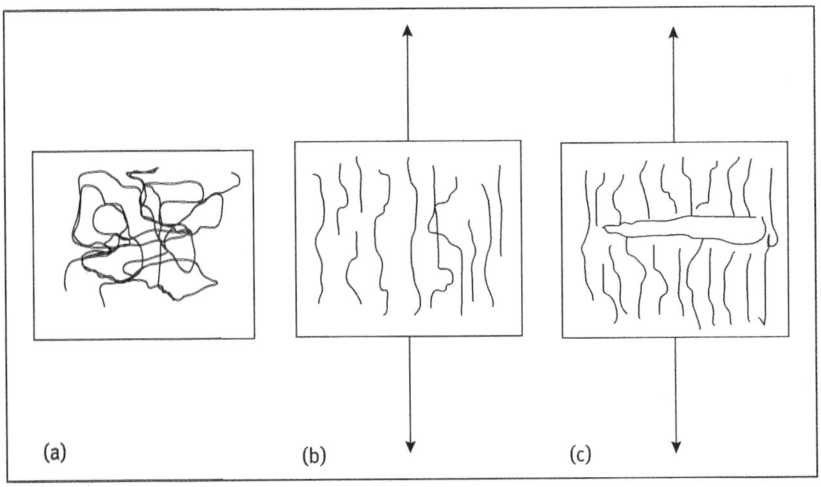

Figure 9.26: a) Unstressed, relaxed molecule structure, b) orientation of molecules due to force and c) destroyed molecules due to additional force.

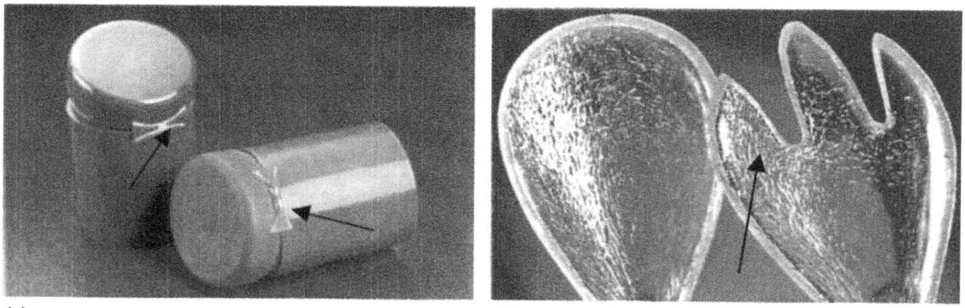

(a) (b)

Figure 9.27: Stress whitening. a) Stress whitening on an integral hinge and b) stress cracks on salad servers (damage visible several weeks after purchase).

N.B.: If aggressive substances are used (e.g., alkali solutions, grease and so on) stress whitening and stress cracks often appear after a prolonged period of machine operation.

Correcting stress whitening/stress cracks

Check and/or change the machine settings, change the mould or moulding compound, start a new cycle and go through Flow Chart 9.13.

9.2.13 Incompletely filled parts

Injection moulded parts with incompletely developed outer profiles are called incompletely filled parts (or short shots) and an example is shown in Figure 9.28. This kind of defect often appears far from the gate if there are long flow distances, or on thin walls (e.g., ribs as shown in Figure 9.29). Due to insufficient mould venting, this defect can also occur in other areas.

Physical cause

There are several physical causes for incomplete filling:
- Injected compound volume is too small (e.g., shot volume is too small).
- Melt flow is impeded due to venting problems.
- Injection pressure is not sufficient.
- Premature freezing of a channel cross-section (e.g., low injection speed or incorrect temperature control in the mould).

N.B.: Incomplete filling due to venting problems does not necessarily cause the diesel effect (Section 9.2.10); therefore, the cause for the defect is often hard to determine.

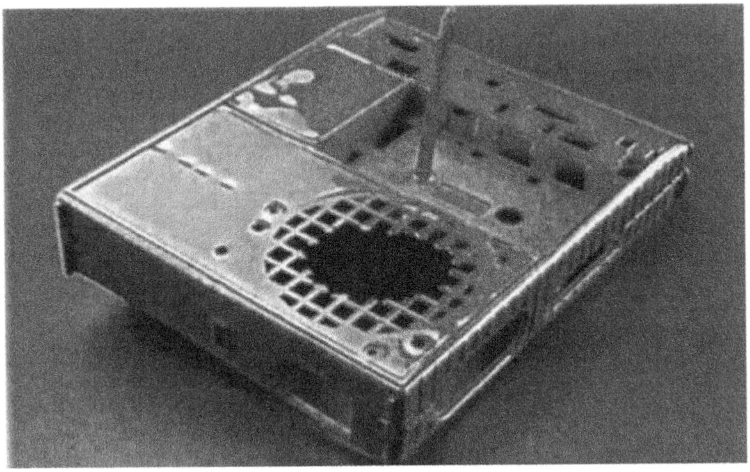

Figure 9.28: Housing with incompletely demoulded lattice.

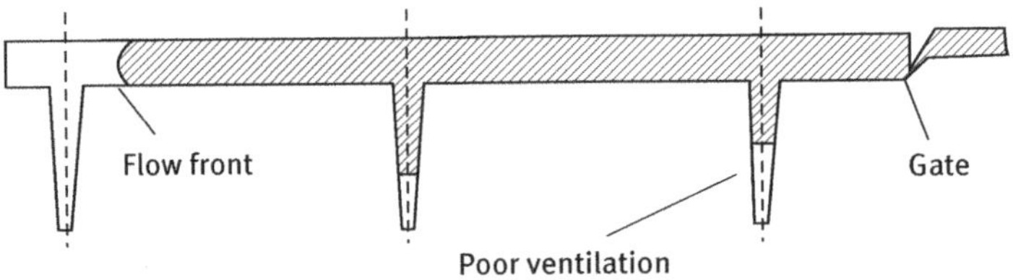

Figure 9.29: Filling problems near thin ribs.

Correcting incompletely filled parts

Check and/or change the machine settings, change the mould or moulding compound, start a new cycle and go through Flow Chart 9.14.

9.2.14 Oversprayed parts (flashes)

Flashes are often created near sealing faces, venting channels or ejectors and look like a more or less developed film-like plastic edge. Fine flashes are not often immediately visible whereas large area thick flashes sometimes stick out several centimetres over the nominal profile, as shown in Figure 9.30.

Physical cause

The possible causes can be divided into four main groups:

- Allowed gap widths have been exceeded (mould tightness is insufficient, production tolerances are too large or damaged sealing faces).

Figure 9.30: Large area overspraying (flash).

- Clamping force of the machine is insufficient or set to low (mould opening force is higher than the clamping force, the mould cannot be kept closed; clamping force deforms platens and the mould).
- Internal mould pressures are too high (shaping pressure at the gap is so high that the melt is even pushed into very small gaps).
- Viscosity of the moulding compound is too low (high internal mould pressures and low flow resistances favour flash formation).

N.B.: Flash formation can develop very quickly (few cycles) and damage the sealing faces (parting surface).

Correcting oversprayed parts (flashes)
Check and/or change the machine settings, change the mould or moulding compound, start a new cycle and go through Flow Chart 9.15.

9.2.15 Visible ejector marks

Ejector marks are depressions or elevations on the ejector side of the moulded part surface. These wall thickness variations can cause gloss differences and depressions on the visible surface of the moulding, as shown in Figure 9.31.

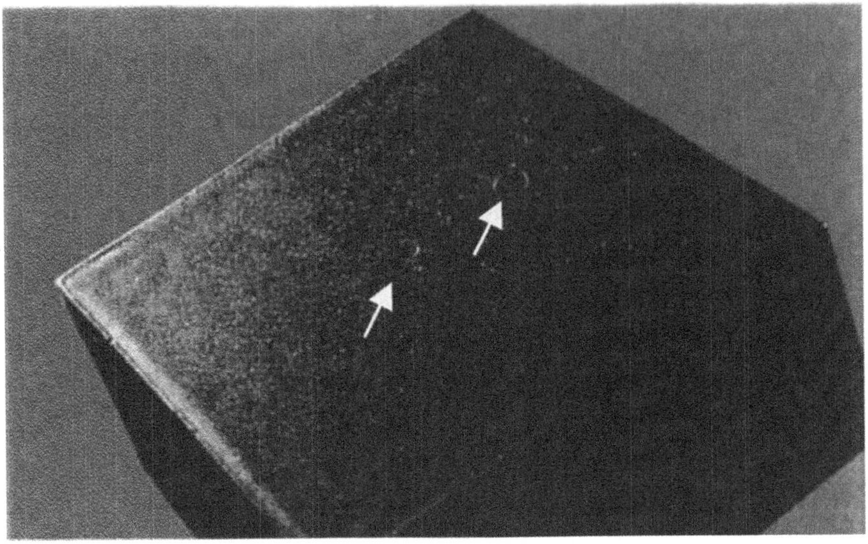

Figure 9.31: Gloss differences near the ejector.

Physical cause

The different possibilities can be divided into four main groups:

- Process-related causes (e.g., premature demoulding or high demoulding forces due to unfavourable machine settings).
- Geometric causes (e.g., incorrect fitting or ejector length).
- Mechanical or strength related (e.g., faulty dimensioning and design of the mould, the moulded part or the demoulding system).
- Thermal causes (high temperature differences between the ejector and mould wall, e.g., Figure 9.32).

Correcting visible ejector marks

Check and/or change the machine settings, change the mould or moulding compound, start a new cycle and go through Flow Chart 9.16.

9.2.16 Deformation during demoulding

Classification of extraction markings, cracks, fractures, overstretched areas and deeply depressed ejectors depends on the degree of damage. Critical defects include moulded parts with undercuts, which are demoulded without movable parts (e.g., slides). Examples of two defect mouldings are shown in Figure 9.33.

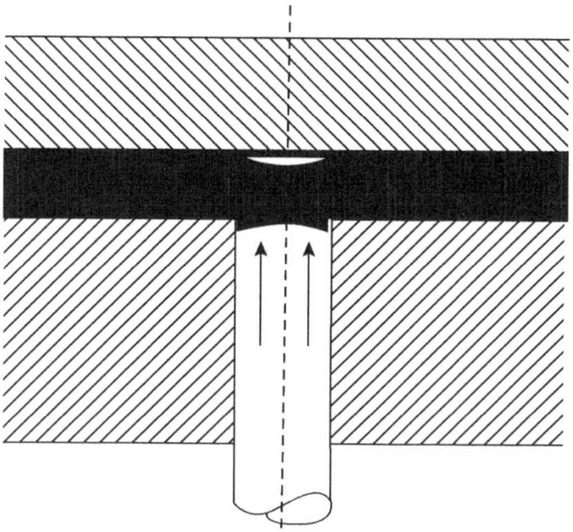

Figure 9.32: Shrinkage near an overheated and poorly fitted ejector.

(a) (b)

Figure 9.33: Deformation during demoulding: a) demoulding grooves on a textured surface and b) deformation due to forced demoulding at an undercut.

Physical cause

The causes of deformation can be classified as follows:

- The forces necessary for demoulding cannot be applied to the moulded part without damaging it.
- The demoulding movement is disturbed.

The amount of demoulding force applied is a crucial criterion and should thus be kept small. In addition to other factors, the shrinkage of the moulded part has a direct impact on the demoulding forces. The shrinkage and demoulding force can be considerably influenced by varying the process parameters, and the geometry of the moulded part is also a very important influencing factor.

In general, low shrinkage is desirable for sleeve and box-shaped parts, since these parts tend to shrink onto the core (solutions are to increase the holding pressure or reduce the cooling time).

Near ribs, the shrinkage retroacts on the demoulding force, because the ribs are being detached from the mould walls (solutions are to decrease the holding pressure or increase the cooling time).

Correcting deformation during demoulding

Check and/or change the machine settings, change the mould or moulding compound, start a new cycle and go through Flow Chart 9.17.

9.2.17 Flaking of the surface layer

If the layers of the material are not homogeneously joined together they start flaking, which can occur at the gate or on the moulded part, and can be either large or very small and thin, depending on the intensity. Examples of both are shown in Figure 9.34.

Figure 9.34: Flaking.

Physical cause

Flaking of surface layers is due to insufficient bonding of adjacent surface layers. The different layers are formed by different flow effects and cooling conditions over the cross-section. Shear stresses and inhomogeneities can reduce the bonding of these layers to such a degree that single surface layers start flaking off as shown in Figure 9.35.

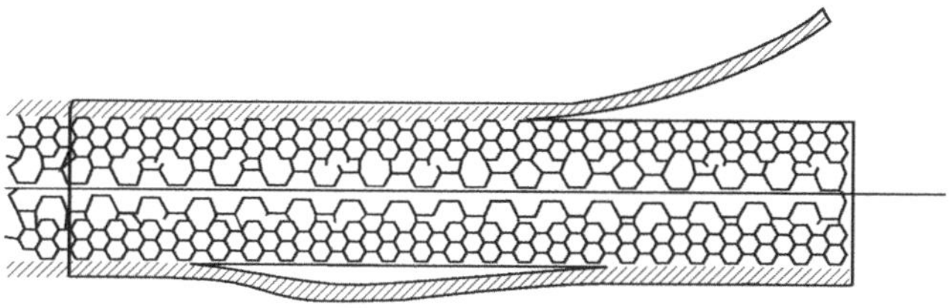

Figure 9.35: Flaking on a cross-section of a moulded part with a different structural formation.

High shear stresses and thermal damage can be caused by:
- High injection speeds.
- High melt temperatures.

Inhomogeneities can be caused by:
- Impurities or other materials among the granules.
- Incompatible dye or masterbatch.
- Moisture in/on the granules.
- Poorly melted moulding compound.

Correcting flaking of the surface layer

Check and/or change the machine settings, change the mould or moulding compound, start a new cycle and go through Flow Chart 9.18.

9.2.18 Cold slugs/cold flow lines

Physical cause

Cold slugs are formed when melt solidifies in the gate or in the nozzle, before the compound is injected, and is transported into the mould with the following shot sequence. If the cold slugs do not melt again, they will cause markings which look like comet tails and can be spread all over the moulded part. The cold slug can also jam a runner, forcing the melt to part. The results are surface defects similar to weld lines, an example is shown in Figure 9.36. Cold slugs are often caused by the wrong nozzle temperature or belated retraction of the plasticising unit. Small nozzle diameters can also have a negative effect. An illustration of the mechanism is shown in Figure 9.37.

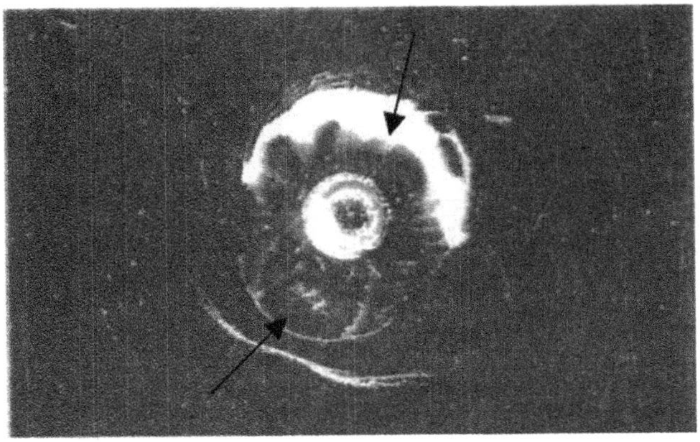

Figure 9.36: Markings caused by a cold slug near the sprue.

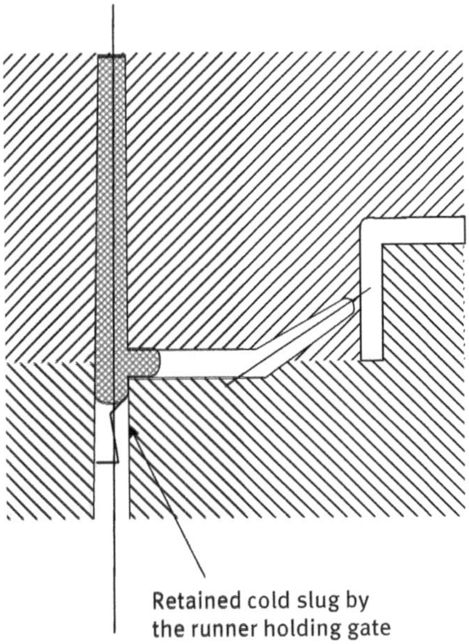

Figure 9.37: Cold slug is transported into the mould by the flow front.

Correcting cold slug/cold flow lines

Check and/or change the machine settings, change the mould or moulding compound, start a new cycle and go through Flow Chart 9.19.

9.2.19 Entrapped air (blister formation)

Physical cause

During the injection, air is entrapped in the melt and appears as a hollow (air blister) on the moulded part. Primarily, there are two factors responsible for this defect:

- Decompression is too high or too fast.
- Plasticising performance is too low.

N.B.: There are two types of hollows, entrapped air and voids. Voids are vacuole hollows, formed by the shrinkage of the moulding compound (see 'sink marks'). Distinguishing between the two is very hard, because of their similar appearance; however, the following hints might be helpful:

- When opening the hollow in a fluid, a void (vacuum) shows no gaseous bubbles.
- Entrapped air defects can be reduced by not using decompression.
- Changing the holding pressure or the holding pressure time has no effect on the size of the hollows.

Moulded parts with hollows are usually not as strong as parts without. Non-transparent parts should be randomly tested by opening them. Examples of trapped air are shown in Figure 9.38.

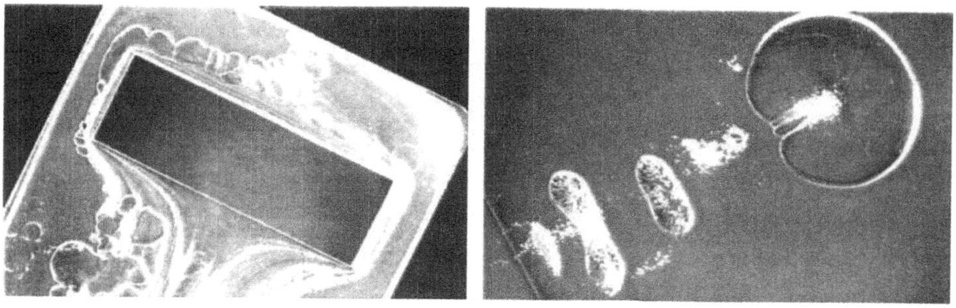

Figure 9.38: Entrapped air.

Correcting entrapped air (blister formation)

Check and/or change the machine settings, change the mould or moulding compound, start a new cycle and go through Flow Chart 9.20.

9.2.20 Dark spots

Physical cause

Black or dark spots appear on the surface due to wear, thermal damage or dirt. Different factors can cause the formation of dark spots or speckled parts:

- Process-related causes, e.g., melt temperature is too high or residence time in the plasticising unit is too long. Incorrect temperature profile in the hot-runner system.
- Mould-related causes, e.g., dirty gate system or wear (dead edges) in the hot-runner system.
- Machine-related causes, e.g., dirty plasticising unit or worn screw and cylinder.
- Polymer or dyeing, e.g., impurities in the granule, high reclaim content or unsuitable dye/masterbatch.

Figure 9.39 shows dark spots resulting from thermal damage to the polymer.

Figure 9.39: Dark spots due to thermal damage.

Correcting dark spots

Check and/or change the machine settings, change the mould or moulding compound, start a new cycle and go through Flow Chart 9.21.

9.2.21 Dull spots near the sprue

Physical cause

Dull spots near the sprue are mainly caused by:
- Small gates
- High injection speeds

Extremely strong orientations of the molecule chains are formed during injection due to high injection speeds, small gate cross-sections and bypasses behind the gate. As there is not enough relaxation time directly behind the gate, the peripheral layers of the melt are frozen while still strongly oriented. Such peripheral layers can only be stretched to a minimum degree and crack under the impact of high shear stresses.

The hot melt inside flows to the mould wall and forms very small notches, as shown in Figure 9.40. The dull appearance is caused by the wide spread reflection in this area.

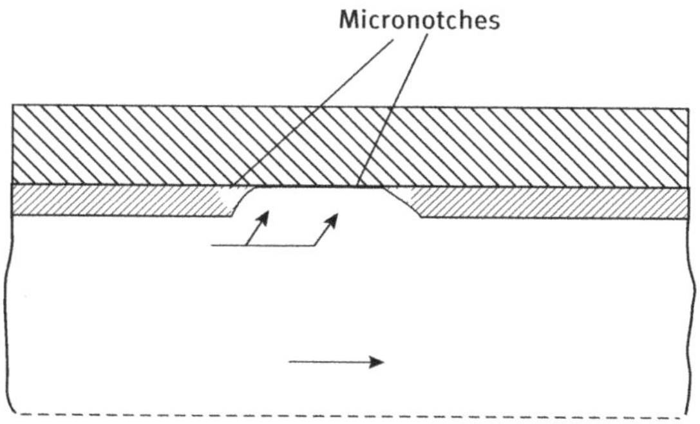

Figure 9.40: Melt flows into the cracked peripheral layer (formation of micronotches).

Correcting Dull Spots Near the Sprue

Check and/or change the machine settings, change the mould or moulding compound, start a new cycle and go through Flow Chart 9.22.

9.3 Data acquisition record

Should surface defects appear during the proving of the mould, the person setting up the machine should try to eliminate or at least minimise the defect by changing the processing parameters. Recording the proving and optimisation process helps to understand the action taken and to evaluate the effects on the quality of the moulded part. Furthermore, the record can be used as a basis for discussion, because it also helps other people to quickly understand the process.

9.3.1 Using the data acquisition record

Recording the moulding process should begin as soon as a defect occurs. In the first part of the record sheet, general data regarding the moulded part, material, machine and mould can be entered, if necessary.

When changing processing parameters, the operator can use the flow diagrams detailed in earlier sections of this chapter. Note that only one parameter per cycle should be changed. Defects can be evaluated using the recommended key for each trial. It can be useful to mark the parts produced during the trials for later reconstruction of process optimisation. This record sheet only contains a selection of the large number of processing parameters available.

9.3.2 Data acquisition record for optimising moulded parts

9.3.2.1 Moulded part data

Name: ....

Moulded part number: ....

Special remarks: ....

Compound data: ....

Trade name: ....

Type: ....

Batch number:..

Type of colouring: ....

Type of dying: ....

Special remarks: ....

Machine data: ....

Type of machine: ....

Machine number: ....

Special remarks: ....

Mould number: ....

Number of cavities: ....

Special remarks: ....

9.3.2.2 Machine settings and defect evaluation

Temperatures (°C):	Basic settings	Test 1	Test 2	Test 3	Test 4	Test 5	Test 6	Test 7
Cylinder temperature (feed zone):								
Cylinder temperature (middle zone):								
Cylinder temperature (front zone): Melt temperature (injected):								
Mould temperature (nozzle):								
Mould temperature (core):								
Mould temperature (defect):								
Hot-runner temperature:								
Back pressure in:								
Metering stroke in:								
Melt cushion in:								

Injection profile:	Basic settings	Test 1	Test 2	Test 3	Test 4	Test 5	Test 6	Test 7
Injection speed 1 in :								
Change-over point 1 in :								
Injection speed 2 in :								
Change-over point 2 in :								
Injection speed 3 in :								
Change-over point 3 in :								
Injection pressure (setting):								
Injection pressure (actual value):								
Holding pressure:								
Holding pressure 1 in :								
Holding pressure 2 in :								
Holding pressure 3 in :								
Holding pressure time :								
Actual values:								
Cooling time:								
Cycle time:								
Defect evaluation:								
Type of defect:								
Evaluation #1:								

(1) Moulded part OK, (2) defect better, (3) defect unchanged, (4) defect worse and (5) new defect created

Temperatures in (°C):	**Test 8**	**Test 9**	**Test 10**	**Test 11**	**Test 12**	**Test 13**	**Test 14**	**Test 15**
Cylinder temperature (feed zone):								
Cylinder temperature (middle zone):								
Cylinder temperature (front zone):								
Melt temperature (injected):								
Mould temperature (nozzle):								
Mould temperature (core):								
Mould temperature (defect):								
Hot-runner temperature:								
Back pressure in :								
Metering stroke in :								
Melt cushion in :								

Injection profile:	**Test 8**	**Test 9**	**Test 10**	**Test 11**	**Test 12**	**Test 13**	**Test 14**	**Test 15**
Injection speed 1 in :								
Change-over point 1 in :								
Injection speed 2 in :								
Change-over point 2 in :								
Injection speed 3 in :								
Change-over point 3 in :								
Injection pressure (setting) :								
Injection pressure (actual value) :								
Holding pressure:								
Holding pressure 1 in :								
Holding pressure 2 in :								
Holding pressure 3 in :								
Holding pressure time:								
Actual values:								
Cooling time:								
Cycle time:								
Defect evaluation:								
Type of defect:								
Evaluation #1:								

Moulded part OK, (2) defect better, (3) defect unchanged, (4) defect worse and (5) new defect created

9.4 Experiences with injection moulded components

The following examples highlight potential errors that can appear on injection moulded components as well as solutions to overcoming them.

9.4.1 Threaded connecting sleeves for ink-drafting apparatus

With this thin cavity three-plate mould arrangement, series injection took place in the parting plane. The filling process and the pressure ratios in the two cavities were thus different. The effect of this was that the quality of the surface of the component was not acceptable. Moreover, very narrow tolerances could not be maintained. These effects could be overcome by correcting the flowpath lengths in the runner (Figure 9.41). Identical flowpath lengths on multiple equipment in the mould are the basic prerequisites for maintaining narrow tolerances.

This basic principle is generally valid in injection moulding technology.

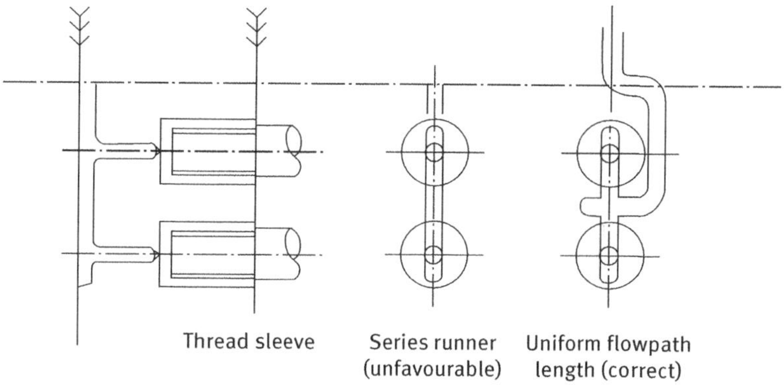

Figure 9.41: Runners for ink-drafting apparatus.

9.4.2 Meter cases

For this meter case, there are dovetail guides on the four side faces. The varying wall thickness thus caused a meeting of flowpaths resulting in an air blister in the vicinity of the gate (Figure 9.42). The following measures were adopted to try and remove the air blister:

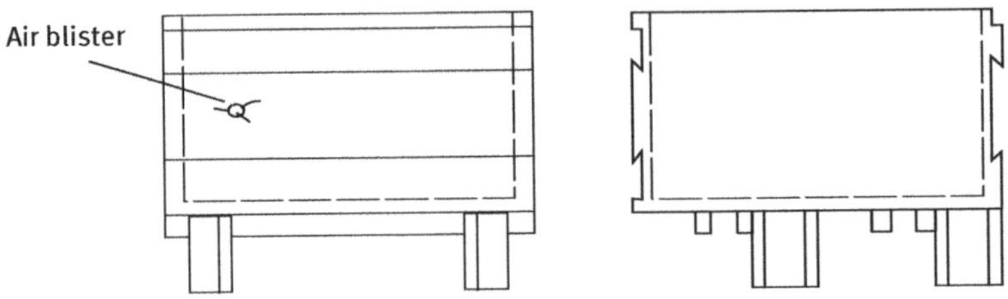

Figure 9.42: Meter case.

1. Reduction of the initial injection speed; however, a hole then appeared at this point, where the compound was no longer coalescing.
2. Raising the compound temperature; initially this was a success, in that, by raising the compound temperature step-by-step, it was possible to reduce the initial injection speed as much as possible. However, after a certain amount of time warping occurred, which can clearly be traced back to very high internal orientation stresses resulting from excessively slow filling.

In order to remove the air blister, even at faster initial injection speeds, the mould clamping force, which had previously been set very high (too high), was reduced until satisfactory running was restored. This example shows that a mould clamping force which is not too high, and which is satisfactorily set, guarantees better air venting.

Moreover, the equipment is kept in better condition as the lower the mould clamping force, the lower the wear.

9.4.3 Wristwatch glass

A glass for a wristwatch displayed convergence points opposite the runner, i.e., very visible joint lines. It was possible to establish clearly, after a filling sequence with this mould, that a uniform flow front was not formed.

When the sample components, as shown in Figure 9.43, were measured later, there were thickness variations of 0.1–0.15 mm. The variations in wall thickness were overcome by supplementary work on the core.
Subsequently, a uniform flow front could be generated and the faults caused by the joint lines are now a thing of the past.

After the fault referred to above had been eliminated, there was still some dissatisfaction with the surface lustre of the watch glass even though the mould inserts were high-gloss polished. The required surface gloss could not be obtained until new mould inserts had been manufactured from non-porous vacuum steel (in accordance with the vacuum ARC refining process).

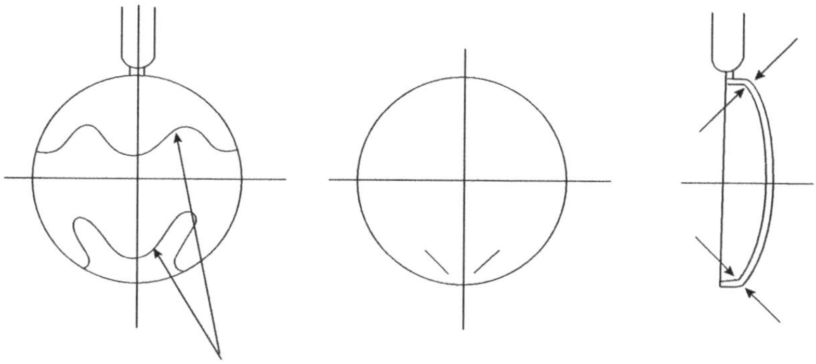

Figure 9.43: Wristwatch glass.

9.4.4 Alarm clock glass

During the manufacture of alarm clock glass, the following problem arose during sampling:

- A uniform flow front was not being formed, due to the edges being some tenths of a millimetre thicker, and an air blister was formed on the side of the component opposite the runner. Since there were electroplated inserts in the core and the mould insert, it was not possible to correct the wall thickness by secondary work. The electroplated inserts would have had to be manufactured again.
- It was decided to remove the air blister by a simpler method shown in Figure 9.44.
- A gas vent was ground, 8.0 mm wide and 0.3 mm thick. The convergence point with the air could now be forced out, and it was then possible to manufacture components of an acceptable quality.

9.4.5 Glass cover for a digital gauge

This component displayed slight sink marks on the face above the finning, especially on the side away from the gate. Injection was carried out using a film gate located at the parting plane. However, the film runner was produced in such a way that the cross-runner to part space was too large; in addition, the runner cross-section was too small.

Moving the cross-runner closer to the component (shorter gate) and better formation of the accumulation base enabled the gate to stay open for longer. Before and after designs are shown in Figure 9.45.

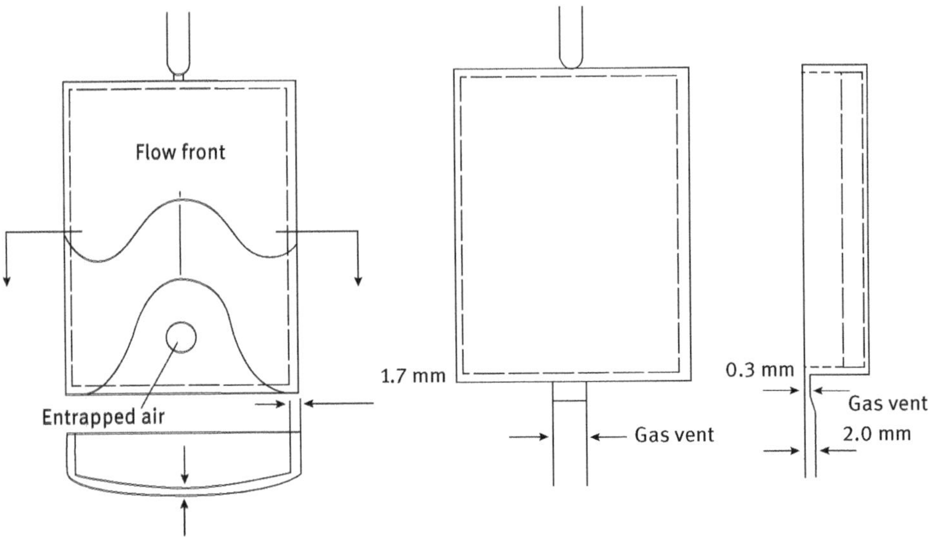

Figure 9.44: Alarm clock glass.

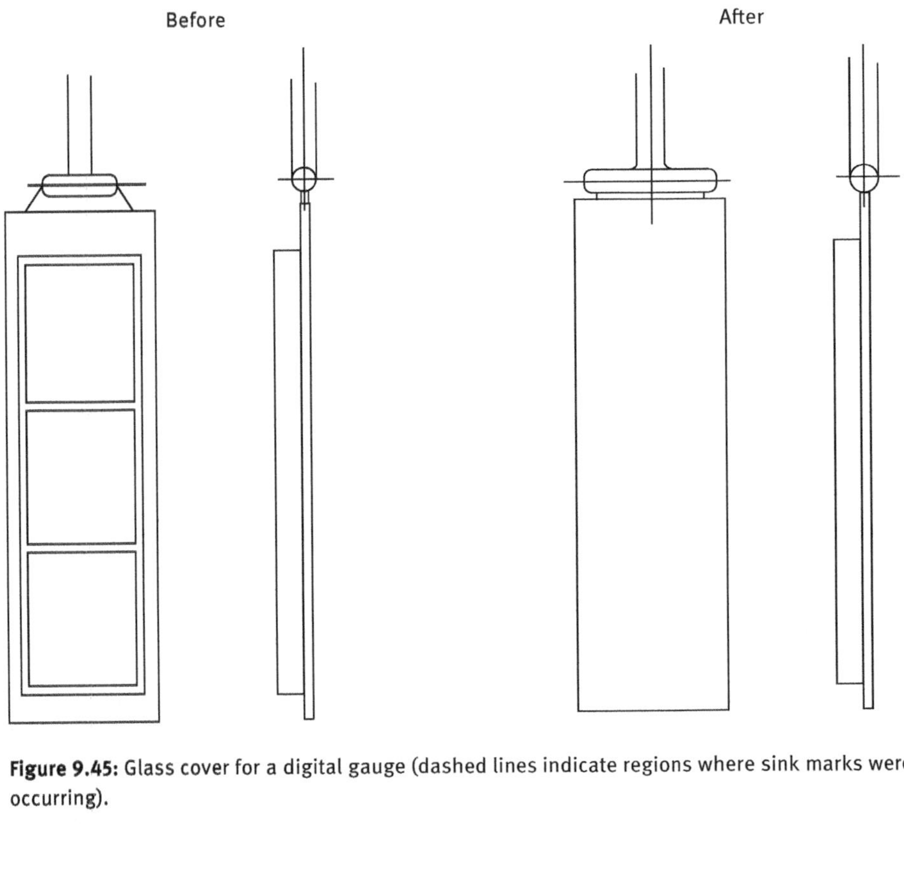

Figure 9.45: Glass cover for a digital gauge (dashed lines indicate regions where sink marks were occurring).

This solution not only eliminated the sink marks, but it was also no longer necessary to select such high melt temperatures and initial injection pressures, which also improved the quality of the component.

9.4.6 Plug boards with insert pins

Despite all the technical skills that injection moulding technology can offer, voids were forming in the centre of the moulding, which was found to be a result of the components being directly connected to the cross-runner Since a gate point of this kind freezes very quickly, the holding pressure was not able to act for a sufficient period.

When using injection moulding technology, it is generally true that gates should start at the centre of the runner or the runner system, so that a plastic flow can form properly from the gate outwards (Figure 9.46).

Direct connections from the runner are therefore highly unfavourable.

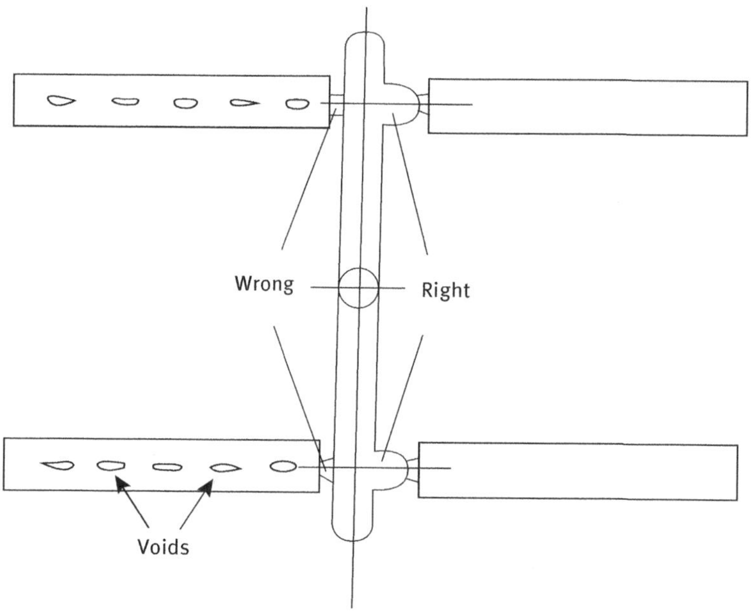

Figure 9.46: Plug boards with insert pins.

9.4.7 Hair slides

It was not possible to manufacture these components without sink marks at the point away from the gate using a sextuple hair slide mould. The root cause lay in the pencil-type transfer from the runner to the gate, where, owing to the lack of an accumulation base, highly viscous material was carried along from the marginal area, thus leading to an premature closing (sealing) of the gate. The sink marks then occurred owing to the lack of holding pressure. However, the sink marks had to be avoided at all costs, as the component was subsequently heat-stamped with film, which did not make contact at the sink marks. Due to the subsequent incorporation of a well-formed accumulation base at the gate, it was possible to manufacture the components without sink marks.

For pin-point gates, the transfer from the runner should always take the form of an accumulation base as it prevents the highly viscous compound at the edge from being carried along the plastic core of the runner and is thus retained for longer (the gate does not freeze until later).

In spite of the gate being the same size, the holding pressure is now effective for longer and the risk that sink marks will form is reduced. Both gating types are illustrated in Figure 9.47.

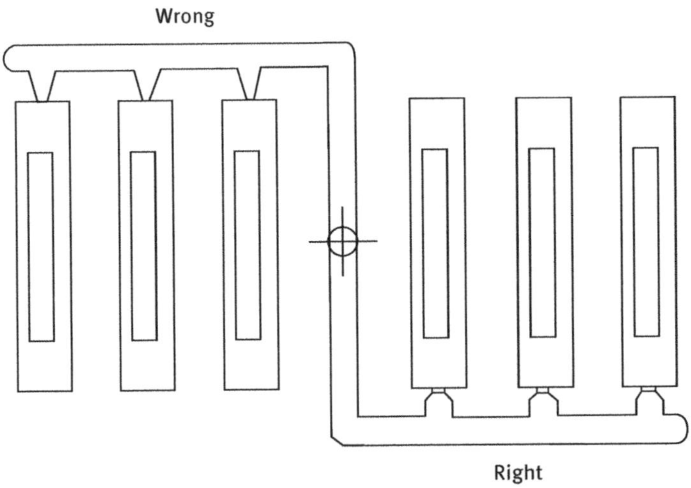

Figure 9.47: Gating in hair slide tool.

For this reason, pencil-form connections should also avoided when using three-plate moulds.

9.4.8 Toothbrush components

These components were injected into the parting plane where a so-called free jet arose through the massive wall thickness because of injection into the free space.

This free jet cools relatively quickly on its top face, and thus can no longer be adequately attached to the compound which follows. As filling goes on, it is compressed together like a snake (sausage injection moulding) and leaves creases behind on the top face of the component.

It was possible to prevent a free jet here by fitting a (sprung) baffle strut, which ensured a flow head of the compound moving away from the gate. When the filling phase ends, the sprung baffle strut is retracted and leaves only a mark behind on the component. The design is illustrated in Figure 9.48.

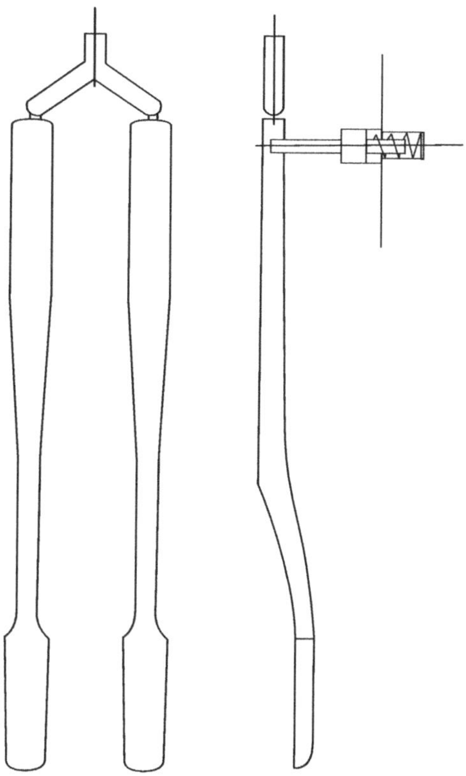

Figure 9.48: Toothbrush component.

Guide values for setting a sprung baffle strut:
- Minimum diameter: 4 mm.
- Distance from gate: 2.5 mm.
- Insertion depth into component: approximately 2/3 of wall thickness.

9.4.9 Screw cap with conical nipple

Screw caps are manufactured in pairs using a three-plate mould.

Free jet formation (sausage injection moulding) occurred during filling as the compartment had a larger cross-section at the connection point, and the jet of compound was therefore initially able to form without hindrance. It was possible to fit a baffle or a sprung baffle strut. It was also possible to avoid free jets by rounding off the sharp transfer from the gate towards the component; thus, the desired flow head is formed when the compound flows in (Figure 9.49).

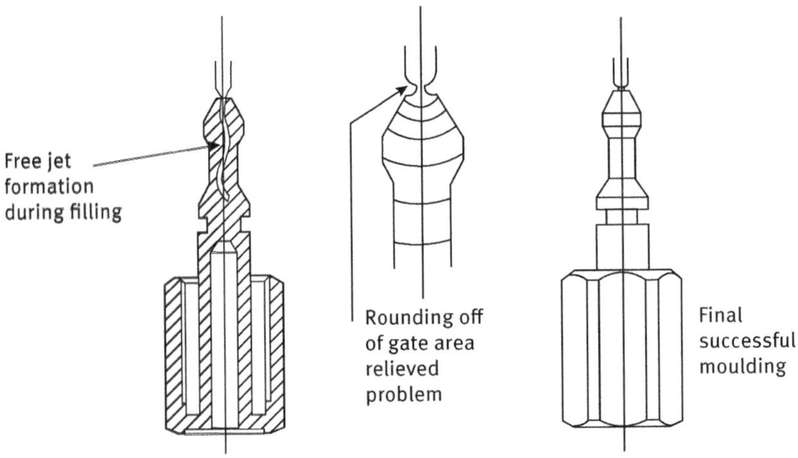

Figure 9.49: Screw cap with conical nipple.

It would be generally favourable to make this edge less sharp, but this cannot be achieved in every case, for example, using tunnel runners, the runner might not be degated at the desired point (directly at the component).

9.4.10 Switch housing

A switch housing was injection moulded using a rod runner. Strong flow line formation took place on the top face of the component and in spite of mould heating, no improvement could be made.

It was not until the transfer from the rod runner toward the component was rounded off that a good top face was produced (Figure 9.50).

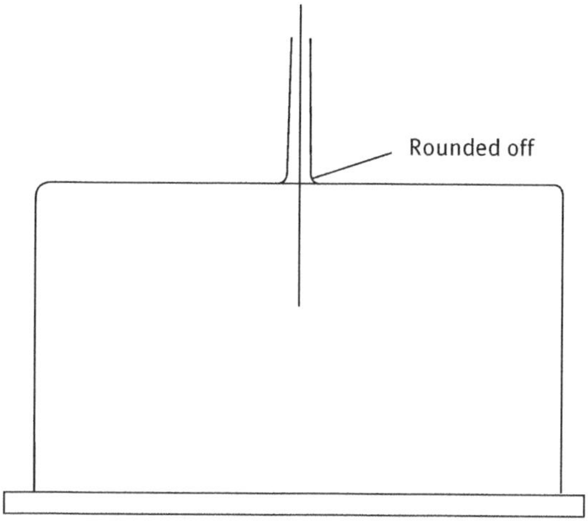

Figure 9.50: Switch housing.

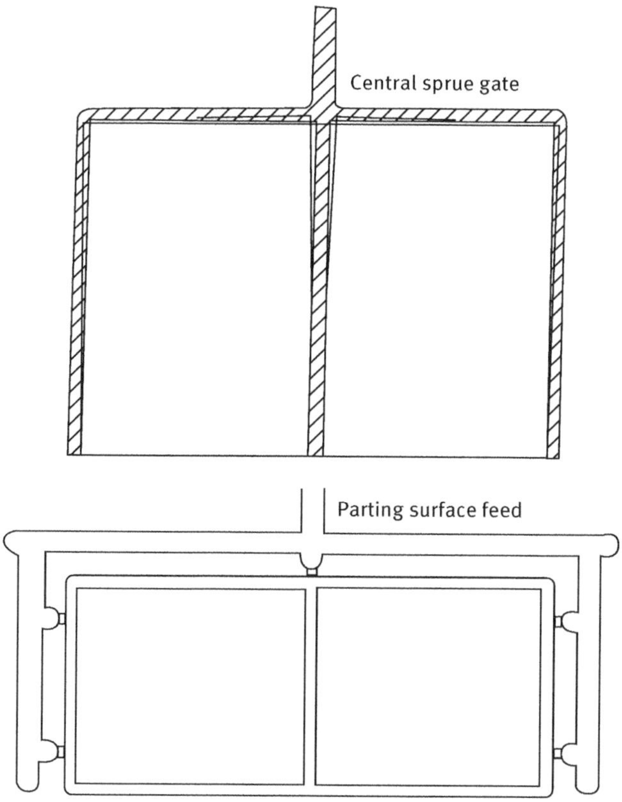

Figure 9.51: Battery housing.

9.4.11 Battery housing

A twin-compartment battery housing was centrally injected using a rod runner. The individual wall thickness and the intermediate wall were approximately 1.5 mm, and the walls were approximately 80 mm high. During injection, a strong wedge effect occurred due to the compound moving forward at the intermediate wall, which is actually in the middle of the component, and although they were manufactured from a single unit, the cores were pressed right against the external mould wall.

To be able to manufacture this component at all, the injection had to be displaced into the parting plane. As the compound now entered the base of cores, the cores were no longer pressed away, but rather were established through the uniform flow front, as shown in Figure 9.51.

In order to extract the enclosed air at the housing base, it was necessary to insert air venting plungers.

10 Advanced processing techniques

10.1 Introduction

Whilst injection moulding (IM) can be considered a relatively young manufacturing technology, many variants of the basic process have been developed. Examples include processes such as injection-compression, gas-assisted, water injection, co-injection, over-moulding and micromoulding, and each of these processes are suitable for specific market applications. An overview of the technologies covered in this chapter is shown in Figure 10.1.

The processes are split into four categories, multimaterial moulding, assisted moulding, micromoulding and additive manufacturing (AM) hybrids, and Section 10.2 onwards will introduce each of these in turn.

Various terminologies are used in multimaterial moulding to describe particular process routes, which have been split into three categories:

- Multicomponent.
- Multi-shot.
- Over-mould.

Assisted moulding techniques include both gas and water. A brief description of each of these technologies, and some of the names commonly used for them in the literature, now follows.

10.2 Multicomponent moulding

10.2.1 Co-injection moulding

Also known as dual injection, sandwich moulding, 2K (or 3K). This type of moulding can be recognised by the production of a defined skin and core structure. The core is fully encapsulated to produce a sandwich-like moulding.

10.2.2 Bi-injection moulding

This involves the simultaneous injection of different materials through different gates as opposed to through the same gate as in co-injection moulding.

https://doi.org/10.1515/9783110654813-010

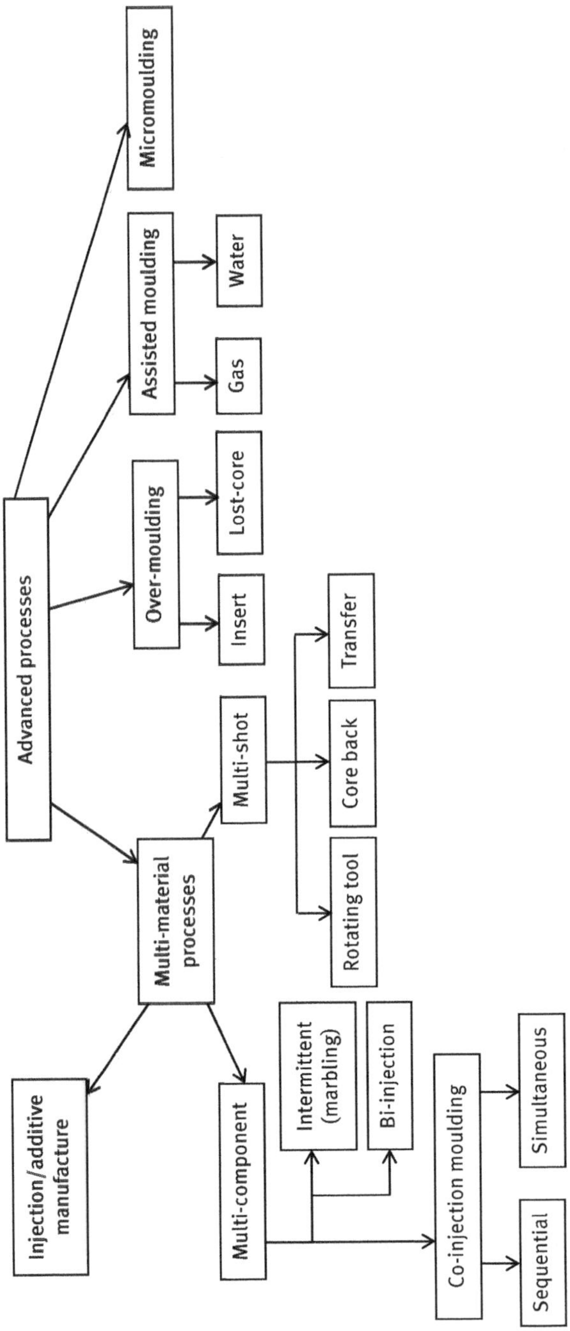

Figure 10.1: Process options for advanced moulding technologies.

10.2.3 Interval injection moulding

Also known as marbleising. This involves the simultaneous injection of different materials through the same gate, with limited mixing, which produces a pattern of either regular or irregular colour distribution, such as marble-type effects.

10.2.4 Injection moulding/additive manufacturing hybrid processes

As AM has matured as a technology, so has interest in hybrid processes that combine the advantages of tool-less manufacture with the efficiency and speed of IM. ARBURG released a freeformer machine in 2013, and there is a considerable amount of ongoing research into developing faster, advanced and more flexible AM techniques. Issues with production speeds, the material feedstocks that can be used, alone or in combination, and the final mechanical integrity of the moulded part (compared with IM) all still need to be overcome. Hybrid process routes combining the benefits of IM and AM therefore offer the most promise in terms of realistic short-term future competition to IM advances. A sample part produced by the freeformer is shown in Figure 10.2.

Figure 10.2: FREEFORMER multimaterial component. Reproduced with permission from ARBURG.

10.3 Assisted moulding

In many ways assisted moulding is analogous to co-injection moulding, as the resultant moulding has a skin and core configuration; however, in this case, the core injection process is used to selectively hollow out parts of the moulding. This process is primarily used for weight savings and cycle time reduction. Gas or water takes the place of the polymer to produce a hollowed out moulding with a reduced section thickness.

Gas injection moulding has no standardised designations. Some common process terminology/trade names include: gas injection moulding technology (GIT), gas-assisted injection moulding (GAIM), gas-assisted injection moulding (GAIN™), gas interior pressure (German: Gas Innen Druck), gas injection process and Kool-Gas™, which is a variation of the standard gas injection technique using precooled gas injection.

Water injection moulding or water-assisted injection moulding (WAIM) is a much more recent technology and less well established.

10.4 Multi-Shot moulding

Multi-shot can be used to describe any process whereby distinct multiple material shots are utilised to produce a single final component, which includes processes whereby preforms are moulded and then transferred to different cavities on the same machine. It can also be used to describe processes where multiple shots are made into the same tool without the tool opening between shots.

10.5 Over-moulding

This technique involves components being placed in an injection mould that are then moulded over with another material. The term over-moulding covers both insert moulding and lost core moulding. This technique is not confined to plastics and over-moulding of metal inserts, such as to produce scissor handles and plastics on ceramics, is commonplace.

10.6 Micromoulding

Micromoulding machines allow very small and precise shots of material (less than a gram) to be delivered into a moulding tool enabling the production of both very small and very thin mouldings, such as tiny gears and fixings. Micromoulding is a specialist area of IM and machines tend to be of a slightly different design than conventional machines, but the process sequence and major components remain the same with a slight variation of the heat transfer dynamics. Due to the size of the moulding and the very thin sections that can be achieved using this method, cooling occurs very rapidly in the tool following injection; furthermore, due to the microcomponents that can be manufactured, novel tooling solutions can be used, such as the use of ceramic inserts (e.g., Figure 10.3).

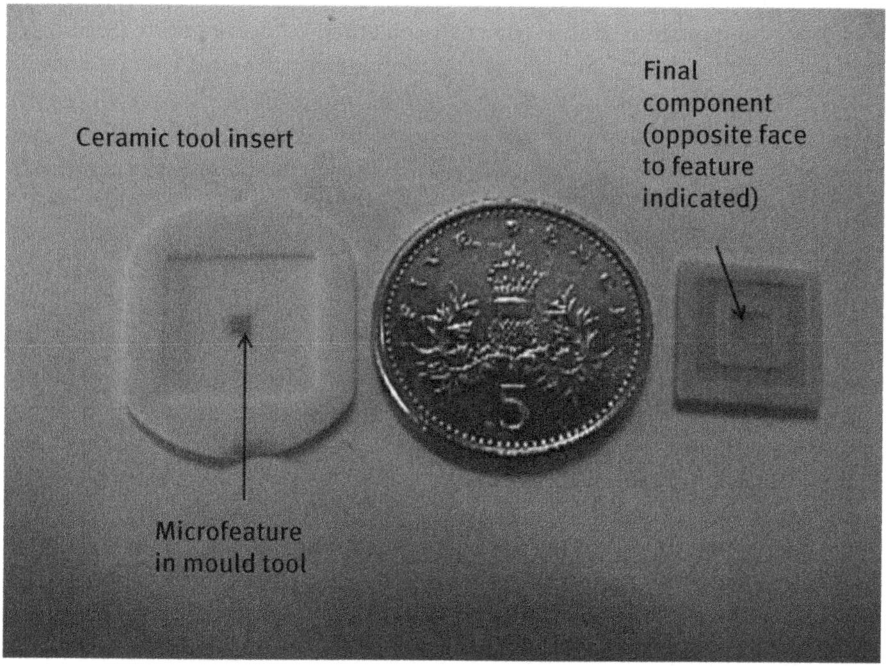

Figure 10.3: Micromoulding: tool insert component (the coin is to allow size comparison).

10.7 Business trends

In 2012, the term 'Industry 4.0' was adopted by the German Government. This strategic aim was to encourage the future manufacturing of highly efficient factories (having lower energy requirements, lower running costs and producing less waste) to link entire supply chains through customised and adaptable manufacturing systems powered by enablers such as the 'internet of things'.

Similar initiatives such as 'smart manufacturing' or 'factory of the future' have followed from other nations, which are all linked by the key themes of the customisation and control of manufacturing processes.

Creating such complex systems for the IM industry means creating a single interface that can link externally to automation systems and control a greater series of operations than conventional machines, whilst also transmitting and receiving information on production. This is entirely suited to the creation of multimaterial mouldings and multifunctional mouldings are a fast growing sector of the IM industry, with all major machine manufacturers offering multimaterial machine capability.

Gas- and water-assisted moulding tend to be less readily available and more specialised, and these technologies very often require licences. The development of gas technology, in particular, has been heavily patented by the industry. (Similarly, microcellular foam moulding (MuCell™) has been heavily licensed.) While not applicable to all moulding designs, assisted moulding can bring a competitive advantage to its user when employing suitable part designs. As well as material savings, these processes can reduce cycle times due to both the cooling effect of the fluids and the reduced section thickness of the polymer.

As well as Automation 4.0, the trend towards multimaterial moulding has been driven by the growing importance of added value manufacturing to give moulders a commercial advantage over their rivals. An increase in the importance of material recycling has also aided the growth of such technologies, where encapsulation enables recyclate materials to be used, in addition, combining multiple processes in one step can produce a range of cost and component benefits, such as the production of lightweight items, or removing postproduction, further assembly or finishing operations.

A further factor pushing the popularity of advanced moulding techniques, especially in Europe and the USA, has been the transfer of a considerable volume of the traditional trade moulding market to cheaper manufacturing sectors in developing countries. To combat the increased competition, trade moulders have sought more technically challenging markets as a means of survival, as much of the traditional standard IM market has been, and continues to be, contracted out. However, the more technically advanced processes can still often (not always) be performed more efficiently and with higher quality by established multimaterial toolmakers and moulding shops.

Whilst German manufacturers were quick to invest in multimaterial moulding technologies, other regions were slower to see the inherent advantages. However, the versatility these processes offer to manufacturers and the current highly competitive moulding market is finally overcoming the resistance generated by the larger capital equipment investment often required to produce more complex mouldings. This is a positive step forward since a well-designed multimaterial component can often repay the initial investment quickly and easily, and can be achieved by combining distinct moulding jobs, such as producing mouldings that then require assembly. By removing the need for the assembly step or perhaps a further finishing stage, a single multimaterial moulding can offer a considerable unit cost reduction. Of equal importance is that a more attractive end product is often the result. Consider, for example, the case of the humble toothbrush. The application of a multimaterial moulding process to this product has transformed this simple and standardised design. The use of combinations of hard and soft feel materials and a myriad of multiple colour combinations have filled endless supermarket shelves with a variety of attractive designs for the consumer, and in the process transformed the toothbrush marketplace. Similarly, consider the ongoing success of in-mould decoration in the packaging industry, where labelling is done as part of the process. This removes the need for finishing stages and enables the production of smart labels that can indicate the freshness of the enclosed foods, or can link to a wireless internet signal to reorder an empty milk carton in the fridge, which all link to future ideals of Automation 4.0 and the 'internet of things'.

A general introduction to each of the technology areas now follows, which will be discussed in more depth in later sections. However, before this, an introduction to some material selection issues will be presented.

10.8 Material selection

When selecting material combinations for multimaterial moulding applications, consideration must be given to the combination of properties required in the final product. In many cases, there must be a certain level of adhesion between the skin and core in order to maintain mechanical integrity, which can be achieved in two ways:

1. The materials are compatible in order to offer some degree of bonding at the interface.
2. A method must be found to mechanically interlock them. In the case of over-moulding and multi-shot this can be achieved with clever usage of material properties and tool design.

Not all parts require adhesion; in fact, in some cases, the requirement may be the exact opposite. If joints are to be produced, it is essential that the mouldings can move freely at the interface, e.g., over-moulding to produce the moving arms and legs on dolls and other similar toys, or to produce ball-and-socket joint mouldings. In cases like these, materials must be selected for their immiscibility to ensure smooth regions of movement. Where adhesion is required, good interfacial bond strength is a prerequisite, otherwise the properties may be governed by a section of significantly reduced thickness. For good adhesion, a certain amount of interdiffusion is required between the melts, which can be achieved when there is high compatibility, or solubility, between the melts.

Tables of compatible and incompatible material combinations are readily available from both machine and materials suppliers, such as the one shown in Table 10.1. However, caution is required when using such tables, as it has been shown that changing from a particular grade of material to another can affect the

Table 10.1: Material compatibility table.

	ABS	ASA	EVA	PA6	PA66	PBT	PC	HDPE	LDPE	PET	PMMA	POM	PP	Modified PPO	PS-GP	PS-HI	SAN	TPU
ABS	+	+	+			+	+	–	–	+	+	–	–	–	*	*	+	+
ASA	+	+	+			+	+	–	–	+	+	–	–	–	*	–	+	+
EVA	+	+	+					+	+				+		+	+	+	
PA6				+	+	*	*	*	*			–	*	–	–	–	+	+
PA66				+	+	*	*	*	*			–	–	–	–	–	+	+
PBT	+	+		*	*	+	+	–	–	+	–	–	–	–	–	–	+	+
PC	+	+		*	*	+	+	–	–	+		–	–	–	–	–	+	+
HDPE	–	–	+	*	*	–	–	+	+	–	*	*	–	–	–	–	–	–
LDPE	–	–	+	*	*	–	–	+	+	–	*	*	+	–	*	–	–	–
PET	+	+				+	+	–	–	+	–	–		–	–	–		+
PMMA	+	+				–		*	*	–	+		*	–	–	–	+	
POM	–	–		–	–	–	–	*	*	–		+	–	–	–	–	–	
PP	–	–	+	*	–	–	–	–	+		*	–	+	–	–	–	–	–
Modified PPO	–	–		–	–	–	–	–	·	–	–	–	–	+	+	+	*	–
PS-GP	*	*	+	–	–	–	–	–	*	–	–	–	–	+	+	+	–	–

PS-HI	*	−	+	−	−	−	−	−	−	−	−	−	−	+	+	+	−	−
SAN	+	+	+	+	+	+	+	−	−		+	−	−	*	−	−	+	+
TPU	+	+		+	+	+	+	−	−	+			−	−	−	−	+	+

−: No adhesion
*: Poor adhesion
+: Good adhesion
ABS: Acrylonitrile-butadiene-styrene
ASA: Acrylonitrile-styrene-acrylate
EVA: Ethylene-vinyl acetate
HDPE: High-density polyethylene
LDPE: Low-density polyethylene
PBT: Polybutylene terephthalate PC: Polycarbonate
PET: Polyethylene terephthalate
PMMA: Polymethyl methacrylate
POM: Polyoxymethylene
PPO: Polyphenylene oxide
PS-GP: General purpose polystyrene
PS-HI: High impact polystyrene
SAN: Styrene-acrylonitrile
TPU: Thermoplastic polyurethane

bond strength. Occasionally, manufacturers may also seemingly disagree on the adhesion properties of materials. Since processing conditions also affect adhesion, experimentation may be required to ascertain optimum conditions for any given material combination. Additives are available to aid the compatibility of materials and not surprisingly, they are called compatibilisers. Through the addition of these materials it is possible to chemically bond some non-adherent materials. Compatibilisers usually contain a third polymer that bonds to, or is soluble in, the two materials. Numerous compatibilisers are commercially available to bond various immiscible materials together, for example, polyamide (PA) 6 and polypropylene (PP) can be bonded by the addition of a maleic anhydride grafted PP. However, compatibilisers tend to be expensive and both the number of chemical sites available at the interface (a result of the number of grafts per chain) and the molten contact time are limiting factors in the final bond strength as there is generally little interfacial mixing.

As well as adhesion, there are other material characteristics that also need to be considered when moulding with materials of different generic families, e.g., the levels of relative shrinkage and thermal expansion values, which may need to be matched or careful consideration given to the requirements before final material selection takes place.

In the case of many co-injection techniques, differences in mould shrinkage and thermal expansion can lead to problems such as sink marks, warpage and residual stresses. With over-moulding techniques, differences in shrinkage or the coefficient of linear thermal expansion (CLTE) can produce high stresses between constrained materials. The result in both cases can be the same, i.e., premature failure.

Consideration of the long-term properties of the various plastic materials making up a multimaterial component is also required, especially if the moulding is put under stress, e.g., during the processing stage if unsuitable processing conditions are used. In service, the moulded part may experience factors such as mechanical stress, chemical attack or high temperature; therefore, all relevant design parameters should be considered as well as those of shrinkage and CLTE previously mentioned. Dependent on the application these parameters may include:

- Time–temperature effect: many properties are temperature dependent. The stiffness, ductility and impact strength may vary considerably with temperature; therefore, testing should be carried out under conditions representative of those found during use.
- Fatigue: dynamic fatigue can occur when stress is applied periodically during use, in applications such as bearings.
- Creep: where parts are exposed to constant loading there may be a change of stiffness over time, which is termed 'creep'.
- Environmental stress cracking: plastics may be embrittled by exposure to water, light, temperature and oxygen, as well as chemical attack.

This section has given an overview of material selection that can be applied not only to multimaterial mouldings but also, in many cases, to standard plastic components. If all relevant process and design criteria are considered at an early stage, conversion to multimaterial processes should be a relatively painless transition. Further material considerations for individual processing techniques, along with current applications and markets, can be found under the relevant technology section.

10.9 Process technology

This section is split into five main areas, the first four focus on multimaterial systems, and the fifth section considers other possibilities such as methods to produce foam materials:
- Multicomponent IM.
- Assisted moulding.
- Multi-shot moulding.
- Over-moulding.
- Others.

Each of these areas will be now be discussed in turn.

10.9.1 Multicomponent injection moulding

This method describes a process whereby plastics are injected into one mould during a single injection cycle and the most common process in this category is co-injection moulding. Other less common variants are bi-injection and intermittent techniques.

10.9.2 Co-injection moulding

Co-injection moulding is a variant of the standard injection process and has been in use since the early 1970s. A number of terminologies are used that can encompass this process, such as sandwich moulding, 2K (two-component) or 3K (three-component), and dual injection or multicomponent. For the purpose of this chapter, co-injection moulding is the preferred terminology. This technique offers the advantages of combining two or more material properties to produce a 'sandwich' structure, which is achieved by making sequential injections into the same mould with one material as the core and another as the skin (Figure 10.4).

The result is the distinctive sandwich structure of skin and core as shown in Figure 10.5.

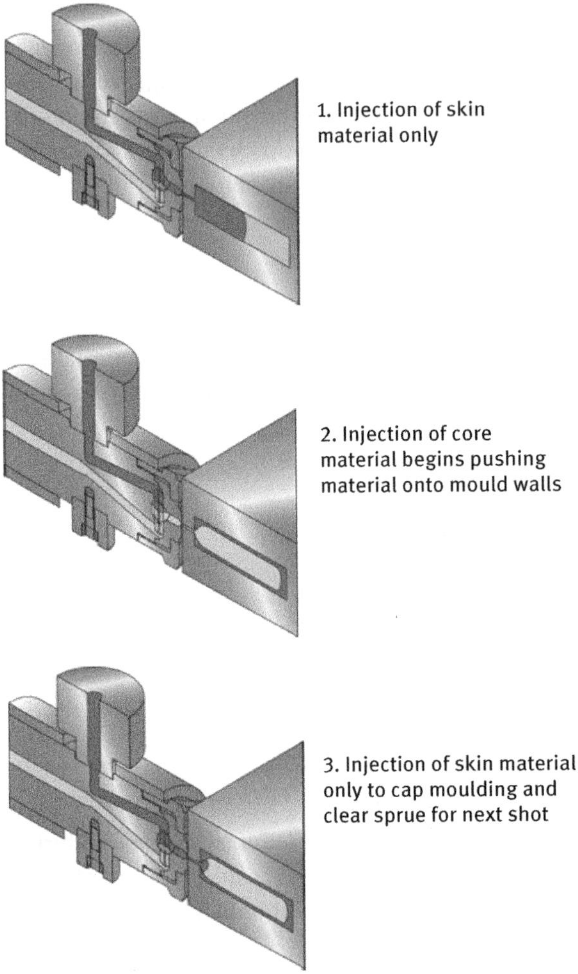

Figure 10.4: Encapsulation of core material by skin during co-injection moulding.

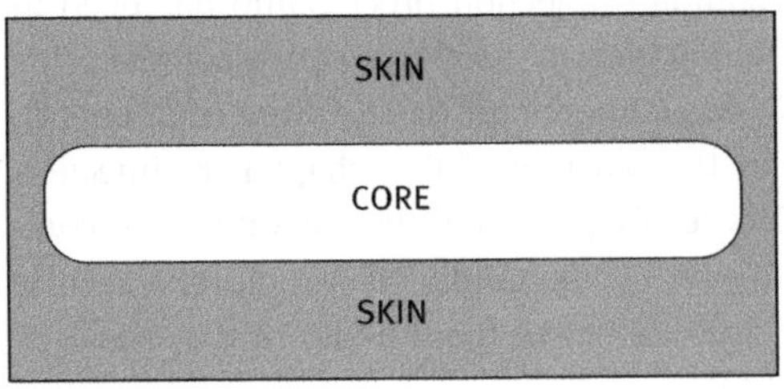

Figure 10.5: Cross-section of the sandwich structure of a co-injection moulding.

Co-injection moulding offers numerous possibilities in terms of a variety of material combinations, some of which are shown in Table 10.2.

Table 10.2: Current applications for co-injection moulding.

Material combination	Properties	Application
Soft feel skin/hard core	High strength core with soft feel skin	Door handles, gear lever
Unfilled skin/core with conductive filler	Electromagnetic interference shielding	Computer housings
Virgin skin/recycled core	Environmentally friendly production, cost saving	Garden furniture, automotive bumpers and fascias
Unfilled skin/reinforced core	High surface finish, structural performance	Automotive door handles
In-mould paint, variable core	No finishing of product required after moulding	Wheel trims
Unfilled skin/foamed core	Good surface finish, low density, high rigidity	Automotive body panels
Pigmented skin, uncoloured core (or reverse)	Reduced pigment cost, aesthetics	Yogurt pot

10.9.2.1 Material selection for co-injection moulding

One of the most difficult technical problems with co-injection moulding is that the core material must be prevented from large-scale mixing with the skin material, in order to retain a consistent skin layer thickness and resultant properties. It must also be prevented from penetrating the skin of the moulding. In addition, there are limitations in the permissible variation of flow characteristics between the two materials. Rheology plays a large part in the relative skin/core distribution; therefore, the viscosity of the materials are of the utmost importance as they affect the process dynamics and resultant core distribution. As a general guide, in order to keep the sandwich configuration and layer thickness consistent, the skin should have the same, or preferably a slightly lower, viscosity than the core. If the skin viscosity is too high, the core melt will flow through the skin and form the surface layer. Core distribution can also be controlled by adjustment of the injection speed and duration, polymer melt temperatures and mould temperature. However, the use of similar moulding temperatures for both materials is recommended, since they are processed simultaneously. The rheology of the materials will be discussed in more detail later in this chapter.

The process does have limitations as there is a need for the skin and core materials to be compatible with each other in terms of adhesion and shrinkage. Adhesion of the layers is necessary to prevent the core material becoming detached from the skin, especially if the moulding is likely to be exposed to mechanical loads; therefore, materials must be compatible, or a suitable compatibiliser used, in the core component. The use of compatibilisers in the core component of the co-injection moulding process was developed and patented by the Rover Group in collaboration with the University of Warwick [1]. Researchers from Warwick also developed and reported methods to mechanically interlock immiscible materials for co- injection moulding [2].

10.9.2.2 Process sequence

During co-injection moulding two compatible melts are injected either sequentially or simultaneously into the mould, thus forming a layered structure. The melt injected first forms the skin, whilst the melt injected afterwards forms the core.

Using two polymers with different properties makes it possible to obtain unique property combinations that are not possible using ordinary IM. A number of these include commercial applications, which are shown in Table 10.2.

The co-injection process was first described and developed by ICI in 1970 and was developed to overcome the surface finish limitations inherent in the structural foam process. Foam mouldings have a rough, irregular surface finish which vary in quality from moulding to moulding. Using a solid skin with a cellular foam core, it is possible to obtain a surface finish equal to a solid part but with added rigidity due to the foam core. There is also the reduction in material cost associated with the foam process as it requires fewer raw materials and is therefore cheaper to produce. This method is often used for products with thickness above 4 mm, for example, some automotive body panels.

10.9.2.3 Co-injection moulding: different techniques

A number of variations of the co-injection moulding process have been developed and can be split into two types, sequential injection and simultaneous injection. These methods will now be introduced and the advantages and disadvantages discussed.

- Sequential Injection: Single-Channel Technique

The single-channel method was patented by ICI in 1970 and was the first commercial co-injection technique. An IM machine with two cylinders is used, one for the skin material and one for the core. The polymer melts are injected sequentially into a mould, first the skin and then the core. A specifically designed valve is used, which allows a first injection of skin material only. At a preset point, the flow is stopped and the core is injected. At the switch point from one extruder to another there is a pressure drop in the mould, which is the major limitation of this method. This switching of polymer flows can cause the flow to stop, giving surface defects such as shadow marks or gloss marks on the mouldings. In terms of skin /core structure, as with all co-injection mouldings, changing IM parameters or the relative material viscosity can be used to control material flow within the process. This process is mainly used for thick sectional parts with foam cores.

– Sequential Injection: Mono-Sandwich Technique

The mono-sandwich technique was developed by Ferromatik Milacron and uses two materials layered in a standard cylinder, as shown in Figure 10.6, which is achieved by melting the skin material in a separate side extruder that plasticates the material. A special hot-runner system then leads the molten material through to the front of the screw in the main cylinder. The resultant melt pressure pushes the screw backwards. When the specified amount of melted material has accumulated in front of the screw, the screw starts rotating and feeds the core material. The injection is then achieved in the same way as for normal IM, i.e., by pushing the screw forward.

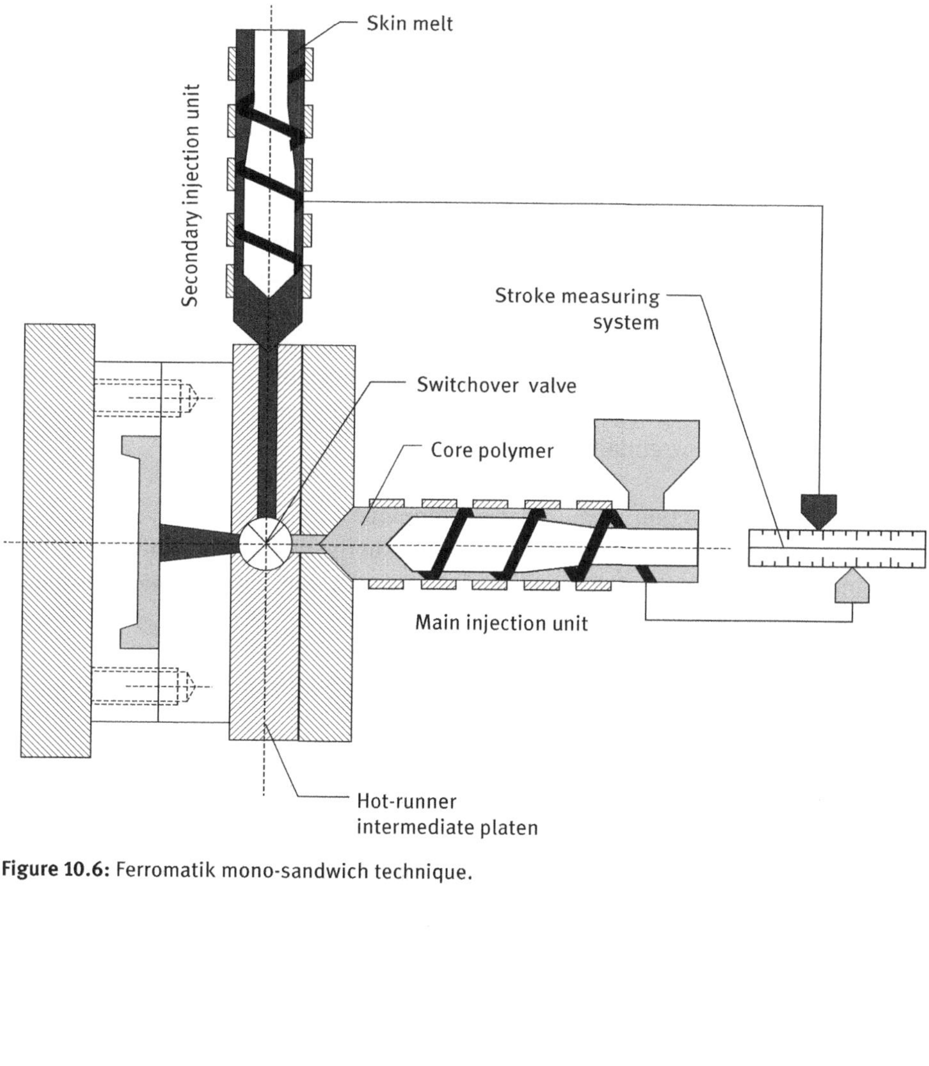

Figure 10.6: Ferromatik mono-sandwich technique.

There is little or no mixing of skin and core melts in the screw as the dynamics of the screw push the material forward rather than mixing the melts together at the material interface. The maximum amount of core material that can be encapsulated will be entirely dependent upon the mould geometry; for simple symmetrical parts this value will be around 65–75% by volume. The main advantage of this method is that a standard IM machine can be converted to a sandwich machine simply by connecting a side extruder to the main injection unit. This method is also particularly good where very thin-walled parts are required. A further advantage reported by the manufacturer is the speed of colour and material change, compared with other sandwich techniques, due to the relatively simple construction. Since the injection is made in the same way as for normal IM, the process control is fairly simple and similar to that of standard injection techniques. The drawback to this method is that by feeding through one injection unit only, there is a lack of the detailed control of the skin/core configuration which is required when moulding complex shapes.

Addmix originated a modified version of the mono-sandwich injection technique, which is similar to the Ferromatik technique but without an ancillary extruder. The second material feedstock is regulated and fed to the screw from a separate hopper, which is controlled by a stroke measurement system, and injection is achieved *via* a special nozzle. In this way the polymers are layered into one injection cylinder. The moulding then proceeds as per conventional techniques.

In summary, sequential injection techniques provide a cheap and useful method of producing sandwich mouldings. It is a technique more suited to simple geometries as the main problem with sequential injection is the lack of control of the skin/core distribution. During melt stream injection, both the skin and core are controlled together using just one velocity profile, therefore, skin thickness cannot be adjusted in various parts of the moulding. To overcome these problems, the simultaneous method was developed, in which skin and core velocities can be controlled separately giving enhanced control.

– Simultaneous Injection: Two-Channel Technique

The two-channel method, developed by Battenfeld in the mid-1970s, includes a phase of simultaneous injection. The process sequence can vary but a typical example would be:

1. Injection of skin material to a preset switch point.
2. Injection of core material begins so both skin and core material flow together.
3. Injection of just core material.
4. Injection of just skin material.
5. Part packing and cooling followed by ejection.

This co-injection moulding process involves the injection of molten plastic into the mould cavity to create the skin layer. After a certain preset time, usually in the region of 0.1–0.3 s, a second plastic that will make- up the core is injected and for a period, there is simultaneous injection of both materials. This simultaneous period of flow is where this process gets its name and also how it differs from the sequential techniques described earlier. The injection of core material pushes the moving layer of skin material against the cavity walls where it cools and solidifies. The final stage of mould filling is injection of the core material only, although sometimes the mouldings are 'capped' with a skin layer to complete encapsulation of the core.

Two injection units are used in this method, which are joined through a specially designed nozzle. In the Battenfeld design, the nozzle is equipped with two separate concentric channels that can be independently, hydraulically opened and closed, which allows the process sequence to be carefully controlled.

A phase of simultaneous injection of skin and core materials avoids the problems inherent in the single- channel technique by maintaining a constant flow front velocity. This can be seen by looking at Figure 10.7 and comparing the pressure profiles and screw velocity of the simultaneous and single-channel methods. In the sequential method, the period of stagnation after injection of skin (A) but before injection of core (B) can be clearly seen, resulting in a drop of cavity pressure and a period where there is no movement of material in the screw. This demonstrates

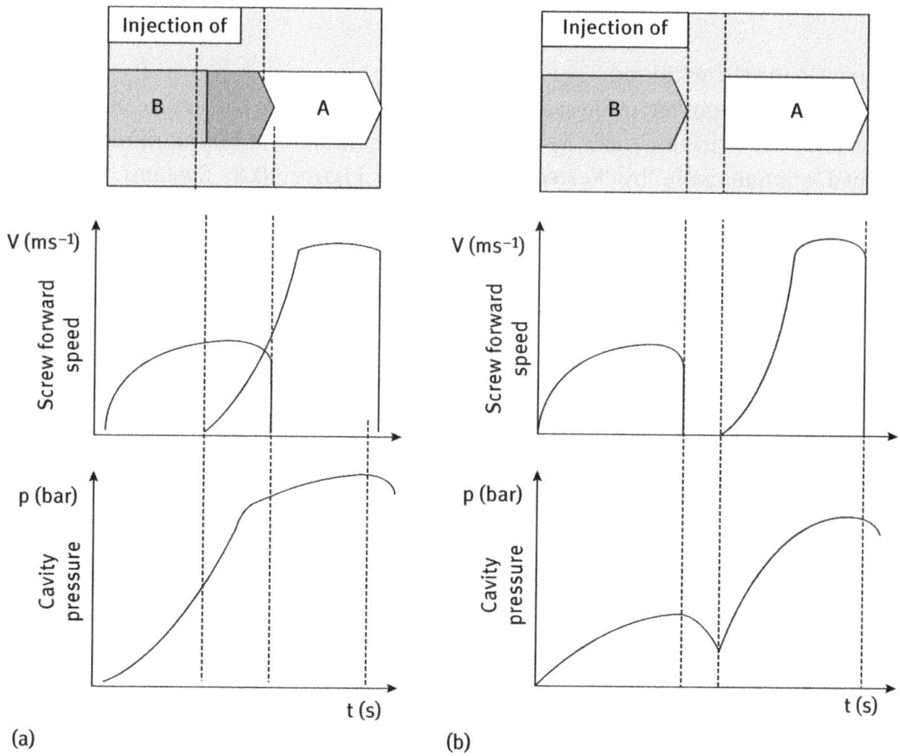

Figure 10.7: Comparison of mould filling of simultaneous (a) and sequential (b) IM.

many of the limitations of sequential injection discussed earlier; however, this pressure drop is less apparent in the simultaneous method.

The length of the simultaneous injection phase depends on both the properties of the material and the mould geometry, with a typical duration of 25% of the injection time for the skin component. Mould filling dynamics dictate the optimum distribution of core material is obtained if the skin viscosity is kept slightly lower than that of the core, which is due to rheological factors that will be explained in Section 10.9.2.5. The separate operation of two injection units, i.e., separate control of the velocity profile of the skin and core, makes it easier to control skin thickness in various parts of the moulding. Additional skin material can also be injected during the holding time to seal the gate area, which also ensures the nozzle is clear of core material and prepared for the next shot.

Again, due to the dynamics of mould filling, material entering the tool and cooling on the walls near the gate can be remelted and flushed away, due to frictional heat generated by the incoming molten flowing melt, which can lead to variations of skin thickness and leave the skin near to the gate region much thinner than on the rest of the moulding. This effect is generally more pronounced on the opposite side of the gate due to the higher shear experienced in this region. In order to overcome this, the three-channel technique was proposed.

- Three-Channel Technique

With the three-channel technique, an extra channel is used in the centre of the gate in order to retain skin thickness in this area. An example of a hot-manifold system developed commercially by Kortec is shown in Figure 10.8. Systems are also

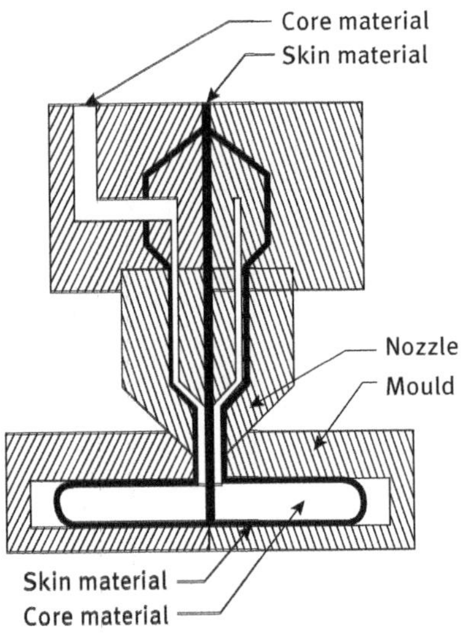

Figure 10.8: Kortec three-channel nozzle technique.

marketed by other companies including Kona, Incoe and Battenfeld. The extra channel can reach the opposite side of the moulding, enabling the two surfaces to be regulated separately and surface thickness controlled. This special design can only be used with a central gate, otherwise the skin/core distribution will be irregular as detailed in Figure 10.10. For other gate geometries or multicavity mouldings, the two-channel or single-channel technique is preferred.

A three-layer technique to combine immiscible material combinations was developed by the Billion Corporation of France. Their solution to polymer incompatibility when using the sandwich IM process employed a third intermediate polymer layer as a binder adhesive, which is analogous to methods used in extrusion blow

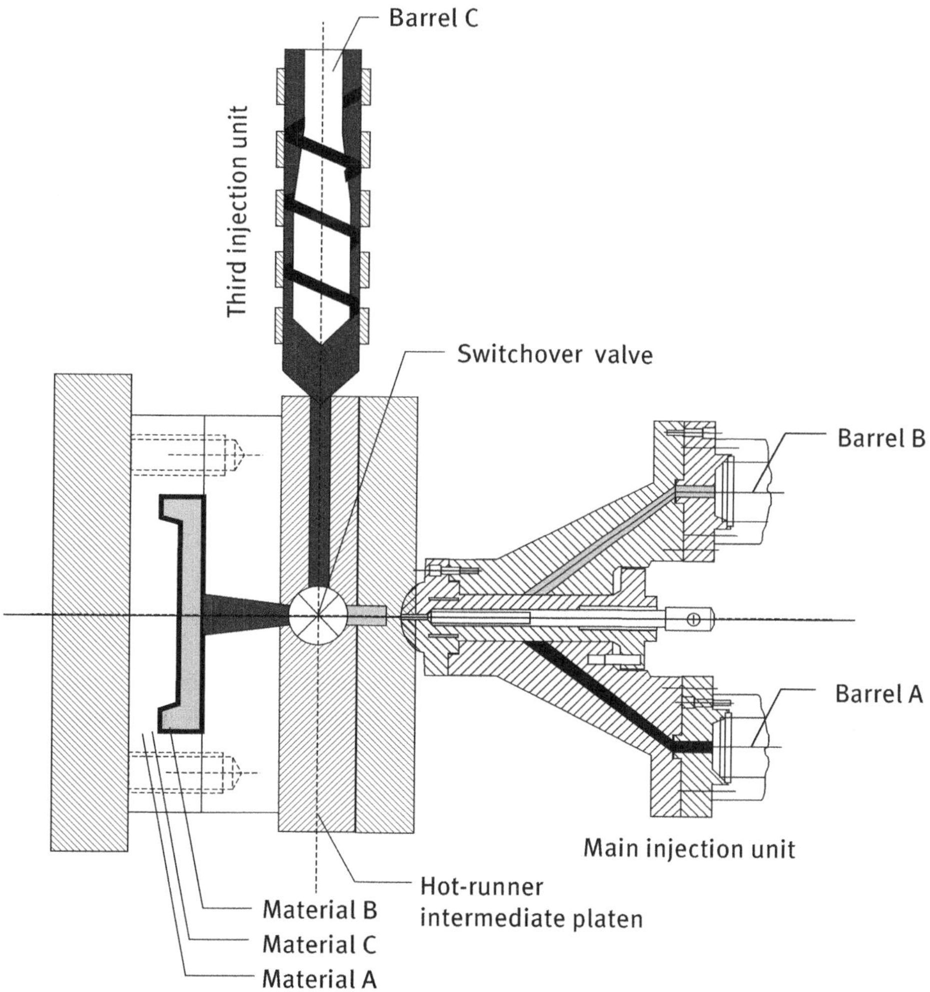

Figure 10.9: Battenfeld three-channel technique.

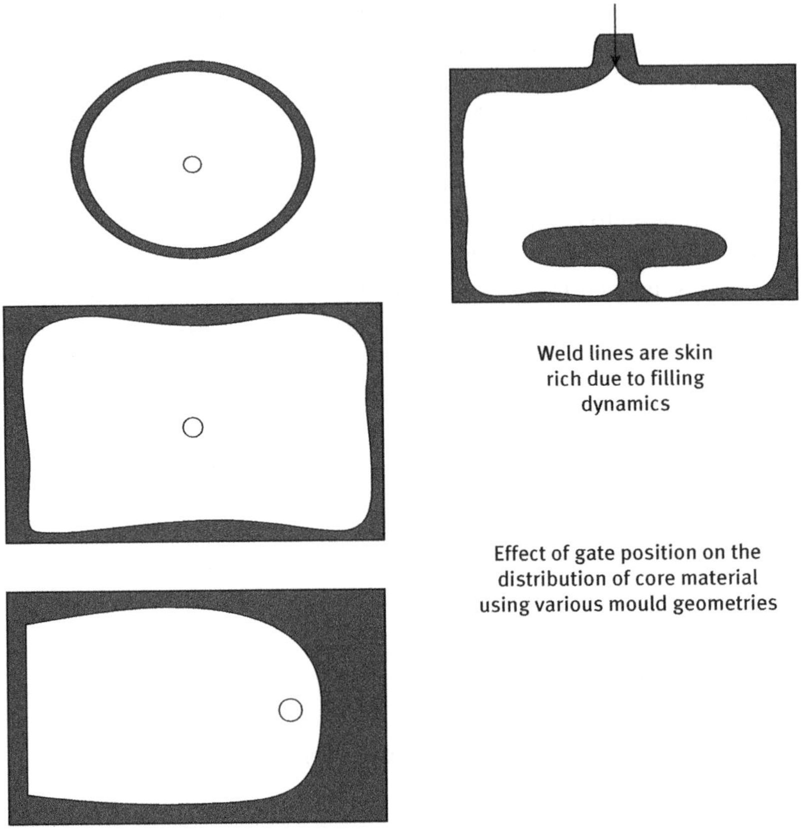

Figure 10.10: Gate position effects.

moulding. However, there are obvious machine cost disadvantages here, as the runner system is complex and a third injection unit is required.

The Battenfeld solution to three-channel moulding is shown in Figure 10.9 and utilises machine configurations also employed in multi-shot techniques, as described in Section 10.11.1. The vertical unit can be used to feed the third material into a combined nozzle system, which is based on their two-layer technology described in Section 10.9.2.

10.9.2.4 Part design and tooling requirements for co-injection moulding

Normal IM tooling can be used for both sequential and simultaneous co-injection moulding providing the following factors are considered. As mould filling and the resultant skin/core distribution is strongly dictated by the gate location, as shown in Figure 10.10, the injection gate must be designed with consideration for the resultant skin/core distribution, otherwise unsatisfactory mouldings may result. If more than one gate is required, or a break in the flow front is produced, the weld line will be skin rich, as the core materials will not completely meet due to the filling dynamics of co-injection moulding. The skin will always pack certain areas before the core material can reach them (Figure 10.10).

10.9.2.5 Rheology and mould filling: why and how co-injection moulding works

For readers who have not seen co-injection moulding in action, it can be very difficult to understand how co-injection moulding can produce the distinctive skin/core structure. It should be remembered that the skin material is injected first to cool and form the skin against the tool wall. The core material then pushes against the skin to cause it to penetrate deeper into the mould cavity. As a result of this process, the core remains encapsulated, providing the viscosities of the skin and core are similar; however, if they are mismatched, this filling pattern is affected. Before this is discussed further, a recap on the basics of rheology and the effects of various parameters on viscosity is required. How a material responds to changes in temperature, shear and pressure can greatly affect processing methods and strategies. Other material properties such as density, elasticity, thermal expansion and thermal conductivity can also affect processing behaviour through, for example, decisions regarding the cooling rates required, shrinkage tolerances and die swell. However, first we will focus on the basics of rheology.

– Rheology Basics

Rheology deals with material deformation and flow and examines the relationship between stress, strain and viscosity. Most rheological measurements relate to simple shear such as shear viscosity and normal stress differences. Material melt flows can be split into three categories, each behaving differently under the influence of shear, as shown in Figure 10.11.

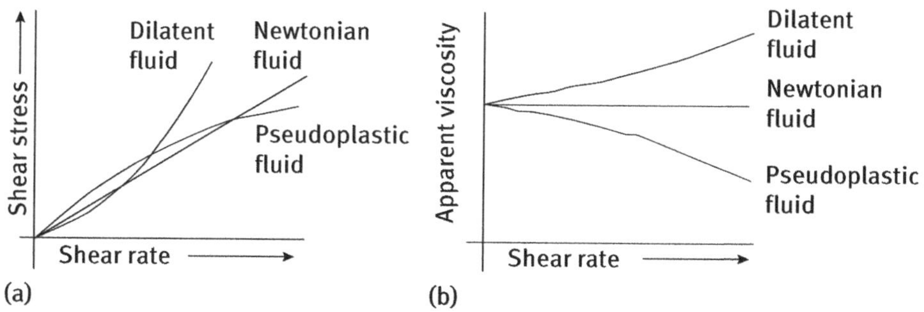

Figure 10.11: Typical stress/shear relationships (a) and apparent viscosity/shear curves (b) for dilatent (shear thickening), Newtonian and non-Newtonian pseudoplastic fluids.

In an ideal Newtonian liquid flow, the viscosity is independent of shear rate. Plastics fall into the category exhibiting shear thinning behaviour, which means they respond to increased shear by a drop in viscosity. However, it can be seen from Figure 10.11 that all melts approximate to Newtonian fluids at very low shear rates. Unfortunately, IM tends to occur at high shear rates. Temperature and pressure also affect the viscosity of the polymer melt; at higher temperatures the viscosity drops, whilst as pressure increases the viscosity increases. A complex picture now emerges, as during IM there are both steep gradients of temperature and pressure. Therefore, during the co-injection process, moulding materials will be subjected to viscosity change throughout the moulding cycle with the skin and core undergoing different shear, temperature and pressure histories.

A further consideration during IM is the response of the skin and core polymer flows to the deformation processes present, which is a complex mixture of shear, elongation and bulk deformations. One way to try to model such effects is to use an element model. A representation of a theoretical polymer element is shown in Figure 10.12 and indicates how stresses can occur in a number of directions. Representations such as these are often used to show the force balance on an element. There are three components of normal stress (xx, yy and zz in Figure 10.12) and six components of shear stress, giving a total of nine stresses (on the three visible planes) using this method of analysis.

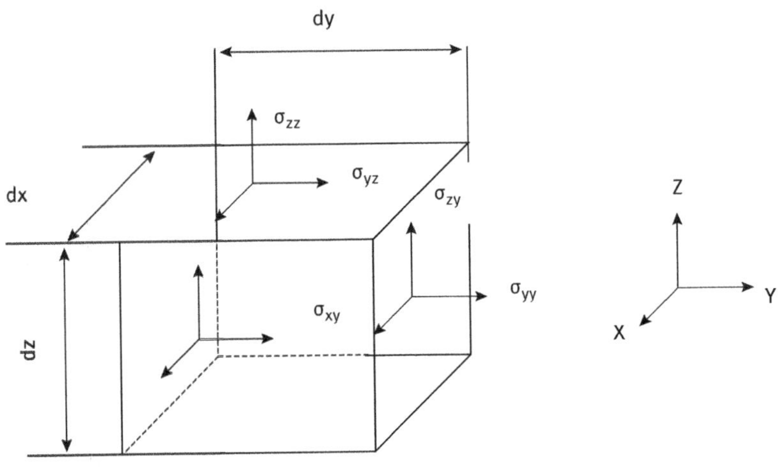

Figure 10.12: Possible stresses on an element.

Shear flow is produced when stress is applied tangentially, extensional flow is the result of normal stress applied to the surface of the material and bulk deformations result when normal stress is applied to all faces.

During shear flow, polymer chains are deformed and orientate in the direction of flow. (The results of this orientation during standard IM were discussed in Chapter 2, Section 2.1.) The tension of the polymer in the flow direction is called the first normal stress. In contrast to this, flow in other directions is relatively small. First normal stress can be defined by the difference between the normal stress component in the flow direction and the normal stress component in the direction of the shear plane. Combinations of some or all these forces are applied to the molten polymer material during the moulding process and are dependent upon the location of the material in the moulding. Analysis using methods such as these, enables material interactions to be better understood.

– Impact of Rheology on the Dual Injection Process

Now some of the factors affecting viscosity have been introduced, the effect of viscosity on skin/core formation can be discussed (Figure 10.13).

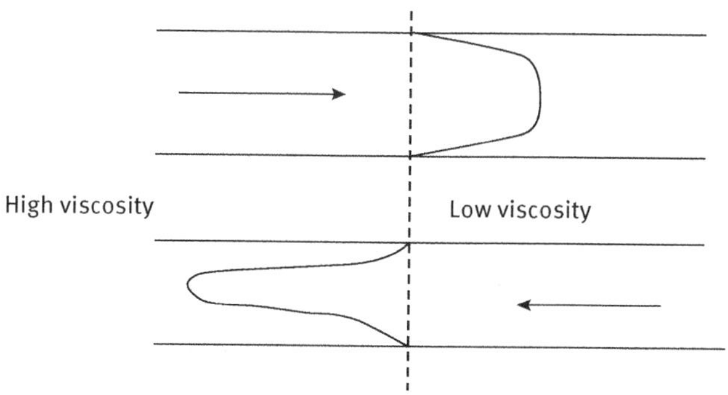

Figure 10.13: Interface behaviour of melts of different viscosity.

High viscosity melts flowing into low viscosity melts will push the low viscosity material in front of it. Low viscosity melts flowing into high viscosity melts will jet through areas of least resistance, giving an effect termed 'melt fingering'. This type of effect can also be seen in GAIM where the core gas component has an effective viscosity of zero. The injected gas follows *via* the path of least resistance, making channels in the hottest, thickest and least viscous parts of the melt stream. Breakthrough of the core component during co-injection moulding can cause unwanted surface defects, whereby the core material is visible in the corners of the mouldings. Hence, a high percentage of co-injection studies have investigated the relationship between the relative viscosities of the materials used within the process and the resultant skin/core distribution to substantiate these findings.

– Previous Studies on the Effects of Viscosity Ratios

It was reported as far back as 1974 that two-phase flows, such as those found in co-injection moulding, are sensitive to differences in the rheological properties of the melts, as different rheological combinations produced different skin/core ratios. Later, the simultaneous system driven by viscosity encapsulation phenomena was proposed, as shown in Figure 10.11.

It has been determined that the most uniform core distributions are achieved at ratios of skin to core between 0.8 and 1.8. Within these ratios about 60% core could be concealed within a square flat plaque moulding without breakthrough, and about 70% core in a disc mould.

Apart from the initial viscosity, the moulding parameters found to have the greatest effect on the skin/core distribution are melt temperature, injection rate and the length of the simultaneous phase. Too short an interval between the injection of skin and core can result in breakthrough, too long and too much skin is forced to the outer edges of the moulding. In terms of other processing effects, it has also been found that, as in IM, there is extensive melt flow in the region of the gate during the postfilling stage of co-injection moulding, which can have implications for the skin thickness around the region of the gate.

From the IM process it is well known that both the temperature of the mould and the speed at which the materials are injected into the mould cavity also determine the end structure. The temperature of the mould will affect the rate at which the skin material freezes, the greater the difference in temperatures of the skin and mould surface, the quicker the material will solidify. The rate at which the skin material is injected will determine how much time the leading edge of the skin material has in contact with the mould surface. A greater speed leads to the skin material penetrating further into the mould and, for co-injection moulding, can help prevent breakthrough. Slowing down the core material injection speed gives a similar, though not exactly the same effect, because as skin injection speeds are increased, especially at low mould temperatures, instabilities in the flow can also increase. These instabilities can result in better mixing at the interfacial region between the skin and core material; however, taken to extremes, these instabilities could also cause breakthrough in the mouldings.

10.9.2.6 Setting optimum parameters

In terms of the interface, the main factor in adhesion strength is the thickness of the skin layer; as the skin thickness increases, the bonding time available for adhesion also increases, resulting in a slower cooling interface. A thinner layer would also experience higher shear from incoming molten material and be more likely to be remelted and swept away into the melt stream.

Therefore, to maintain consistent skin thickness across a co-injection moulding, viscosity and the effects of other moulding parameters need to be controlled. It has been found that injection speed is the key parameter to obtain uniform skin distribution. During co-injection moulding, lower speeds have been found to be most effective and also minimise disturbances in laminar fountain flow, especially in the gate region, compared with higher speeds. Using materials of different generic families, be they immiscible or compatible, is likely to cause a different elastic response to shear and stresses within the mould at the skin/core interface, especially in high shear tooling. This fluctuation in stress will affect the mixing dynamics within the interfacial area and in some cases may cause defects on the surface of the skin/core interface. However, with the development of being able to combine immiscible materials *via* compatibilisation or mechanical techniques, it is likely that commercial interest in such material combinations will prompt further advancements. There are two obvious areas for this growth; firstly, the recycling potential in of co-injection technology, i.e., using mixed material combinations and secondly, the possibility of combining distinct manufacturing processes *via* one in-mould operation. An example of this is the production of in-mould primers for PP to improve paintability or adhesion, which removes the need for the pretreatment stages currently used. Examples of current applications for co-injection techniques now follows.

10.9.2.7 Co-injection moulded application case studies

- Preforms for Blow Moulding

Injection blow moulding is commonly used to produce containers such as PET carbonated drinks bottles. During this method, preforms are injection moulded and then inflated in a separate operation. Core bars are used to transfer mouldings from the moulding station to the inflation station. Multilayer extrusion blow moulding has been practised for some time with nine, or even more, distinct layer configurations possible. Multilayer injection blow moulding is a slightly newer technique.

- Thermoplastic Olefinic Elastomer Fascias and Bumpers

Automotive applications need to be cost-effective and potentially recyclable to meet End-of-Life Vehicle Directive requirements (see Section 10.9.2.8). Co-injection offers not only the potential to incorporate recyclate into the core but also the utilisation of postindustrial painted recyclates; one such example is the Ford P207 fascia which, to comply with part recyclability, contains recycled material. Once the initial investment in machinery is overcome it is also an economical approach in terms of both material utilisation and recycling potential.

10.9.2.8 Recycling and legislation

The thermoplastic olefin fascia is an example of the recycling potential inherent in the co-injection moulding technique. Environmental legislation affects a number of sectors of the plastic industry and while this legislation varies globally, in Europe, packaging manufacturers must consider the directive 'Packaging and Packaging Waste Directive, 94/62/EC', automotive manufacturers and suppliers must consider the 'End-of-Life Vehicle Directive, 2000/53/EC', and the manufacturers of electronic and electrical equipment must comply with the demands of the directive on 'Waste Electrical Electronic and Equipment Directive, 2002/96/EC'. Other similar legislation is in place in countries such as the USA and Canada. What this means in real terms is that many current and future plastic applications will demand consideration of recycling capability. On a global scale, if IM alone was considered, the quantity of material requiring recycling is considerable. In this respect, co-injection moulding is an ideal solution as recyclate materials can be buried in the core and with the exception of transparent materials and tooling of complex geometries or multiple gating, standard IM process can easily be adapted to co-injection moulding. With this in mind, it is therefore highly surprising that this method is not more widely utilised.

10.9.2.9 Discussion and conclusions

Co-injection moulding provides processing routes which enable property combinations that are, in general, not possible with conventional IM. However, there are a number of factors that have tended to limit the commercial uptake of co-injection moulding technologies:
- High capital cost – co-injection machines are around 40% more expensive than an ordinary IM machine.
- Restrictions of co-injection moulding with regard to mould geometry: producing parts with sharp corners, changes in wall thickness, ribs or bosses are difficult due to the problems of distributing skin and core materials.
- Weld lines – at weld lines there is only skin material, and special solutions, like overflow channels, are necessary if the core material is needed here.
- There is a complicated and poorly understood relationship between rheology and process parameters. Interrelationships exist between the melt temperature, viscosity, and temperature differences between the skin and core melts.
- Restrictions with regard to material combinations. Even for compatible materials, the choice of a certain grade could influence adhesion and result in inferior properties of the moulded product. Processing conditions can affect the rates of interdiffusion of the skin and core, which can subsequently affect the properties of the final component. Again, the effects are complicated and not well understood.

However, with the increased need to meet environmental recycling legislation and targets, in addition to an increased understanding of the process, it is surprising that uptake of co-injection technology has remained low.

10.9.3 Bi-injection

During bi-injection, materials are processed simultaneously at different points into a tool. In this regard, it shares features of co-injection (simultaneous injection of two materials but through one nozzle) and core back moulding (where injection is at two points but in sequence). This method is discussed in detail later.

This moulding process has a shorter cycle time than core back moulding and can achieve the usual requirements of multi-shot, such as multicolour or hard/soft combinations. The knit-line is also stronger due to the higher temperature at the interface when the flows meet; however, the materials do not maintain good separation and definition at the interface, which can be a problem in potential applications for this technique.

10.9.4 Interval injection moulding

This method produces marbling-type effects and is generally, though not always, confined to mouldings of different coloured materials of the same type, which can produce a random or regular colour distribution pattern as required. It works by injecting the materials in either a simultaneous or sequential manner from a combined nozzle, as shown in Figure 10.14, and is similar to methods used to produce co-injection mouldings. In this case however, the injection does not produce a skin/core configuration, the material flows together and small-scale mixing operations take place but not at a level to completely mix the two materials. The two injection units are coupled using a special interval unit, inside which the mixing nozzle is located, as shown in Figure 10.15.

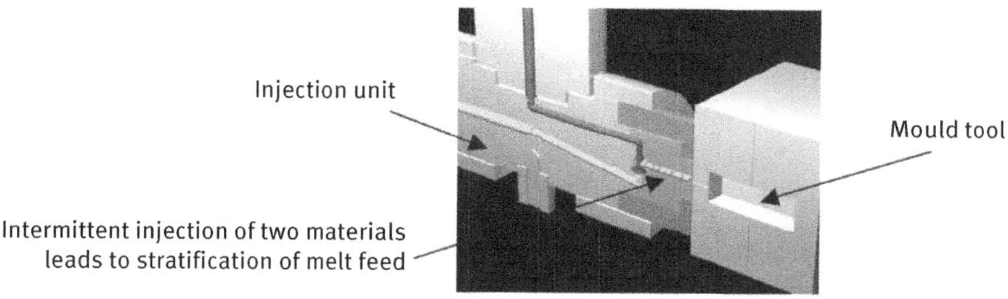

Figure 10.14: Interval IM.

The resulting colouration is controlled by the mixing dynamics of the materials in question and the shear they are subjected to, for example, by the size and position of the sprue. Injection can be simultaneous or set to an alternating cycle to give a pulse-type effect. The size of the injection steps and the speed of injection will also alter the colouration of the resultant moulding. Applications tend to be mainly for cosmetic applications such as covers and boxes, and applications such as buttons where the colouration can be used to decorative effect (e.g., Figure 10.16).

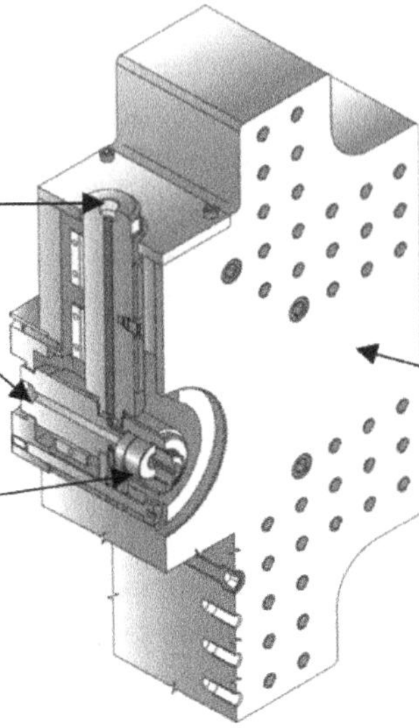

Figure 10.15: Unit for interval moulding.

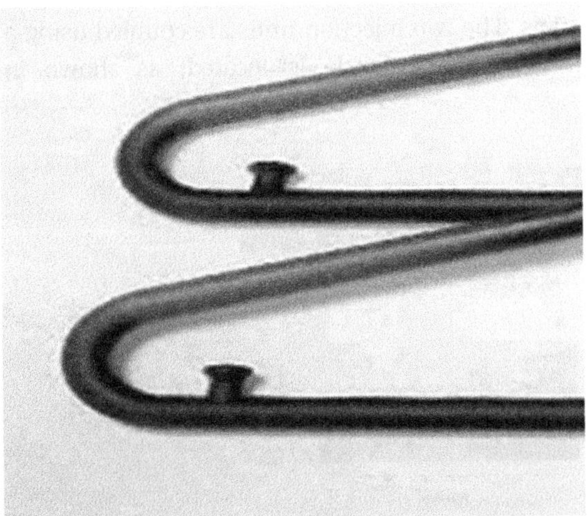

Figure 10.16: Marbling effect on a moulding.

10.10 Assisted moulding

10.10.1 Gas injection moulding technology

The characteristic feature of GIT is the filling of a mould with two different materials. Plastic materials form the first component and the second component consists of a gas, generally nitrogen. The two components do not mix. All standard IM machines which are equipped with a device for introducing gas are suitable for the GIT process. The injection of the gas may be performed by a machine nozzle or by a separate injection module in the mould.

10.10.2 Process technology

The process sequence begins with the injection of the first component – the plastic. A dose of approximately 70 ± 20% of the volume of the cavity is proportioned and injected.

Shortly before the conclusion of the material injection phase, the gas injection phase begins. The brief overlapping of the material injection and gas injection phases is intended to prevent a speed-break of the melt front and the switchover marking is governed by this. The best location for introducing the gas is in areas with large melt accumulations. The gas fills the cavity and forces the melt forward; design-specific cavities in the moulded part are the result.

Once the cavity is completely filled, the gas holding pressure phase begins and is applied until the part is dimensionally stable (maximum gas pressure is 400 bar). The gas pressure during this time is constant throughout the entire canal. Due to the relatively low gas pressure, there are correspondingly low interior pressures in the mould, in turn, necessitating low clamping forces of the IM machine.

After the initial solidification of the melt, the gas pressure is reduced. This takes place either by allowing the gas to escape into the ambient atmosphere, or by recovering a certain proportion (up to 90%) through the machine nozzle or the mould nozzle – the process is illustrated in Figure 10.17.

10.10.3 Patent situation

Development in GIT has been restricted to some extent by the number of patents that exist relating to this process. If a process that is protected by a patent is employed, expensive licence fees must sometimes be paid. This section is written without regard for the patent situation, which must be clarified if considering adoption of this technology.

10.10.4 Advantages and disadvantages of gas injection moulding technology

Depending on the mould construction, the following advantages can be realised:
- Greater range of configuration options in mould design.
- Material savings (generally 20–30%).

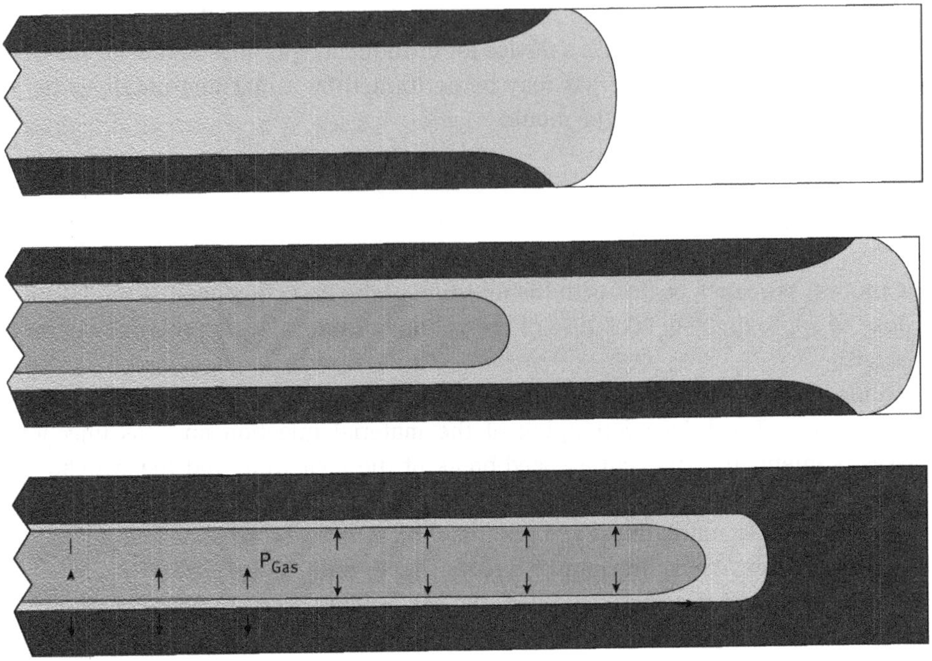

Figure 10.17: The gas IM process.

- Cycle time reduction for moulded parts with thick walls.
- Increase in mechanical rigidity with equal weight.
- Moulds without sink marks.
- More uniform shrinkage, lower residual stresses, significantly less distortion.
- Reduced clamping force.
- Implementation of long flow paths.
- Better surface in comparison to foamed parts.
- Simpler mould construction possible in some cases.
- Possibility of integrating thick and thin areas.

The most significant disadvantages of GIT are:
- Additional costs for gas-pressure generation equipment and pressure-regulator modules, gas, machine nozzles or mould nozzles, licence fees.
- Frequent jetting.
- Empirical derivation of setting parameters.
- Greater weight deviations.
- Hole at the injection point.
- Strength/tightness at sealing is frequently inadequate.
- Welding seams are typically more frequently visible.
- Injector must be cleaned more often in some cases.

10.10.5 Process Variations in the application of gas injection moulding technology

10.10.5.1 Gas injection through the machine nozzle

Here, gas is introduced directly through the machine nozzle. The process sequence is: the hydraulic GIT needle shut-off is opened, the plastic material is injected and the material cushion for the sealing process remains in the screw prechamber, as illustrated in Figure 10.18a.

The hydraulic GIT needle shut-off nozzle is closed, nitrogen is introduced, the melt core is displaced and the pressure holding phase is active. After the conclusion of the pressure holding phase, the recovery phase of the nitrogen gas begins. There is a brief retraction of the nozzle for pressure reduction – this stage of the process is shown Figure 10.18b.

Finally, the hydraulic GIT needle shut-off nozzle is opened and the material cushion is injected for sealing, as shown in Figure 10.18c.

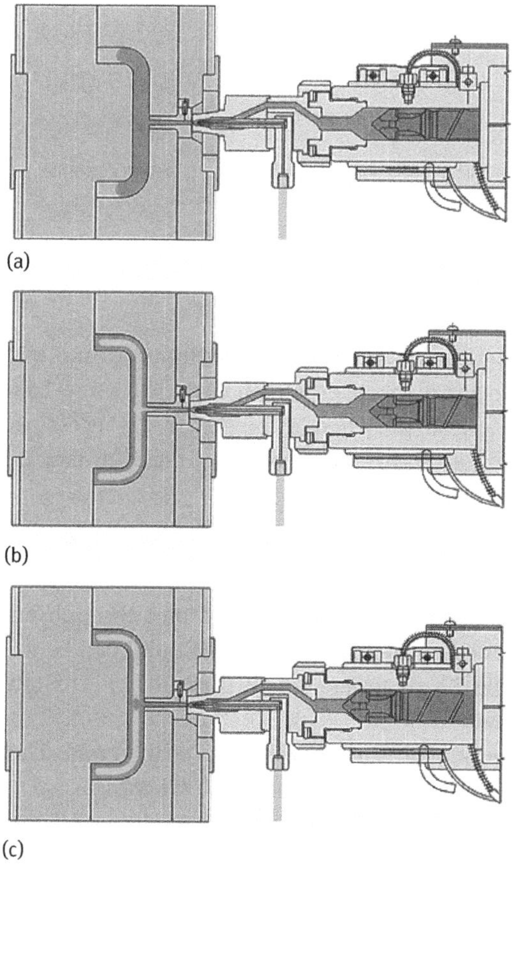

(a) (b) (c)

Figure 10.18: Gas injection through the machine nozzle.

10.10.5.2 Gas injection through an injector module in the mould

This can be carried out in two ways; the gas can be introduced either through the sprue or directly into the moulded part. The hydraulic needle shut-off nozzle is opened and the mould material is completely injected. No material cushion remains behind – this stage of the process is shown in Figure 10.19.

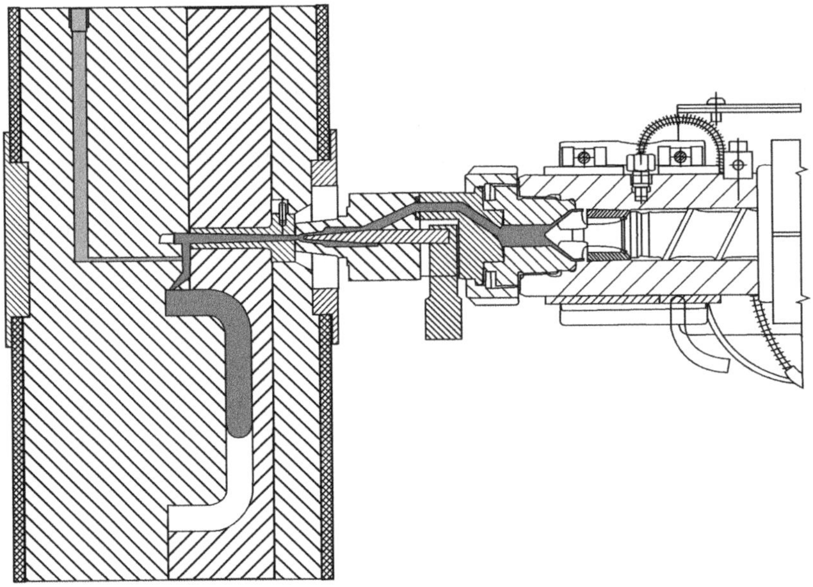

Figure 10.19: Gas injection through an injector module in the mould.

The hydraulic needle shut-off nozzle is closed and nitrogen is introduced into the mould through the injector module. The gas displaces the melt core. The screw can proportion doses as the holding pressure is provided through the gas pressure phase. Upon conclusion of the gas holding phase, the recovery of nitrogen is activated – this process stage is shown in Figure 10.20.

10.10.5.3 Gas injection in the moulded part

During this process, gas is introduced directly into the moulded part through an injector module in the mould. In this process, the gas introduction point can be located anywhere on the moulded part. The disadvantage here is that the hole created in the part cannot be automatically sealed.

In this process, it is possible to work with the nozzle open at intervals. The mould material is completely injected and no material cushion is left behind.

Nitrogen is introduced into the mould through the injector module and the holding pressure is realised through the gas holding phase. The delayed dosing

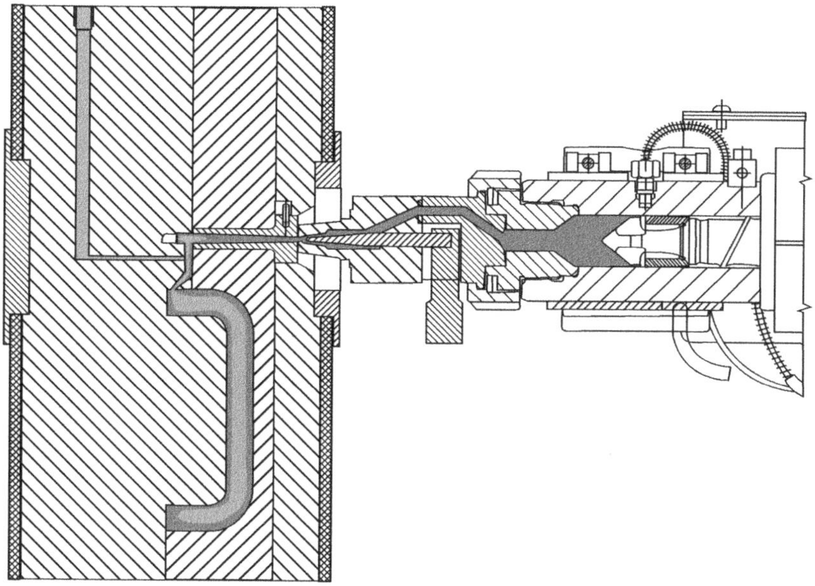

Figure 10.20: Gas injection through an injector module in the mould.

phase runs until the sprue is sealed. Upon conclusion of the gas holding phase, nitrogen recovery is activated and the screw can simultaneously begin dosing again. The process sequence is shown in Figure 10.21.

10.10.5.4 Melt blow moulding technology

Melt blow moulding technology is divided into two subprocedures, melt back pressure technology and melt extrusion technology in a secondary cavity. The advantage of these processes is the avoidance of switchover markings on the surface of the moulded part.

10.10.5.5 Melt back pressure technology/gas injection in the moulded part

Using melt back pressure technology, the entire moulded part is first filled with material (Figure 10.22a). Nitrogen is then introduced at the flowpath end, which forces the plastic melt back into the plasticising cylinder (as shown in Figure 10.22b). With the hydraulic needle shut-off nozzle open, the mould material is injected (the moulded part is filled completely). The hydraulic needle shut-off nozzle remains open, the gas is injected and the screw is withdrawn. After the completion of gas injection, the hydraulic needle shut-off nozzle is closed. The gas holding phase and recovery of the gas follow and the screw can simultaneously begin dosing again (Figure 10.22).

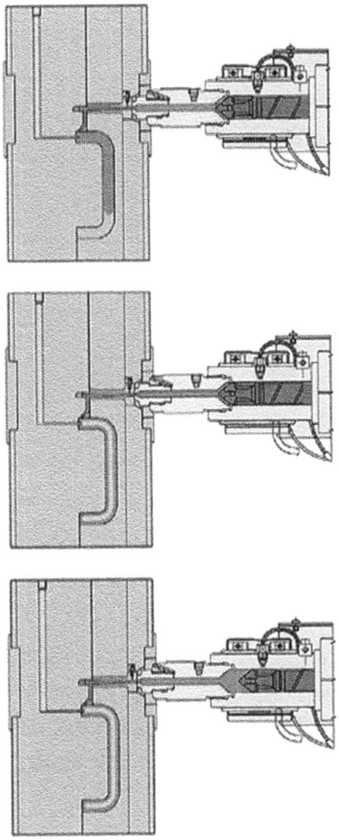

Figure 10.21: Gas injection in the moulded part.

10.10.5.6 Melt extrusion technology/gas injection in the moulded part

Using melt extrusion technology, the cavity is first completely filled with material (Figure 10.23a). Concurrent with the start of gas injection, one or more secondary cavities, into which the excess melt may be displaced, are opened (Figure 10.23b).

The process is carried out using an open nozzle. The material is injected, the moulded part is filled completely and the second cavity is closed. A hydraulic slide gate opens the secondary cavity and gas injection begins concurrently, with the excess material forced into the secondary cavity. The gas holding sequence runs to completion and the recovery of gas follows the gas holding phase. The screw can simultaneously begin dosing again (Figure 10.23).

10.10.5.7 Core pull technology

In core pull technology, the cavity is first completely filled and the moulded part is subjected to melt back pressure. Before or during the introduction of gas, an

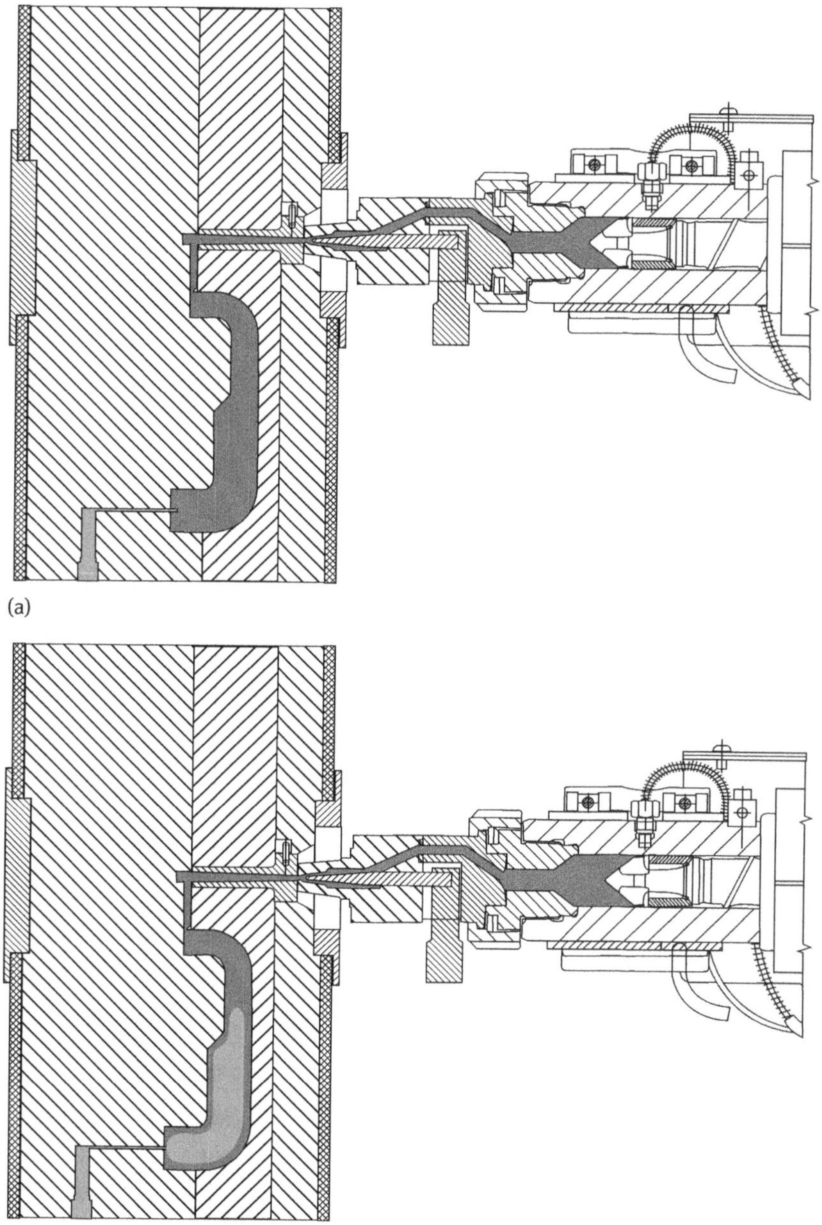

(a)

(b)

Figure 10.22: The melt blow moulding technology process.

additional displaced volume is created in the main cavity by the withdrawal of one or more of the cores.

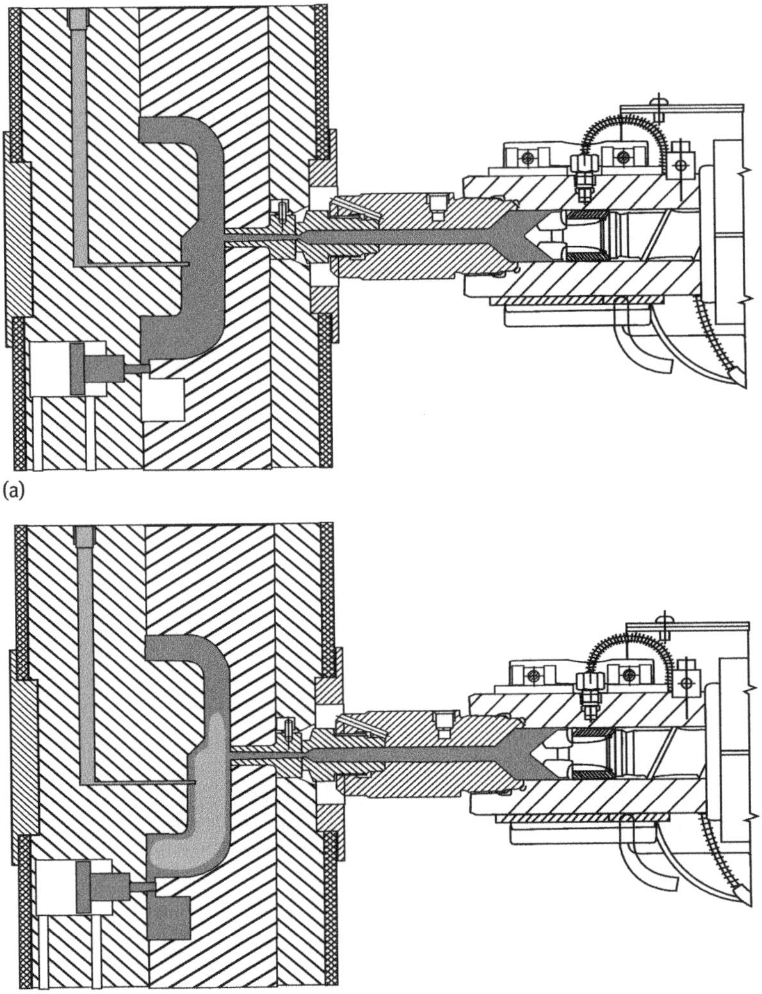

(a)

(b)

Figure 10.23: Melt extrusion GIT.

Using core pull technology, the process operates with an open nozzle. The core is inserted, the material is injected and the moulded part is completely filled. The core is withdrawn from the moulded part hydraulically, thus creating additional volume. The mould material is completely injected and the injection of gas into the moulded part occurs simultaneously. The gas holding phase then begins and once completed, is followed by gas recovery. After the gate is sealed, the screw can begin dosing again (Figure 10.24).

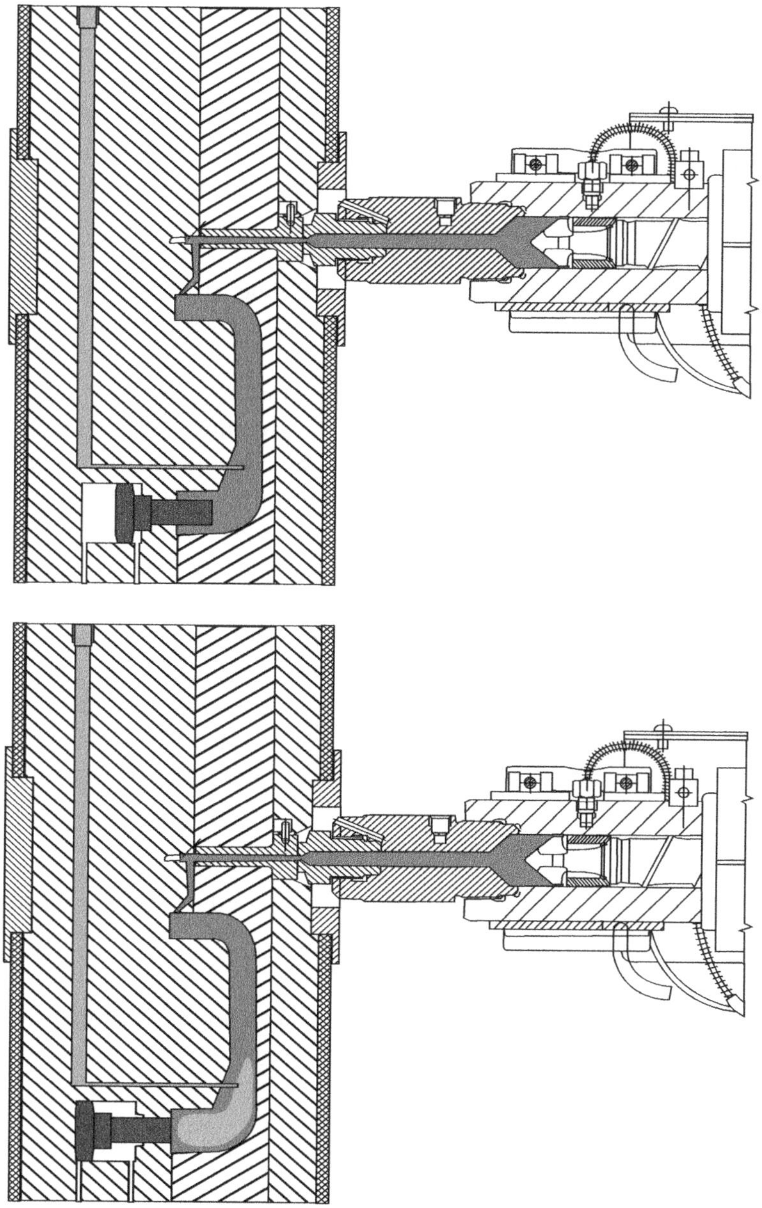

Figure 10.24: Core pull technology GIT.

10.10.5.8 Systems technology for the implementation of gas injection moulding technology

- Gas Pressure Regulation
  Gas pressure regulation consists of the pressure regulation module and the electrical control unit. The pressure regulation module reduces the established system pressure to the desired gas pressure and electrical control of the pressure regulation module is achieved through the machine control unit or an external memory with programmable controls. The switchover from melt injection to gas injection can take place either as a function of time, stroke or pressure. The hydraulic pressure of the IM machine or the internal pressure in the mould can be used for the pressure-dependent impulse.
  Two different design concepts exist for the control unit:
  - Stationary unit: the stationary unit consists of the pressure regulation modules and an electrical control unit which is integrated into the machine controls. This process option is implemented *via* the control unit and external operating elements are not necessary. The simple and direct entry of all necessary parameters *via* the user monitor of the machine control unit is especially advantageous with this system. In addition, the GIT device can be integrated into the monitoring function of many modern machines.
  - Mobile control unit: the mobile control unit consists of pressure regulation modules and a separate electronic control unit with a manual programming device. The link to the machine is created *via* an interface. The mobile gas pressure control unit can be quickly installed on any other IM machine using the GIT interface.
- Gas Injection Moulding Technology: nozzle Systems

Machine nozzles and mould nozzles are available for the introduction of gas into the moulded part.

- Machine Nozzle
  After the cavity is filled with the plastic melt, gas is also injected through the same machine nozzle, i.e., the injection gate and the gas opening are at the same location of the moulded part.
- Mould Nozzles
  The advantage of mould nozzles, in particular, is that the melt injection gate and the gas injection point do not have to be at the same location. As a result, several gas introduction points may be arranged on the mould piece to meet specific requirements; hence, separate gas runners can be controlled independently, allowing special requirements to be fulfilled. Mould nozzles are thus distinguished by their great flexibility which is advantageous in mould designing and modification. With regard to the manner of installation, a distinction is made between fixed and movable mould nozzles:

- Fixed mould nozzles: these mould nozzles are installed in a fixed position in the direction of mould ejection. They can be easily placed in the mould since they have especially small dimensions.
- Movable mould nozzles: when mould nozzles cannot be installed in the direction of mould ejection for design reasons, they must be installed with movement capability (the nozzle can then be inserted into the cavity or withdrawn from it). The system for moving the mould nozzle can be operated using nitrogen.

10.10.5.9 Configuration guidelines for gas injection moulding technology

- General

  Rod-shaped parts, such as handles, are especially suited for the GIT process. These parts are characterised by very high wall strength allowing:
  - The possibility of producing plastic parts in one process run, which have very high wall strength and no sink marks.
  - Material savings of up to 50%.
  - Significantly shorter cycle times resulting from shorter cooling times due to the reduced wall thickness.
  - Two-piece parts which previously had to be glued or welded together after the IM process can be manufactured in one process run.

- Cross-sections

  In the design of the cross-section of moulded parts which are to be produced *via* the GIT process, a factor to be considered is the cross-section formed by the gas pocket should be as round as possible. For this reason, an attempt should be made to achieve a circular outer contour in order to avoid uneven material accumulation. Design guides are shown in Figure 10.25.

  Since rounded external contours are frequently impossible in practical applications, the following configuration guidelines should be observed:
  - Implement external contours as close to round as possible.
  - Avoid sharp edges, round liberally instead.
  - Avoid material accumulation at the corners.
  - Provide a uniform wall thickness for the entire moulded part.

  In the configuration of right-angled cross-sections, it is possible that the melt will not be displaced at the narrow sides. Observe these rules of thumb for this situation:
  - Maximum moulded part width = 3 to 5 times the height of the moulded part.
  - Minimum moulded part length = 5 times the height of the moulded part.

- Bends and Curves

In the area around bends there is the possibility of material accumulation on the exterior surface (as shown in Figure 10.26), as well as reduction of the wall thickness on the interior surface. For this reason, the following should be observed:

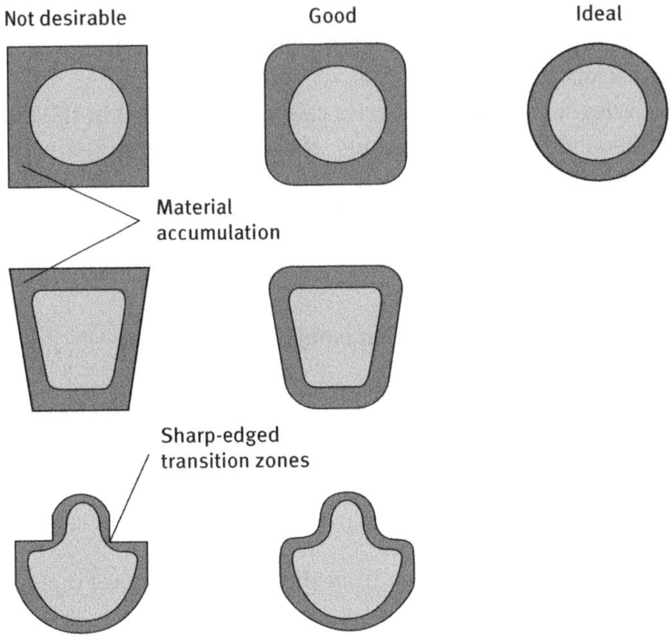

Figure 10.25: Cross-sections in the GIT process.

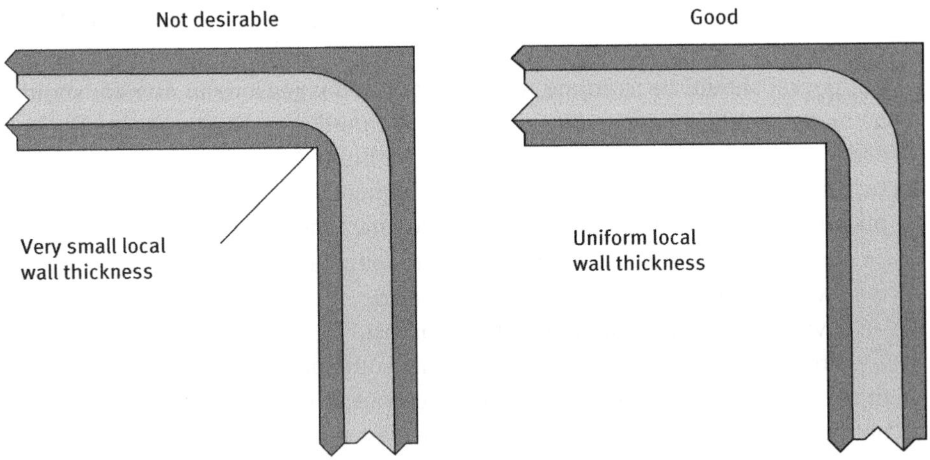

Figure 10.26: Local wall thickness at a 90° angle.

  - Avoid sharp-edged areas.
  - Select the greatest possible bend radius.
- Cross-section Changes

Significant changes in the cross-section frequently have to be made within a part. In this instance, it should be remembered that the melt flow and the gas flow are guided from the larger to the smaller cross-section, which leads to the desired flow of plastic in the cavity and hence to a uniform distribution of the melt. As a reciprocal effect, uniform distribution of the melt creates a uniform distribution of the gas.

- Gas Introduction/Injection Location The injection position and the position of gas introduction are critical factors. To achieve wall thicknesses which are as uniform as possible, it is extremely important to fill the cavity homogeneously with the melt so that the gas bubble can be specifically guided as designed.

The following basic rules must be observed when selecting the location of gas introduction:

- Implement gas and melt injection from the front side of the moulded part in a latitudinal direction whenever possible.
- Plan for only one flowpath and one cavity whenever possible.
- Do not establish the injection location in a visible area or in an area of mechanical stress of the moulded part.
- Position the introduction of gas in the vicinity of the material injection so that the gas bubble can follow the melt flow and only expand in one direction.
- Observe the force of gravity, fill the moulded part from the bottom to the top in order to achieve the desired swell flow.

If it is necessary to establish the gas introduction point at a different location to the material injection point of the moulded part, then the following point should be observed:

- The injection point lies exactly in the middle of the moulded part.
- The flowpaths are balanced as precisely as possible.
- The mould piece is configured symmetrically.

10.10.6 KoolGas™

KoolGas™ is a further development of GAIM, which was originally developed at the University of Warwick, UK. In this process, cryogenically cooled gas is used to form the moulding core and to provide internal cooling. It has several important advantages over standard GIT techniques, including reduced cycle time, smooth bores, balanced thermal stresses and improved morphology across the moulding wall. It can also be used to produce thinner walls than conventional methods. The equipment requirement is similar to conventional GIT except that a special cryogenic cooling module is required to cool the gas to suitable temperatures.

10.10.7 Water-assisted injection moulding process

The WAIM process, or water injection technology, was developed by the IKV (Institute for Plastics Processing) in Germany in 1998. Although this was not the first research work in this field, significant barriers to success in previous work had to be overcome. The aim was to replace nitrogen gas with water in order to reduce cooling times; however, the use of water injection presented technical challenges, i.e., to generate both high pressures and high flow rates as well as defining a single hollow section. With these problems now overcome, WAIM has become a viable fluid injection technique of increasing commercial interest.

The process itself is very similar to GIT except during the water injection stage, one or more hydro-pumps are used to inject the water, at temperatures between 10 and 80 °C and up to pressures of 350 bar, into the cavity containing the preinjected polymer. Like GIT, the melt is displaced into the low pressurised areas completely filling the cavity. The water flows through the hollow body to provide cooling until the part has solidified sufficiently to allow ejection. Finally, the water is removed either by gravity-induced draining or by feeding compressed air through one of the nozzles.

The WAIM process can be operated using a short-shot technique but current WAIM applications tend to employ the full-shot technique with the subsequent displacement of the melt into overflow channels to produce the hollow shape.

The main advantage of WAIM is the reduction of cycle time, which is due to the higher specific heat of water compared with nitrogen gas. Cycle time reductions of up to 70% have been claimed using WAIM, as opposed to conventional GIT. This technology has generally found applications in rod-shaped mouldings with large diameters (typically above 40 mm in diameter).

Because of viscosity differences between gas and water (water is much more viscous), larger injection nozzles are required for water injection in order to achieve suitable flow rates. Therefore, the size of the nozzle places a restriction on its potential applications.

Investment costs are similar to those associated with GIT, including items such as a water pressure generator, pumps and nozzles. There are also safely issues associated with this method as both operators and moulds need to be protected from the high pressure water jets and water damage.

There are a number of machine manufacturers who now offer water injection systems, for example, Battenfeld offer AQUAMOULD™, Engel offer WATERMELT™ and Ferromatik Milacron offer the Aquapress system, to name but three.

10.11 Multi-shot moulding

Multi-shot moulding has been around for over 50 years and is used as a method of placing materials either side by side (abutting), one on top of the other with in an overlap or superimposition of one shot onto another. Special tooling and machinery is required to achieve this and common examples of such mouldings are keypads with the numbers made of one colour and the letters moulded-in using another colour. The advantage of this technique is elimination of the printing processes which would otherwise be required to mark the keypads. Multicolour automotive tail lights are also made using this technology. Another common application of this technology is to combine hard and soft materials to produce a 'soft feel' component. Handles, such as on doors or toothbrushes, are common products benefiting from this technology.

Multi-shot moulding techniques are well established and their growth has been advanced by the development of thermoplastic elastomer (TPE) materials, enabling rigid and flexible material combinations, such as those described previously, to be employed. These can also be seen in a variety of other applications from automotive seals to bras.

A number of moulding methods can be employed to produce multi-shot mouldings; however, whilst mouldings can be produced by a variety of methods, those produced from the same material, but *via* the multicolour multi-shot process, enjoy the highest market share.

Multi-shot techniques produce not only multicolour but also multimaterial mouldings. The most common methods are the use of tool rotation, the core back technique and transfer tools. The latter technique can be used to move shots from one cavity to another, which is very similar to over-moulding. These processes will be described in detail in Sections 10.11.2–10.11.4. Some applications are given in Table 10.3.

Table 10.3: Applications for multi-shot mouldings.

Material combination	**Properties**	**Application**
Same material/ different colours	Aesthetics	Buttons, mobile phone casings, toothbrushes, automotive light casings
Soft feel/hard feel	Increased customer appeal through 'soft touch' properties	Various handles, toothbrushes, cameras, screwdrivers
Transparent/ coloured	Viewing panel incorporated into moulding	Light casings
Very flexible TPE, hard substrate	Sealing properties	Lids with moulded on seals
LSR, hard substrate	Sealing properties	Seals

LSR: Liquid silicone rubber(s)

In addition to these applications, the multi-shot process can also be used to produce power tool cases, battery cases and domestic appliances (e.g., kettle handles, vacuum cleaners, lawn mowers and electric toothbrushes).

10.11.1 Machine technology

Multi-shot processes, as the name implies, require multiple shots of material to make a single component and a separate injection unit is required for each material. In addition, special tooling and equipment is required to mould the multiple shots. Multi-shot capability can either be built into the injection tool or controlled *via* the IM machine. To facilitate the multi-shot process, multiple injection units can be arranged to feed machines in a number of ways, as Section 10.11.1.1 will explain.

10.11.1.1 Injection unit configurations

Multiple injection units can be arranged around the clamping units as combinations of horizontal and vertical units in piggybacked or right angle configurations (e.g., Figure 10.27).

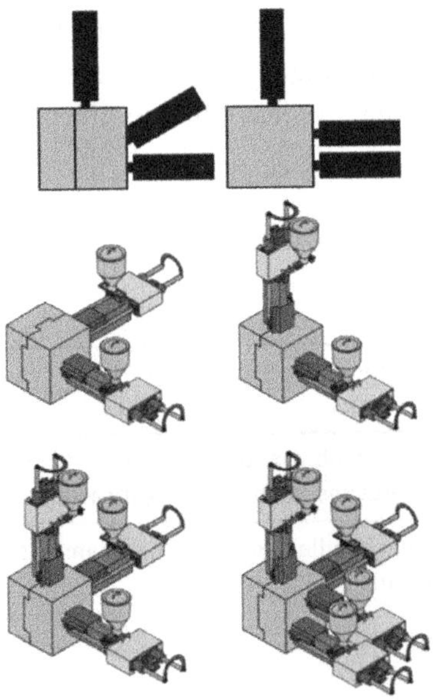

Figure 10.27: Possible arrangements of injection units.

Using a vertical injection unit can save space and hence is favoured for multi-shot moulding at the split- line. To change a mould, the units can be slid towards the nozzle. When the vertical position cannot be used, perhaps due to lack of factory height, the second unit can be positioned at right angles. The position of the unit can be adjusted both horizontally and vertically, although the latter option is available as an option rather than as standard, by some manufacturers. Again as in the vertical unit, it can be moved to the nozzle side to change the mould.

Using the piggyback method, like the vertical method, the units can be angled above the main unit which saves floor space however, the piggyback arrangement requires less ceiling height. The additional unit can be moved in two ways, it can be attached to the main unit but with the nozzles thermally isolated or with a separate cylinder which moves independently of the main cylinder.

- *Plastication Design*
  In terms of the injection screw configurations within the injection units, these tend to have become standardised at 22D, which refers to the ratio of screw length (L) to its diameter (D) and is called the L/D ratio. For faster production and high-quality mixing this is often increased up to 26D. Specialised screws for hygroscopic materials, which enable venting, screws for increased mixing of materials and configurations for temperature-sensitive materials, such as polyvinyl chloride (PVC), are all commonly available from suppliers.

- *Machine Type*
  When robots are required to control the insertion and removal of parts, machinery design can be an issue. The use of tie-bar-less machines as opposed to the conventional tie-bar models is usually a contentious subject with advantages and disadvantages inherent in both machinery designs. However, when it comes to multimaterial technology, tie-bar-less machines can offer distinct advantages including the ease with which robots can be utilised and the larger mould mounting area these machines offer, which can often mean a smaller capital outlay on machinery as a smaller machine may be purchased.

Now machinery issues have been briefly overviewed, individual technologies for the multi-shot process will be introduced. These are:
- Core back moulding.
- Rotary method.
- Transfer moulding.

10.11.2 Core back moulding

Core back moulding is referred to in various literature as composite IM or multi-shot, and is a tooling controlled process. The manufacturer Battenfeld uses the trade name Combiform for this process.

Core back moulding, thought of simplistically, is one tool taking multiple shots within a single machine cycle. It allows different areas of the tooling to be opened or closed for specific material feeds, which is achieved through the use of moving slides or inserts, as illustrated in Figure 10.29.

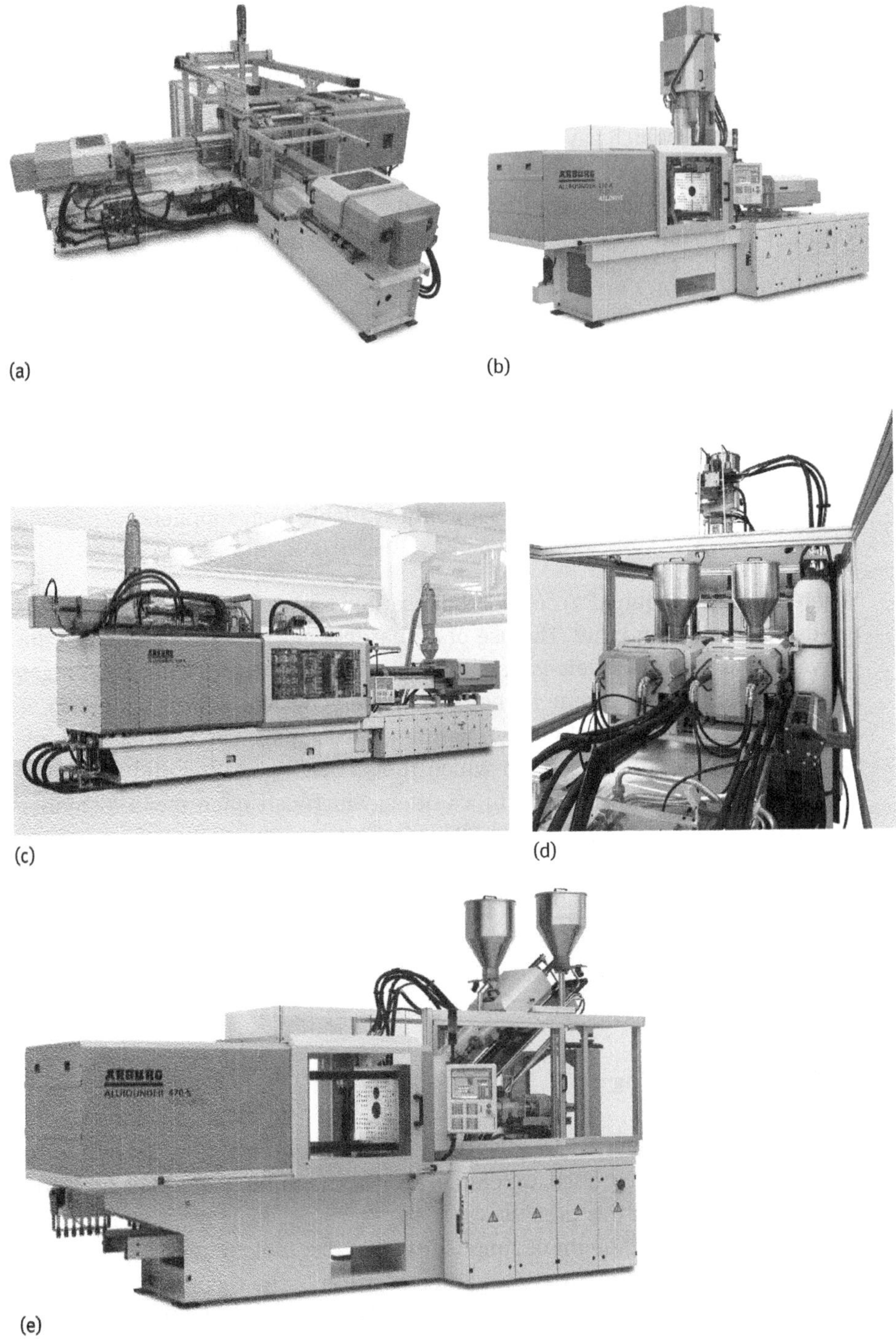

Figure 10.28: Possible arrangements of injection units.

In Figure 10.29a, the insert is closed, which restricts the first injected material to this area of the cavity. When the insert is opened, as shown in Figure 10.29b, the second material can feed into the newly opened cavity area and flow into the material already injected, resulting in the multi-shot component. In this way, slides and inserts are used to block and control the cavity openings to particular material feeds.

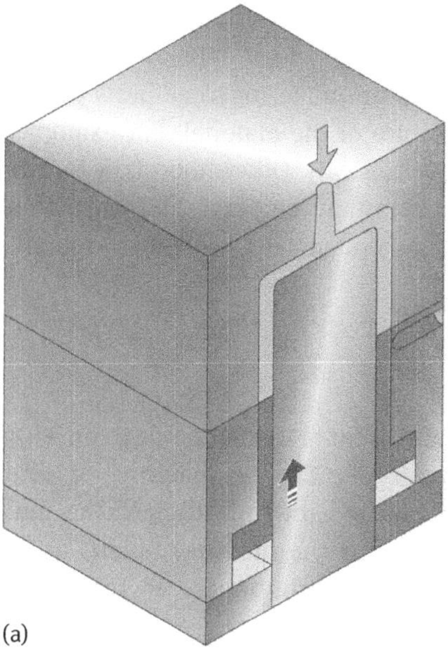

(a)

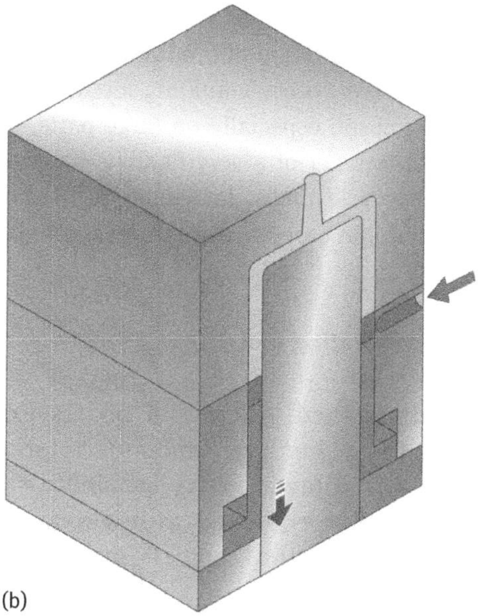

(b)

Figure 10.29: Core back moulding.

For example, a two-component process, such as shown in Figure 10.29, may consist of the following stages:

1. The first material is injected into the cavity.
2. Using a core puller to activate a slide seal, a further area of cavity is revealed.
3. The second material is injected into the cavity.
4. The completed multi-shot component is ejected.

The main advantages of this method, when compared with other multi-shot techniques, are the ability to complete the process without opening the mould or transporting the preform. However, increasing the number of components beyond two significantly increases the cost of tooling due to the increased intricacy required. The machine must also have the necessary means to actuate all the slides in the tooling, which is usually achieved by adding the core pullers to the injection machine specification, though other methods of actuation may also be used.

As with all multimaterial injections, attention must be paid to the compatibility of the melts. The use of the core back technique enables greater bond strengths to be achieved, compared with other multi-shot techniques, as the time between injection of the first melt and injection of the second material can be optimised. However, the sequence of injecting the first material and then the second material is longer than in other multi-shot techniques, which proceed in parallel. Therefore, for components that can be produced by both techniques, a detailed analysis of the economic implications of the process routes may be required to determine the most appropriate method of production.

However, other moulding methods may be inappropriate for some parts due to tooling costs, especially on large parts, or the nature of the material. For example, robotic transfer of a very flexible material may be difficult. Other important factors may be cycle times, for example, the cooling times may be very short on a thin moulding. In cases such as these, a core back tooling system may be the best option.

10.11.3 Rotating tool

In this method, the mould rotates through 180° for a two-shot part or 120° for a three-shot part, and this rotational capability can be machine or tool based. An integral rotary capability can be designed into the tool or the machine can be equipped with a rotary attachment to the moving platen, as in Figure 10.30; the choice usually comes down to economics. If rotational

Figure 10.30: Rotating platen.

capability is to be used regularly, it is cheaper to have it on the machine than to continually buy more expensive tooling. A rotary platen must have an accurate indexing device to control the rotation and the stroke needs to be both fast and cushioned to prevent damage. The platen must also have facility to mount ejector pins. An example of a mounted electric unit is shown in Figure 10.31. A tool is mounted on the indexing unit and moulding proceeds as shown in Figure 10.32.

Figure 10.31: Electric indexing unit.

The process operates in parallel so at any stage there is a shot being produced by each cavity, which makes the overall cycle time per moulding shorter than the core back technique described earlier. Generally, the moulding produced in the first cycle should only melt on the very surface, which gives the good material separation required but still forms an adequate bond; this does, however, require good control of the process.

10.11.4 Transfer moulding

In this method, instead of rotating the mould, a robot is used to transfer the moulding to the next cavity where it can then be over-moulded (Figure 10.33). In this example, the robot will move both the lower moulding to the upper

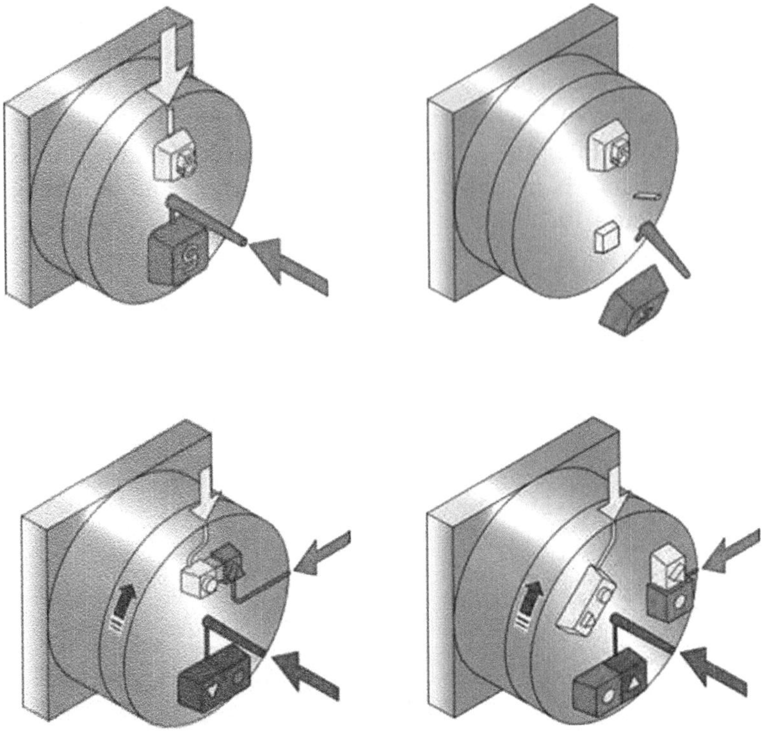

Figure 10.32: Two-component (top) and three-component (bottom) multi-shot moulding.

cavity and the complete moulding from the upper cavity as the tool opens after each cycle. Like rotary methods, moulding proceeds in parallel with a moulding produced in each of the cavities during any one cycle. Therefore, the cycle time will be dependent upon the moulding requiring the longest moulding time. As in rotating methods, a good bond is required whilst maintaining a distinct separation of materials. A high level of accuracy is required when placing the insert in order to obtain good definition and registration on the final component. A means must therefore exist to hold the preform accurately in place before the second material is injected onto it.

Transfer moulding is not restricted to one machine as robotics can be used to move the preform to a second machine; however, this would involve investment in both another machine and a second injection tool. If multiple layers are required, this method can be used, for example, to produce a four-layer moulding; this could be achieved by producing two layers in one machine, with robotic transfer to a second machine whereby two further layers could be added. The use of robots is increasing in IM operations due to several advantages of using automation, which include reduced cycle times and the ability

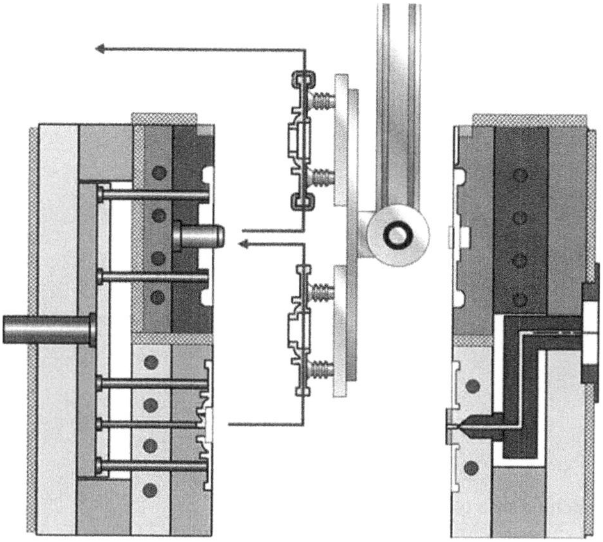

Figure 10.33: Transfer tool with robot.

to control the process *via* the machine interface. The robots can also ensure surface scratches or damage are kept to a minimum if an aesthetic component is in production.

Some example layouts of the production cells of two machines are shown in Figures 10.34–10.36.

In this layout (Figure 10.34), the use of only one robotic system is required to produce a two-component moulding, thereby reducing the tooling costs required for the production of two-component parts. Existing conventional injection moulds can be used and therefore lower investment costs are required. In this case, the machines can also be operated separately as needed.

In Figure 10.35, two robots are employed in a connected cell but, as in Figure 10.34, there is still flexibility in the use of standard machines and tools that can be operated independently as required.

Figure 10.36 also shows two machines each with an independent robotic system but with a separate processing station module capable of operating robots in horizontal or vertical movements.

10.11.5 Multi-shot using a single injection unit

The mono-sandwich technique for co-injection moulding was described in Section 10.9.2.3 to which the reader should refer for machinery details. This technique can also be used for over-moulding by using the core back technique, again

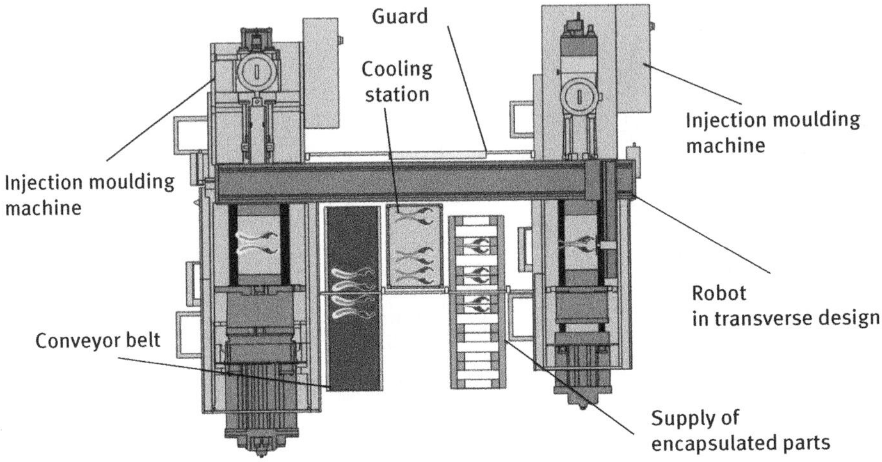

Figure 10.34: Interlinking of shared machine by a robot.

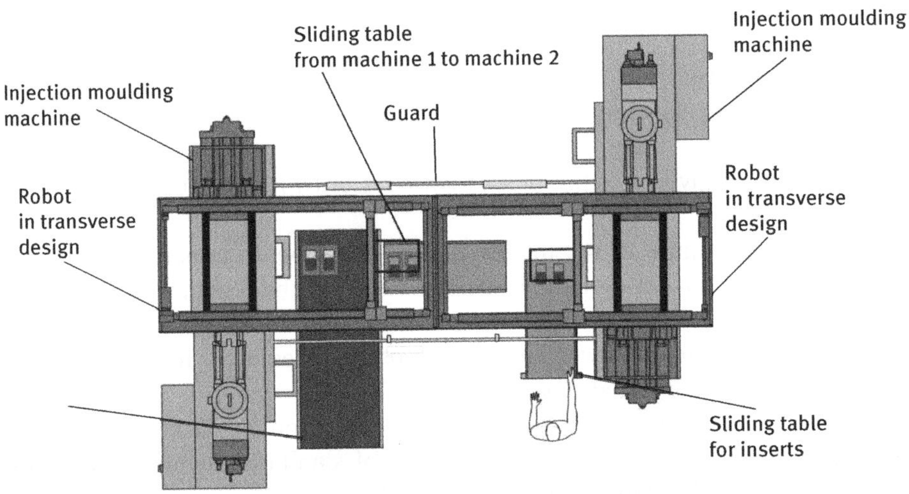

Figure 10.35: Connection of two parallel production cells with a sliding table.

described earlier. In this technique, termed the mono-sandwich 5 process, additional information is required by the runner system in order to release different valves as necessary. Once the melt is layered, the first component is injected, the valve is then switched within the mould to expand the cavity and the rest of the shot is injected to create a multi-shot component.

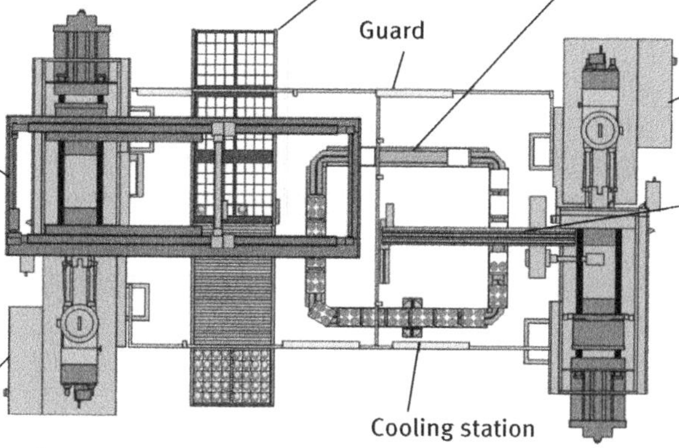

Figure 10.36: Connection of two production cells with a circulation system.

10.11.6 Materials for multi-shot moulding

10.11.6.1 Material selection for multi-shot moulding

This section will introduce the factors to be considered during material selection when using the multi- shot moulding method, as well as providing some introductory information on two classes of materials which may not be familiar to the non-specialist moulder, namely TPE and LSR. This section will begin however by considering adhesion.

Tables of compatibility have already been introduced, such as the one shown in Table 10.1, which describe the properties of different materials, including miscibility, and therefore likely adhesion. Other characteristics may also be considered, for example, possible modifications through alloying or the addition of compatibilisers. The thermal properties are also important, i.e., both softening and melting temperatures and ranges, due to the requirements of limited melting to form a bond but ensuring the individual materials are clearly separated. However, changing the melt and mould temperature and injection speed of the moulding machine will influence bond strength. The molten contact time will also drastically affect the final bond strength and the separation achieved. If the first shot becomes molten at too deep a layer, it will flow when the second material is injected onto it and consideration of this factor will also influence what type of multi-shot technique should be employed. The testing of bond strength can be achieved by peel testing or by standard tension testing as required.

10.11.6.2 Material process order

The previously mentioned melting and softening characteristics can affect the potential process order. Since the material injected first is already well on its way to cooling when the second material is injected, it is better to mould the material with the lowest melting temperature first rather than *vice versa*. This is where one technique may become more preferable over another for certain material combinations, for example, core back moulding can use less than a full cycle between material injections. This means injection of the second component can occur before full cooling has taken place, therefore increasing the potential of achieving adhesion in some cases, which is not possible with the rotary or transfer methods.

10.11.6.3 Using thermoset materials

The combination of thermoplastic/thermoset materials is a further area of development – there are two possibilities for this process:

1. The thermoplastic is moulded and placed in the thermoset mould.
2. The thermoset is moulded and cured and placed in a thermoplastic mould.

With the first of these processes, the choice of thermoplastic will be dependent upon the temperature required in the thermoset mould. Very few materials may be suitable if high cure temperatures are required.

Using the second option, it may prove difficult to achieve a bond between a fully cured thermoset and the molten thermoplastic, as there is no interdiffusion. However, clever process design, making use of both material properties, such as shrinkage, and tooling to promote mechanical interlock, may remove the need for adhesion. This is an area of ongoing commercial development in both Germany and the UK and is likely to hit commercial exploitation in the near future.

The core back technique offers an increased opportunity to achieve adhesion between materials and allows the thermoplastic to contact the thermoset before the part has fully cured. However, this method presents problems, as controlling the mould temperature and the differing thermoset/thermoplastic flow fronts would be extremely difficult.

Thermoset flexible/rigid combinations can be achieved by combining LSR with thermosets. The similar processing temperature requirements and process control is greatly enhanced using this combination, providing the mould is designed with adequate consideration of flows.

10.11.6.4 Liquid silicone rubber

LSR can also be combined with some thermoplastics and is an area of considerable commercial interest, with one current application in the automotive industry. General Motors produce a thermoplastic/LSR multi-shot air intake manifold for the

Northstar model [3], other potential applications include shower heads and water-resistant mobile phones.

The reason for this interest is the added versatility offered to soft/hard combinations using LSR materials, as they are thermally stable and show little change in properties with temperature, which can be of particular interest at elevated temperatures where the use of TPE may become limited.

LSR can also undergo rapid cure cycles, and as postcure is not necessary, the finished parts cost less than those produced using other thermoset technologies.

Material grades which are injection mouldable can be formulated as two-component systems and cure at temperatures of around 170–230 °C. Like moulding with thermoplastic materials, the cycle times are part dependent, but cycle times of 15–60 s are typical. Because mix-meter pump systems are used to feed the injection machines, raw material handling of LSR materials is relatively simple. More details on LSR materials can be found in Chapter 7.

10.11.6.5 Thermoplastic elastomers

The properties of TPE that have made them so commercially successful are their low modulus and flexibility. The ability to recover from stress and return to their original shape makes them suitable for applications such as sealing rings. These types of applications comprise the traditional markets of thermosetting rubber materials, which TPE materials are now replacing. In addition, TPE have the added advantage of recyclability over thermoset materials, which is especially important in markets where environmental legislation is in place.

Whilst newer material types have emerged over recent years, primary TPE types can be categorised into two generic classes, block copolymers (styrenics, copolyesters, polyurethane and PA) or TPE blends and alloys (thermoplastic polyolefins and thermoplastic vulcanisates). These TPE types are known as two- phase systems as essentially, a hard thermoplastic phase is coupled mechanically or chemically with a soft elastomer phase, resulting in a TPE that has the combined properties of the two phases.

TPE usage has increased significantly in recent years and is expected to continue to rise. Worldwide, consumption was estimated at 1,400,000 metric tonnes/year in 2000.

Some examples of common TPE are shown in Table 10.4. As well as adhesion considerations, the suitability of a TPE will also depend upon properties such as its hardness and compression ratio. These properties may not be familiar to readers with a background in thermoplastics, therefore a brief description of each is included.

Hardness can be defined as the resistance of the material to indentation and is usually measured on a Durometer using a Shore Hardness scale. TPE tend to be rated on a Shore A scale, the softest materials range from around 3 Shore A and the hardest to 95 Shore A. Out of interest, thermoplastic materials are measured on a

Table 10.4: Examples of TPE and substrates.

Type	Elastomer description	Subgroup	Possible substrates
TPE-O	Polyolefin blends	PP/EPDM PP/EPDM crosslinked	PP PA (modified)
TPE-V	Polyolefin alloys	Various	Polystyrene, ABS, PET, ASA, ABS/PC blends
TPE-S	Styrene	SEBS Styrene-butadiene- styrene SEBS/PPE	PP PA (modified)
TPE-A	Polyamide	PA12 based PA6 based	PA
TPE-E	Polyester	Polyester ester Polyether ester	PA*(polyester ester only) PET, PBT
TPE-U	Polyurethane	Polyester urethane Polyether ester urethane Polyether urethane	PA, ABS, POM, PC, PBT, PVC

EPDM: Ethylene propylene diene monomer
SEBS: Styrene-ethylene/butylene-styrene
PPE: Polyphenylene ether

different, Shore D scale, which gives an indication of the difference in their hardness properties. TPE material grades are available with a range of hardness levels, however, a factor to bear in mind is that hardness is also a function of the thickness of the material and the substrate beneath. In multi-shot applications, the thinner the layer of elastomer on the substrate, the harder it will feel due to the effect of the harder substrate beneath it. It is for this reason that standards, such as the American Society for Testing and Materials, specify a material thickness of 0.250 inches. The compression set is often specified for sealing-type applications and is a measure of the deformation after compression for a specified deformation, time and temperature.

10.11.7 Multi-shot application case studies

Now the major areas of the multi-shot process have been described, two recent applications of this technology by UK-based companies will be described.

10.11.7.1 Trio knob

An interesting example of a relatively early UK multi-shot application from Sifam, now SMT multi- shot, Devon, UK, details the cost savings that can be achieved using multi-shot technology. The original costings are from 1996, but the relative

savings involved are still applicable today. The three production methods are as follows:
1. Traditional: moulding two components (a knob and a coloured cap), assembly and printing.
2. Two-shot: knob and cap moulded together, eliminates assembly but still requires printing.
3. Three-shot: part redesigned so that print effects are instead moulded in, thereby eliminating assembly and printing.

It can be seen from Table 10.5 that despite the major increase in tooling costs, the following factors reduced the job cost:
- Utilising only one injection machine instead of two.
- Keeping the cycle time the same due to sequential moulding whereby the three cavities are each moulding in any one cycle.
- Removal of assembly and printing costs.

Table 10.5: Run cost comparison for Trio Knobs.

Production method	Traditional	Two-shot	Three-shot
Tool cost	£8,500	£12,000	£19,000
Cycle time (secs)	21 + 21	21	21
Machine cost (100,000 pieces)	£1,241	£657	£695
Materials	£345	£365	£385
Assembly/print	£1,650	£825	£0
Job cost	£3,236	£1,847	£1,080

Moulding was carried out using a three-cavity tool, moving the rotating side of the mould tool by 120° after each injection cycle. At any one stage there are three mouldings in the tool; a first single-shot, a two-shot intermediate moulding and a complete three-shot component, which is then ejected. Since all three cavities fill simultaneously, the cycle time is fixed to the cavity requiring the longest fill time.

10.11.7.2 Stanley screwdriver

Stanley Tools, Engel and Burnett Polymer Engineering, Northampton, UK, developed a four-shot process to manufacture screwdriver handles. A nylon core is used over which two different coloured layers of PPE are then moulded. These layers give the final part both aesthetic appeal and commercial branding. A final TPE is added for a soft feel grip. Adhesion is achieved in the layers by clever use of material properties and tooling design, with bonding occurring *via* mechanical interlock. This imaginative design results in an incredibly strong component giving improved impact resistant and torque. An application of this type requires extreme accuracy in location of the various components, to ensure definition of the end product and avoid damage to the tooling. This product highlights what can be achieved by the successful collaboration between manufacturer, moulder and machinery supplier.

10.11.8 Limitations to multi-shot

Perhaps the most important issue with this process is the cost of tooling. Obviously this varies with the part design and complexity of tooling required. More complex rotating mould tools may also be more prone to breakdown than standard tools. Another problem associated with tooling concerns machine size. Over-dimensional, bulky moulds require larger machines, increasing both machine costs and space requirement due to the larger machine footprint. One way around this is to use tie-bar-less machines, and the advantages of this machine configuration in relation to multi-shot have already been discussed. The use of two injection machines in a transfer process also involves large investment as two machines and tools are required.

Accurate positioning of the mould cavities is crucial. Moving and rotating the tool to achieve precision positioning of parts, often at very high speeds to optimise cycle times, can create wear and inaccurate registration, and also places restrictions on cycle times due to the time required to carry out this change of positioning.

In methods using the same mould and multiple cavities, process restrictions may limit bonding strength when multiple materials are employed, as only one tool temperature can be employed. The choice of tool temperature may be compromised from the optimised bonding temperature.

10.12 Over-moulding

In this process, a component termed a preform is placed into the tool of an IM machine. A second material is then moulded onto or around the preform. Two methods fall under this category: insert moulding and lost core moulding.

10.12.1 Insert moulding

Insert moulding with plastics is a two-step process whereby a first preform component is placed into the open mould cavity. Injection then proceeds as with traditional moulding methods by injection of a molten plastic onto the preform. This process is not limited to two material components and the resultant mouldings can be built up layer-by-layer until the required number of layers is achieved. It is also not limited to plastic and a number of materials have been successfully employed as inserts, such as wood veneers (Figure 10.37), textiles, metals or functional films.

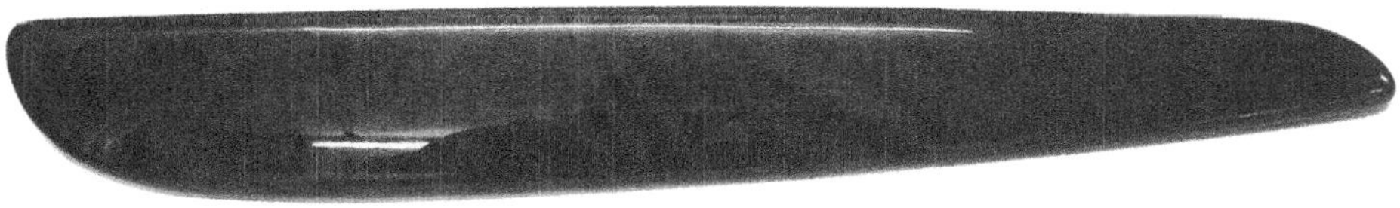

Figure 10.37: Example of wood veneer moulding produced by insert moulding onto ABS [5].

Inserts can be loaded by hand or by robots, and must be accurate in both their dimension and their placement into the over-moulding tool to prevent tool damage and provide accurate registration of one material on another. A mechanism must also exist to hold them in place within the tool. In this way, this process has similar requirements to that of in-mould lamination techniques and transfer moulding, which is commonly used to decorate plastics with films or foils, and is detailed in a specialist Review Report [4].

10.12.2 Lost core moulding

The lost core technique, like insert moulding, is often used in combination with metals as well as plastics. This technique produces a hollow component, similar to those produced by techniques such as extrusion blow moulding or GAIM and overcomes disadvantages inherent in both these processes. It enables high dimensional accuracy, unlike extrusion blow moulding, as well as a defined interior surface which is not possible using GAIM. For the manufacture of plastic components, the core is first produced either from a low melting point metal alloy, usually tin bismuth, or a soluble plastic material. The core is then inserted either by hand or robot into a tool and over-moulded. The core material can then be melted or dissolved out and the final component cleaned of any residual waste material. The advantages and disadvantages of this technique are detailed in Table 10.6.

Table 10.6: Process attributes of lost core moulding.

Advantages	Disadvantages
Mould design is relatively simple, Complex geometry is possible, High-quality surface finish Good, seam-free interior surface Weight saving High cavity pressures are permissible	Expensive High unit cost Long development times Core materials can be expensive

Applications for this technology include nylon air inlet manifolds for car engines, canoe paddles, tennis rackets, BMX bike wheels and hot water heating pumps.

10.13 A final word on energy efficiency

As energy efficiency has become an increasingly important factor in process and machine selection for some applications, as well as in the continuing debate on hydraulic powered versus electric powered machines, a brief introduction to measurements of the energy efficiency of the IM process is given here.

In an alternating current, circuit voltage (V) and current (I) are both time-dependent. Energy efficiency is measured by effective power that is:

$$\text{Momentary effective power } P(t) = \text{voltage } V(t) \times \text{current} I(t) \tag{10.1}$$

This effective power (Pt) relationship with voltage (V^t) and current (I_t) is shown in Equation 10.2: Therefore capacitive power in the AC circuit:

$$P_t = V_t \times I_t \times \cos \varphi \tag{10.2}$$

This is entirely different to the actual installed power and concepts should not be confused. The relationship between power and energy is shown in Table 10.7.

Table 10.7: Power and energy comparisons.

Electrical power	Electrical energy
Equation: $P = V \times I \times \cos \varphi$	Equation: $E = P \times t$
Unit: watt (W) or kilowatt (kW)	Unit: kilowatt-hour (kWh) = kW × h
1,000 W = 1 kW	Also often used: watt-second (Ws)
Power is a momentary value, as is, e.g., voltage	Energy is the amount of power over time

Energy measurement therefore requires the evaluation of all (entire machine) non-sinusoidal voltages and currents, and suitable measuring devices must be used that allow the automatic determination of the important values of:
- Phase shift cos φ.
- Supply frequency.
- Energy requirement over a defined time period.
- Average power.

In order to do this it is necessary that the machine and process are already thermally balanced, i.e., already running at suitable parameters and no changes are made, and the measurement is carried out over a sufficient period to ensure accuracy. It is important like for like measurements are made if any comparisons are to be drawn from the results. The final relationship between energy requirements are shown in Equation 10.3:

$$\text{Specific energy requirement (kWh)/kg} = \frac{\text{average power consumption (kW)}}{\text{material throughput (kg/h)}}$$

$$= \frac{\text{average power consumption kW} \times \textit{cycle time}(s)}{\text{shot weight kg} \times 3,600} \quad (10.3)$$

To enable realistic economic comparisons of machines or processes to be made, repeatable measurement methods must be used to monitor the specific energy requirement.

10.14 The future?

A whole host of advanced IM technologies exist, a number of which are beyond the scope of this chapter. Whilst this chapter is not a conclusive guide to all the currently available technologies, the developments of note include micromoulding, which enables tiny components to be accurately and cost-effectively manufactured, and is of especial interest to the medical industry. The use of supercritical fluids to assist in moulding is also a growing field. These materials (generally supercritical carbon dioxide or nitrogen), offer a huge number of potential applications in the fields of polymeric foam production, for viscosity reduction (and therefore melt temperature reduction), as contaminant removal from recyclates and as compatibilising agents. Other multimaterial IM variations also exist and the rapid advances in AM can also not be ignored. It is likely that in future factories, commercially successful moulders will require both the understanding and adaptability to utilise a wide range of the available process options. It is also likely that developments will continue to extend beyond the use of just plastic materials. Future components may be hybrids of many different materials, such as plastics, metals and ceramics, all created as a single component in automated factories wired to the 'internet of things' and connecting the entire supply chain from cradle to grave.

Tooling costs will be key to many future developments in this area, due to the prohibitive costs often involved in tooling for multiple materials. Developments in both the mouldability and adhesion characteristics of both thermoset and thermoplastic materials, and machinery improvements in terms of enhanced speeds and control are also likely to further enhance processing and material combination options. With such potential inherent in these processes it is likely that the future limitations of this technology may only be defined by the limits of the designer's imagination.

References

1. R.A. Easterlow, P.D. Stidworthy, R.J. Coates and G.F. Smith, inventors; Painted Plastic Articles, assignee; EP0816066, 1998.

2. V. Goodship and K. Kirwan, *Plastics, Rubber and Composites*, 2001, **30**, 11.
3. K. Baraw in *Proceedings of the Annual Conference*, Composites Institute, Society of the Plastics Industry, 1997, S 15-C/1.
4. J.C. Love and V. Goodship, *Decorating Plastics*, Rapra Review Report, Smithers Rapra, Shawbury, UK, 2002.
5. K. Makenji, V. Goodship, S. Buckley and E. Evered, *Progress in Rubber, Plastics & Recycling Technology*, 2006, **22**, 4, 225.

Further reading

1. J. Avery in *Gas-Assist Injection Molding: Principles and Applications*, Hanser Publishers, Munich, Germany, 2001.
2. L. Shau-Tarng in *Polymeric Foams: Innovations in Processes*, Technologies, *and* Products, CRC Press, Baco Raton, FL, USA, 2016.
3. V. Goodship, B. Middleton and R. Cherrington in *Design and Manufacture of Plastics Components for Multifunctionality: Structural Composites, Injection Molding and 3D Printing*, Elsevier, Atlanta, GA, USA, 2016.

Abbreviations

3D	Three-dimensional
ABS	Acrylonitrile-butadiene-styrene
AM	Additive manufacturing
APE	Aromatic polyester
ASA	Acrylonitrile-styrene-acrylate
ASTM	American Society for Testing and Materials
BgVV	Bundesinstitut für Gesundheitlichen Verbraucherschutz Und Veterinärmedizin
BMA	Ultralow abrasion
CA	Cellulose acetate(s)
CAB	Cellulose acetate butyrate
CLTE	Coefficient of linear thermal expansion
CP	Cellulose propionate
CPVC	Chlorinated polyvinyl chloride
CS	Compression set
DAP	Diallyl phthalate
ECTFE	Ethylene chlorotrifluoroethylene
EP	Epoxide/epoxy resins
EPDM	Ethylene propylene diene monomer
ETFE	Ethylene-tetrafluoroethylene
EVA	Ethylene-vinyl acetate
FDA	US Food and Drug Administration
FEP	Tetrafluoroethylene hexafluoropropylene
GAIM	Gas-assisted injection moulding
GIT	Gas injection moulding technology
HDPE	High-density polyethylene
HF	High frequency
HV	Hot vulcanisation
IJM	Injection moulding machine(s)
IKV	Institute for Plastics Processing
IM	Injection moulding(s)
ISO	International Organization for Standardization L/D Length/Diameter
LCP	Liquid crystal polymer(s)
LDPE	Low-density polyethylene
LIM	Liquid injection moulding
LLDPE	Linear low-density polyethylene
LSR	Liquid silicone rubber(s)
MF	Melamine formaldehyde
MP	Melamine-phenol formaldehyde
MW	Molecular weight
MWD	Molecular weight distribution
NR	Natural rubber
PA	Polyamide(s)
PAA	Polyacrylic acid
PAE	Polyacrylic ester
PAEK	Polyaryl ether ketone(s)
PAI	Polyamide imide
PAR	Polyacrylate(s)

https://doi.org/10.1515/9783110654813-011

PBT	Polybutylene terephthalate
PC	Polycarbonate
PE	Polyethylene
PEC	Polyester carbonate
PEEK	Polyether ether ketone
PEEKK	Polyether ether ketone ketone
PEI	Polyetherimide
PEK	Polyether ketone
PES	Polyether sulfone
PET	Polyethylene terephthalate
PE-U	Thermoplastic elastomer (urethane grade)
PF	Phenol formaldehyde
PFA	Perfluoroalkoxyalkane
PKV	Highly wear resistant
PM	High-chromium, powder-metallurgical steel
PMMA	Polymethyl methacrylate
POM	Polyoxymethylene
PP	Polypropylene
PPA	Polyphthalamide
PPE	Polyphenylene ether
PPO	Polyphenylene oxide
PPS	Polyphenylene sulfide
PS	Polystyrene
PS-GP	General purpose polystyrene
PS-HI	High impact polystyrene
PSU	Polysulfone
PTFE	Polytetrafluoroethylene
PU	Polyurethane
PVC	Polyvinyl chloride
PVDF	Polyvinylidene fluoride
RT	Room temperature
RTV	Room temperature vulcanisation
SAN	Styrene-acrylonitrile
SB	Styrene-butadiene
SBS	Styrene-butadiene-styrene
SEBS	Styrene-ethylene/butylene-styrene
T_g	Glass transition temperature
T_m	Melt(ing temperature)
TPE	Thermoplastic elastomer(s)
TPU	Thermoplastic polyurethane
UF	Urea formaldehyde
UP	Unsaturated polyester
UV	Ultraviolet
WAIM	Water-assisted injection moulding

Appendix: Trade Names, Specific Weight and Suppliers of Some Plastic Materials

Abbreviation	Trade names	Specific weight (g/cm3)	Plastic type
ABS	Terluran (15), Novodur (15), Lustran (15), Cycolac (20), POLYLAC (35)	1.03–1.07	Thermoplastic
ASA	KIBILAC (3), GELOY (20)	1.07	Thermoplastic
CA	Cellidor A, S, U (4), Tenite (9)	1.26–1.32	Thermoplastic
CAB	Cellidor B (4), Tenite Butyrat (9)	1.16–1.22	Thermoplastic
CP	Cellidor CP (4)	1.19–1.23	Thermoplastic
DAP	Raschig DAP (18), Supraplast (22)	1.51–1.78 filled	Thermoset
EP	Epoxidor (18)	1.7–2.0 filled	Thermoset
ETFE	Tetzel (7), NEOFLON (5)	1.7	Thermoplastic
EVA	Freenflex (23), Wacker EVA (24), Supraplast (22), Elvax (7)	0.92–0.95	Thermoplastic
FEP	Teflon FEP (7), NEOFLON (5)	2.14–2.17	Thermoplastic
HDPE	Rigidex (15), Eraclene (23)	0.94–0.96	Thermoplastic
LDPE	Lucalen (19), Riblene (23), Lupolen LD (19)	0.91–0.93	Thermoplastic
MF	Meloplast (18), Resopal (20), Spuraplast (22)	1.5–2.0 filled	Thermoset
PA6 PA66	Akulon (6), Ultramid (2), Durethan (16), Zytel (7), Grilon (10)	1.13 1.14	Thermoplastic
PA11 PA12	Grilamid (10), Vestamid (13), Rilsan (31)	1.04 1.02	Thermoplastic
PA amorphous	Trogamid (13)	1.12	Thermoplastic
PBT	Crastin (7), VALOX (20), ARNITE (6)	1.29	Thermoplastic
PC	Makrolon (4), Lexan (20), WONDERLITE (3)	1.20–1.24	Thermoplastic
PC/ABS	WONDERLOY (3)	1.10–1.17	Thermoplastic
PCTFE	NEOFLON (5)	2.1–2.12	Thermoplastic
PAEK	AvaSpire (21)	–	Thermoplastic
PF	Resinol (18), Supraplast (22)	1.25–2.0 filled	Thermoset
PET	Arnite (6), Rynite (7)	1.37 crystalline 1.34 amorphous	Thermoplastic
PI	EXTEM (20)	1.42	Thermoplastic, thermoset

https://doi.org/10.1515/9783110654813-012

(continued)

Abbreviation	Trade names	Specific weight (g/cm3)	Plastic type
PIB	Oppanol (2)	0.91–0.93	Thermoplastic
PMMA	Degalan (13) –additive, Diaklon (14), Plexiglas (13), Elvakon (14)	1.18	Thermoplastic
POM	Ultraform (2), Delrin (7), Hostaform (12), Celcon (12)	1.41–1.43	Thermoplastic
PP	Dexflex (19), INEOS (15)	0.9	Thermoplastic
PPO	Rarely used alone as cannot be melt processed, but as component of NORYL (20) – see PPE	1.06 (*modified grade*)	Thermoplastic
PPE	NORYL (20) (these are modified grades)	1.06–1.10	Thermoplastic
PPS	Ryton PPS (17), Xtel (17)	1.34	Thermoplastic
PS	INEOS styrenics (15), Lacqrene (1)	1.04	Thermoplastic
PES	Udel (21), Ultrason (2)	1.24–1.37	Thermoplastic
PTFE	Teflon (7), Polyflon (5)	2.14–2.20	Thermoplastic
PUR	Desmopan (4), Hytrel (7), Elastollan (2)	1.14–1.26	Thermoplastic
PVC soft	Vestolit (11), Vinnol (24), EVICOM (15)	1.2–1.35	Thermoplastic
PVC hard	EVICOM (15), Vestolit (11), Vinnol (24)	1.38–1.94	Thermoplastic
PVDF	Solef (21), Kynar (1)	1.76	Thermoplastic
SAN	Luran (15), Lustran (15), KIBISAN (3)	1.08	Thermoplastic
SBS and SEBS	Dryflex (8), Mediprene (8)	1.04	Thermoplastic
SI	Raschig silicones (18), ELASTOCIL (24)	1.86–1.68	Thermoplastic, thermoset
TPU	Elastollan (2), Europrene (23)	–	Thermoplastic
UF	Ralupol (20)	1.17–2.0 filled	Thermoset

Supplier key:
(1) Arkema
(2) BASF
(3) CHEMIE
(4) Covestro
(5) Daikin Industries
(6) DSM
(7) DuPont
(8) Elasto
(9) Eastmann

(10) EMS-GRIVORY
(11) Mexichem
(12) Celanese
(13) Evonik Industries
(14) LUCITE International
(15) INEOS
(16) Laxness
(17) Chevron Phillips
(18) Raschig
(19) Lyondellbasell
(20) SABIC
(21) Solvay
(22) Sud-West-Chemie
(23) Versalis
(24) Wacker

Index

https://doi.org/10.1515/9783110654813-013

Printed and bound by CPI Group (UK) Ltd, Croydon, CR0 4YY
13/09/2026
14960857-0001